IET SECURITY SERIES 18

Physical Layer Security for Wireless Sensing and Communication

Other volumes in this series:

Volume 1	**Information Security: Foundations, technologies and applications** A. Awad and M. Fairhurst (Editors)
Volume 2	**Engineering Secure Internet of Things Systems** B. Aziz, A. Arenas and B. Crisp
Volume 3	**Mobile Biometrics** G. Guo and H. Wechsler (Editors)
Volume 4	**User-Centric Privacy and Security in Biometrics** C. Viuelhauer (Editor)
Volume 5	**Iris and Periocular Biometrics** C. Rathgeb and C. Busch (Editors)
Volume 7	**Data Security in Cloud Computing** V. Kumar, R. Ko and S. Chaisiri (Editors)
Volume 8	**Hand-Based Biometrics: Methods and Technology** M. Drahanský (Editor)
Volume 9	**Authentication Technologies for Cloud Computing, IoT and Big Data** Y.M. Alginah and M.N. Kabir (Editors)
Volume 10	**Nature-Inspired Cyber Security and Resiliency: Fundamentals, Techniques and Applications** E.M. El-Alfy, M. Eltoweissy, E.W. Fulp and W. Mazurczyk (Editors)
Volume 12	**Voice Biometrics: Technology, trust and security** C. García and G. Chollet (Editors)
Volume 14	**Privacy by Design for the Internet of Things: Building accountability and security** A. Crabtree, H. Haddadi and R. Mortier (Editors)
Volume 16	**Machine Learning, Blockchain Technologies and Big Data Analytics for IoTs: Methods, technologies and applications** A. Kumar Tyagi, A. Abraham, F. Khadeer Hussain, A. Kaklauskas and R. Jagadeesh Kannan

Physical Layer Security for Wireless Sensing and Communication

Edited by
Hüseyin Arslan and Haji M. Furqan

The Institution of Engineering and Technology

Published by The Institution of Engineering and Technology, London, United Kingdom

The Institution of Engineering and Technology is registered as a Charity in England & Wales (no. 211014) and Scotland (no. SC038698).

First published 2022

The Institution of Engineering and Technology
Futures Place
Kings Way, Stevenage
Hertfordshire, SG1 2UA, United Kingdom

www.theiet.org

British Library Cataloguing in Publication Data
A catalogue record for this product is available from the British Library

ISBN 978-1-83953-527-7 (hardback)
ISBN 978-1-83953-528-4 (PDF)

Typeset in India by MPS Limited
Cover Image: Yuichiro Chino/Moment via Getty Images

Contents

Foreword

The fifth generation (5G) of mobile communication signaled a paradigm shift in wireless networks by introducing a variety of services satisfying different performance requirements rather than concentrating on increasing the achievable data rates. As a result, enhanced mobile broadband (eMBB), massive machine-type communication (mMTC), and ultra-reliable low-latency communication (uRLLC) services were introduced to support a large number of devices and critical applications. Sixth-generation (6G) cellular communication standards intend to broaden this variety of applications and use cases even further, laying the groundwork for a human-centric digital society. This vision of the digital society encompasses work, healthcare, education, industry, entertainment, banking, and transportation – of which communication is just one (albeit critical) part.

Given the diverse use cases in next-generation networks, it is clear that wireless systems will not only be used for communication but also to acquire information about the environment and/or users using sensing techniques to improve the overall network performance. However, the very fact that radio signals can be used to gain environmental awareness renders them prone to being eavesdropped on, manipulated, jammed, and exploited by malicious nodes/entities. Although commonly utilized, conventional cryptographic security is not designed for networks with the diversity of capabilities as seen in the current (or future) wireless systems since it requires certain computational capabilities at all nodes. As such, these methods are unable to scale well with the decentralized and heterogeneous networks that are becoming more prevalent. On the other hand, physical layer security (PLS) offers a potential supplementary solution to guarantee the confidentiality, integrity, authenticity, and availability of communications by leveraging the dynamic aspects of the wireless environment. This book focuses on core concepts related to PLS and its application with respect to future wireless communication networks. The summary of each chapter is as follows:

- **Chapter 1** starts off the book by envisaging what the next-generation networks could look like and highlights the need for secure wireless transmissions. A general overview of cryptography, its limitations, and the ability of PLS to solve these issues is then presented, thereby laying down the foundation for more detailed PLS discussions in the forthcoming chapters.
- **Chapter 2** introduces information theory, followed by the basics of PLS via the established information theory framework. Then, a classification of developed PLS techniques considering main approaches and attacks is presented. This

chapter aims to be a technical introduction to PLS, its techniques, and performance evaluation approaches.

- **Chapter 3** expands the scope of PLS to safeguarding the entire wireless environment. A generalized framework that encompasses the different PLS approaches is introduced. This framework splits the PLS fabric into observation and modification planes. In line with this, different domains of PLS are discussed under these plane concepts, enabling a vision of future PLS mechanisms for next-generation wireless systems.
- **Chapter 4** revisits the role of the wireless channel on PLS. The chapter first delves into the basic channel exploitation approaches for PLS. Owing to the fact that not every channel feature can be suitable for PLS applications, the chapter establishes and discusses the main criteria that must be considered while selecting a certain channel feature for PLS. These criteria are established by considering not only the traditional eavesdropping attack but also other security threats such as spoofing, impersonation, and jamming attacks. Afterward, the chapter explores new channel features that have been observed in the recent 5G and expected in the future generations of networks from a PLS perspective and highlights research directions.
- **Chapter 5** presents the analog domain PLS mechanisms in wireless communication networks based on the transceivers' hardware, known as physical layer authentication (PLA). First, PLA and its metrics are introduced. Then, the PLA techniques using radio frequency (RF) hardware components are deeply discussed, where PLA attributes are extracted and used jointly or separately to provide an accurate authentication process. Finally, new unique attributes are exploited for PLA in 5G communication systems including beam pattern and channel sparsity. To this end, some critical discussion is raised to explore the research challenges of PLA mechanisms for next-generation wireless networks.
- **Chapter 6** presents an upcoming archetype incorporating context-awareness for PLS. Context or situation awareness indicates a level of knowledge of the underlying phenomenon occurring at a given time and/or location. The variety of wireless devices, their capabilities, and limitations, along with their applications or usage areas render the classical one-size-fits-all approach to security infeasible and inefficient. Wireless sensing and the increasing implementations of artificial intelligence (AI)/machine learning (ML) in wireless networks have facilitated the possibility of tailored security mechanisms based on device capabilities, quality of security (QoSec) requirements, applications, and environments. To this end, the radio environment map concept, an enabler of contextaware security, the context-aware PLS framework, and select implementations from the literature are examined.
- **Chapter 7** presents an overview of the inherent security characteristics provided by different waveforms and the control channel signals against different security attacks. When these two fail or when higher security levels are required, PLS enters the picture. To this end, the signal domain physical modification for PLS, which refers to the procedures that modify the transmitted signal based on the

parameters extracted from the observation plane to ensure secure communication, is then discussed.

- **Chapter 8** continues the examination of the modification plane by discussing the secure resource allocation (RA) and scheduling approaches and their popular methodologies. By carefully allocating the available resources based on the PLS idea, information security can be efficiently achieved. In this context, performance indicators and fundamental RA and scheduling-based PLS optimization challenges are highlighted. Later, the significance of PLS-based scheduling in wireless network downlink transmission is discussed. Finally, some potential issues with PLS-based scheduling and RA are highlighted.
- **Chapter 9** highlights how spatially distributed nodes in the network can be used to thwart eavesdropping, jamming, and spoofing attacks through the modification plane. In this context, the role of relays, base stations (BSs), and reconfigurable intelligent surface (RIS) is considered in cooperative communication, coordinated multipoint (CoMP), and smart radio environment paradigms, respectively.
- **Chapter 10** focuses on the usage of PLS in Internet of things (IoT) networks in an interrogative manner. In order to do this, first, the general structure of IoT networks is introduced, then the unique features of the IoT from the PLS perspective along with the possible challenges on the way are examined. The existing PLS techniques that are exploited for IoT networks are investigated, and the future directions of PLS in IoT networks are specified.
- **Chapter 11** discusses various security challenges incurred by sensing and joint radar and communication (JRC) systems, followed by prospective PLS solutions to counter these vulnerabilities. Starting with a brief introduction of wireless sensing and critical use-cases, this chapter explains exploratory, manipulation, and disruptive attacks on sensing information with relevant PLS techniques to realize secure sensing systems. Afterwards, security threats and possible solutions are delineated for JRC systems including dual-function radar communication (DFRC) and radar communication coexistence (RCC) scenarios.
- **Chapter 12** discusses RF and free-space optical (FSO) communications in non-terrestrial networks (NTNs), a key enabler of ubiquitous connectivity in future generation networks, from a PLS perspective. To this end, first NTNs are introduced and security concerns are highlighted. Then, the secrecy performance of RF eavesdropping in a high altitude platform station (HAPS)-aided satellite communication system, where the HAPS node relays the signal from the low earth orbit (LEO) satellite to the ground station (GS) is investigated. Different system models and numerical results to show the PLS performance of FSO eavesdropping in NTN systems are presented. The chapter is concluded with open challenges and future research areas.
- **Chapter 13** examines PLS for cognitive radio networks (CRNs). CRNs are susceptible to several sorts of physical layer attacks; thus, PLS has lately been employed for investigating and enhancing the security of CRNs. In light of this, this chapter explores the main physical layer attacks that threaten CRN users along with the primary countermeasures against them. Integration of PLS and CRNs with other technologies and applications is also presented.

- **Chapter 14** discusses the utilization of ML techniques in PLS. After providing some background on ML, this chapter highlights selected use cases for PLS, including signal relation-based physical layer authentication (PLA), identification of multiple RF impairments, cognitive radio security, and IoT security. ML is used in these scenarios due to its ability to jointly optimize the different parameters encountered. Following this, major challenges for practical ML approaches are highlighted, such as complexity, theoretical analysis, standard requirements, and cost.
- **Chapter 15** discusses the implementation of quantum technology, which is advancing quickly in communications networks. In particular, Quantum Key Distribution (QKD) and Quantum Random Number Generation (QRNG) have become key applications for cryptography and physical layer security. In QKD, security is guaranteed by the rules of physics and is, in theory, impenetrable by an interceptor, in contrast to classical cryptography and network security. To ensure system security, unusual quantum physics postulates and features like superposition, entanglement, and quantum measurement uncertainty are specifically utilized. The laws of quantum physics also ensure that QRNG generates real randomness which is unpredictability based on historical data and uniform statistical distribution. In this study, we review prominent aspects of QKD and QRNG, related standards, protocols, and hardware components. In addition, a thorough analysis of the current QKD protocols and QRNG techniques is also provided.

About the editors

Hüseyin Arslan is a professor at Istanbul Medipol University, Turkey, where he is the Dean of the School of Engineering and Natural Sciences. He has also worked as a part-time consultant for various companies and institutions including Anritsu Company and The Scientific and Technological Research Council of Turkey. Hüseyin conducts research in wireless systems with an emphasis on the physical and medium access layers of communications. His current research interests cover 5G and beyond radio access technologies, physical layer security, interference management (avoidance, awareness, and cancellation), cognitive radio, multi-carrier wireless technologies (beyond OFDM), dynamic spectrum access, co-existence issues, non-terrestrial communications (high altitude platforms), joint radar (sensing) and communication designs. He has been collaborating extensively with key national and international industrial partners and his research has generated significant interest in companies such as InterDigital, Anritsu, NTT DoCoMo, Raytheon, Honeywell, and Keysight technologies. In addition to his research activities, he has also contributed to wireless communication education and developed a number of courses at the University of South Florida including a unique Wireless Systems Laboratory course (funded by the National Science Foundation and Keysight technologies) where he taught the theory and practical aspects of wireless communication system with the most contemporary test and measurement equipment. He has served as general chair, technical program committee chair, session and symposium organizer, workshop chair, and technical program committee member for several IEEE conferences. He is a member of the editorial board for the *IEEE Communications Surveys and Tutorials* and the *IEEE Sensors Journal*. He has also served as a member of the editorial board for *IEEE Transactions on Communications, IEEE Transactions on Cognitive Communications and Networking (TCCN),* and other scholarly journals. He holds a Ph.D. degree in electrical engineering from the Southern Methodist University (SMU), Dallas, TX, USA.

Haji M. Furqan is currently a post-doctoral researcher in the Communications, Signal Processing, and Networking Center (CoSiNC) at Istanbul Medipol University, Turkey. He received his B.E. and M.Sc. degrees in electrical engineering from COMSATS Institute of Information Technology (CIIT), Islamabad, Pakistan in 2012 and 2014, respectively. He received his Ph.D. from Istanbul Medipol University, Turkey. His research focuses on physical layer security, REM security, context aware security, cognitive security, MIMO, OFDM, V2X, and 5G and beyond technologies.

List of acronyms

1D	one-dimensional
3GPP	Third-Generation Partnership Project
4G	fourth generation
5G	fifth generation
5GB	fifth generation and beyond
5G-NR	5G New Radio
6G	sixth generation
ABC	ambient backscatter communication
ACK	acknowledgment
ADC	analog-to-digital converter
ADoA	angle difference of arrival
AES	Advanced Encryption Standard
AF	amplify-and-forward
AFE	analog front-end
AI	artificial intelligence
AMZI	Asymmetric Mach Zender interforemeter
AN	artificial noise
AoA	angle of arrival
AoD	angle of departure
AP	access point
ARQ	automatic repeat request
AUROC	area under the ROC
AWGN	additive white Gaussian noise
B5G	beyond 5G
B92	Bennett 1992
BB84	Bennett Brassard 1984
BBM92	Bennett Brassard Mermin 1992
BBU	baseband unit
BCQI	best CQI
BER	bit error rate
BLER	block error rate
BlueFMCW	blue frequency-modulated continuous-wave
BS	base station
CB	coordinated beamforming
CCSK	cyclic code shift key
CDD	cyclic delay diversity

CDF	cumulative distribution function
CDMA	code division multiple access
CE	central entity
CF	compress-and-forward
CFO	carrier frequency offset
CFR	channel frequency response
CIR	channel impulse response
CJ	cooperative jamming
CL	centralized learning
CNN	convolutional neural network
CoMP	coordinated multipoint
COW	coherent one way
CP	cyclic prefix
CPNN	conventional prepossessing neural network
CPU	central processing unit
CQI	channel quality indicator
CR	cognitive radio
CRN	cognitive radio network
CS	coordinated scheduling
CSI	channel state information
CSS	chirp spread spectrum
CTF	compute-and-forward
CU	communication user
CVQKD	continuous-variable QKD
D2D	device-to-device
DAC	digital-to-analog converter
DARPA	Defense Advanced Research Projects Agency
DAS	distributed antenna system
DC	difference convex
DCTF	differential constellation trace figure
DD	delay diversity
DDoS	distributed denial of service
DE	dispersion entropy
DES	Data Encryption Standard
DF	decode-and-forward
DFRC	dual function radar communication
DFT	discrete Fourier transform
DI	device independent
DL	deep learning
DMCS	discrete modulation coherent state
DNN	deep neural network
DoD	Doppler delay diversity
DoF	degree of freedom
DoS	denial of service

DPS	dynamic point selection
DSRC	dedicated short-range communication
DSS	Digital Signature Standard
DSSS	direct sequence spread spectrum
DUT	device under test
DVQKD	discrete-variable QKD
E91	Ekert Protocol
ECC	elliptic-curve cryptography
EE	energy efficiency
eICIC	enhanced inter-cell interference coordination
ELPC	extremely low-power communication
EM	electromagnetic
eMBB	enhanced mobile broadband
eNB	evolved NodeB
ERLLC	extremely reliable and low-latency communication
ESC	ergodic secrecy capacity
ETSI	European Telecommunications Standards Institute
FBMC	filter bank multi carrier
FC	fusion center
FDD	frequency division duplex
FEC	forward error correction
FeMBB	further-enhanced mobile broadband
FFT	fast Fourier transform
FH-MIMO	frequency hopping multiple-input and multiple-output
FHSS	frequency hopping spread spectrum
FIR	finite impulse response
FL	federated learning
FMCW	frequency-modulated continuous-wave
FSO	free-space optical
GAN	generative adversarial network
GFDM	generalized frequency division multiplexing
GG02	Grosshans and Grangier Protocol 2002
GMCS	Gaussian modulated coherent state
gNB	gNodeB
GNSS	global-navigation satellites system
GPS	global positioning system
GPU	graphic processing unit
GSM	Global System for Mobile Communications
HAPS	high altitude platform station
HARQ	hybrid automatic repeat request
HetNet	heterogeneous network
HST	high-speed train
I	in-phase
IA	interference alignment

ICI	inter-cell interference
ICIC	inter-cell interference coordination
IDS	intrusion detection system
IEEE	Institute of Electrical and Electronics Engineers
IFFT	inverse fast Fourier transform
i.i.d.	independent and identically distributed
IoT	Internet of Things
IQI	in-phase/quadrature imbalance
ISAC	integrated sensing and communication
ISI	intersymbol interference
ISO	International Organization for Standardization
ITU	International Telecommunication Union
JRC	joint radar and communication
JT	joint transmission
KDR	key disagreement rate
KGR	key generation rate
KPI	key performance indicator
LAN	local area network
LDHMC	long-distance and high-mobility communication
LDPC	low-density parity-check
LEO	low earth orbit
LiDAr	light detection and ranging
LoRa	long-range
LoS	line-of-sight
LPI	low probability of intercept
LSTM	long–short term memory
LTE	long-term evolution
LTE-A	LTE-Advanced
M2M	machine to machine
MA	multiple access
MAC	medium access control
MAQ	multi-bit adaptive quantization
MBM	media-based modulation
MCS	modulation and coding scheme
MDI	measurement device independent
MDI-CVQKD	continuous variable measurement device independent QKD
MDI-QKD	measurement device independent QKD
MIMO	multiple-input multiple-output
MISO	multiple-input single-output
MITM	man in the middle
ML	machine learning
MMFN	multi-module fusion neural network
mMIMO	massive multiple-input multiple-output
mMTC	massive machine-type communication

mmWave	millimeter wave
MoA	molecular absorption
MoS	molecular scattering
MPC	multipath component
MRT	maximum ratio transmission
MTL	multi-task learning
MUI	multi-user interference
NACK	negative acknowledgment
NIST	National Institute of Standards and Technology
NLoS	non-line-of-sight
NN	neural network
NOMA	non-orthogonal multiple access
NOMA-IoT	non-orthogonal multiple access-Internet of things
NP	non-deterministic polynomial
NR	New Radio
NTN	non-terrestrial network
OFDMA	orthogonal frequency-division multiple access
OFDM	orthogonal frequency division multiplexing
OOBE	out of band emission
OSI	open systems interconnection
PA	power allocation
PAPR	peak-to-average power ratio
PBCH	broadcast channel
PD	probability of detection
PDCCH	downlink control channel
PDF	probability density function
PDMA	path division multiple access
PDP	power delay profile
PDSCH	downlink shared channel
PE	permutation entropy
PER	packet error rate
PF	proportional fair
PFA	probability of false alarm
PHY	physical
PL	pathloss
PLA	physical layer authentication
PLL	phase-locked loop
PLS	physical layer security
PMD	probability of missed detection
PMI	precoding matrix indicator
PN	pseudo-random
PNS	photon number splitting
PRACH	random access channel
PRNG	pseudo random numbers generator

PSD	power spectral density
PSS	primary synchronization signal
PU	primary user
PUCCH	uplink control channel
PUEA	primary user emulation attack
PUSCH	uplink shared channel
Q	quadrature
QAM	quadrature amplitude modulation
QBER	quantum bit error rate
QED	quantum electrodynamics
QKD	quantum key distribution
QKDN	quantum key distribution networks
QoS	quality of service
QoSec	quality of security
QPSK	quadrature phase shift keying
QRNG	quantum random number generators
RA	resource allocation
RAN	radio access network
RB	resource block
RCC	radar communication coexistence
RCI	regularized channel inversion
RE	resource element
REM	radio environment map
REMo	radio environment monitoring
RF	radio frequency
RFID	radio frequency identification
RI	rank indicator
RIS	reconfigurable intelligent surface
RL	reinforcement learning
RN	relay node
RNG	random number generator
RNN	recurrent neural network
ROC	receiver operating characteristic
RR	round robin
RRDPS	round robin differential phase shift
RRH	remote radio head
RRM	radio resource management
RSA	Rivest Shamir Adleman
RSS	received signal strength
RSSI	received signal strength indicator
RSU	road side unit
RTI	radio tomographic imaging
RTT	round trip time
SARG04	Scarani Acn Ribordy Gisin 2004

SC	secrecy capacity
SC-FDMA	single carrier frequency division multiple access
SDO	Standards Developing Organizations
SDR	software defined radio
SE	spectrum efficiency
SECOQC	Secure Communication based on Quantum Cryptography
SHA	Secure Hash Algorithm
SIMO	single-input multiple-output
SINR	signal-to-interference-plus-noise ratio
SIR	signal-to-interference ratio
SKG	security key generation
SL	supervised learning
SLS	sector level sweep
SMS	short message service
SNR	signal-to-noise ratio
SON	self-organizing network
SOP	secrecy outage probability
SPAD	single photon avalanche photo diodes
SPD	single photon detectors
SPDC	spontaneous parametric down conversion
SR	secrecy rate
SS	spread spectrum
SSS	secondary synchronization signal
STAP	space–time adaptive processing
SU	secondary user
SVD	singular value decomposition
SVM	support vector machine
SW	spherical wavefront
TAS	transmit antenna selection
TCP	transmission control protocol
TDD	time division duplex
TDES	Triple Data Encryption Standard
TDoA	time difference of arrival
TF	twin field
TF-QKD	twin field QKD
THz	terahertz
ToA	time of arrival
TP	transmission point
TRNG	true random number generator
TTI	transmission time interval
UAV	unmanned aerial vehicle
UD	unidimensional
UE	user equipment
ULA	uniform linear array

umMTC	ultra-massive machine-type communication
uRLLC	ultra-reliable low-latency communication
USL	unsupervised learning
UWB	ultrawide band
V2V	vehicle-to-vehicle
V2X	vehicle-to-everything
VA	virtual anchor
VANET	vehicular ad hoc network
VLC	visible light communication
VR	visibility region
Wi-Fi	wireless fidelity
WLAN	wireless LAN
WPE	wireless propagation environment
WSN	wireless sensor networks
WSS	wide sense stationary
WSSUS	wide sense stationary uncorrelated scattering
ZF	zero forcing

Chapter 1

Wireless communication networks and the need for security

Muhammad Sohaib J. Solaija[1], *Haji M. Furqan*[1], *and Hüseyin Arslan*[1]

Wireless systems are becoming an increasingly important part of our daily lives. The proper operation of many crucial applications and use cases, including those in healthcare, finance, e-commerce, transportation, industrial automation, etc., depends on secure and reliable communication. Conventional cryptographic security mechanisms are widely used, but they are unable to scale with increasingly heterogeneous and decentralized networks. On the other hand, by utilizing the dynamic features of the wireless environment, physical layer security (PLS) offers a promising complementary solution to ensure the authenticity, confidentiality, integrity, and availability of legitimate communications. In this chapter, we first try to paint a picture of the next-generation networks before emphasizing the need for secure wireless transmission in future networks. Later, we highlight the limitations of cryptographic methods, and the ability of PLS to address them is provided. We conclude the chapter by briefly discussing a recently developed unified framework for PLS in future networks.

1.1 Introduction to next-generation wireless networks

Wireless communication systems have evolved dramatically over the last few decades. This is evident from the last, i.e., fifth-generation (5G) cellular networks where the emphasis has shifted towards connecting *things* rather than just *people*. As a consequence, voice calling and Short Message Service (SMS) have long since been replaced (for the most part) with broadband applications. In fact, according to Ericsson's mobility report [1], it is estimated that by 2027 mobile subscribers would rise to 9.1 billion with 93% (as opposed to 84% today) of them being broadband customers. This is catered to by the enhanced mobile broadband (eMBB) service under the 5G umbrella, which also promises to support ultra-reliable low-latency communication (uRLLC) and massive machine-type communication (mMTC) applications.

[1]Department of Electrical and Electronics Engineering, Istanbul Medipol University, Turkey

This paradigm shift toward connected everything is expected to be more pronounced in the next-generation of wireless networks with sixth generation (6G) and the consequent adoption of services such as further-enhanced mobile broadband (FeMBB), extremely reliable and low-latency communication (ERLLC), long-distance and high-mobility communication (LDHMC), ultra-massive machine-type communication (umMTC), and extremely low-power communication (ELPC) [2,3]. As such, 6G envisions a digital society centered around the various facets of human life including education, banking, trading, entertainment, manufacturing, healthcare, etc. Wireless communication then becomes a critical enabler for such a society since ubiquitous, reliable, fast, robust, and secure connectivity is necessary for supporting the aforementioned parts of society. Figure 1.1 provides a glimpse of how the future networks would look like in terms of deployment scenarios, enabling technologies, and applications.

It follows intuition that in order to satisfy the increasingly diverse user/application requirements, diverse enabling technologies and approaches are needed. As Figure 1.1 illustrates, some of these enablers include incorporation of higher frequency bands, integrated sensing and communication (ISAC), reconfigurable intelligent surface (RIS)-aided smart radio environments, non-terrestrial networks (NTNs), etc. Their heterogeneous nature, in turn, renders the optimization of network operation quite challenging. As [4] points out, this imposes three requirements

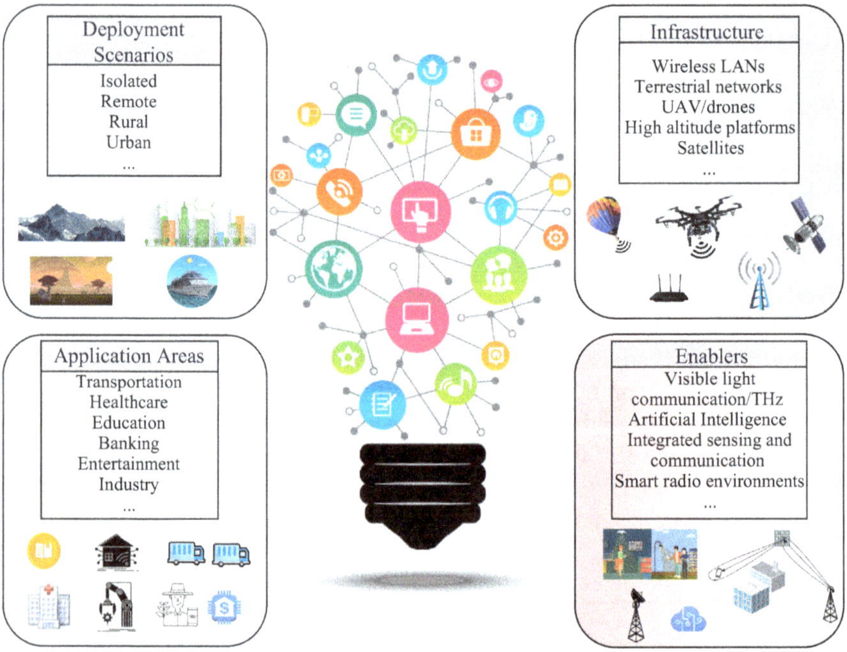

Figure 1.1 A glimpse of the next-generation communication systems capable of supporting the digital society of the future

from the network itself, namely awareness, flexibility, and intelligence. Here awareness refers to the knowledge of the network infrastructure, application requirements, user patterns, and propagation conditions; flexibility* refers to the capability of the network to use its resources (in time, frequency, space, power) in a dynamic manner depending on the instantaneous requirements; and intelligence is the ability to decide on what might be the optimum decision in any given situation.

While the network usually has considerable information about the users' behavioral patterns and available network infrastructure, more information can be extracted about the propagation medium and surroundings using radio signals. As these signals traverse the propagation media, they experience different phenomena such as reflection, refraction, diffusion, scattering, etc. that can be used to increase awareness of the radio environment [5]. The flexibility, on the other hand, is enabled by the introduction of more controllable transmission parameters. For instance, in the case of 5G, multiple numerologies allow the adoption of PHY layer parameters such as subcarrier spacing and cyclic prefix (CP) duration to cater to varying requirements like latency or massive connectivity [6]. Given the (expected) increased heterogeneity of 6G in terms of deployment scenarios, frequency spectrum, application requirements, and device capabilities, the number of transmission parameters (and associated optimization problems) would increase manifold. Consequently, artificial intelligence (AI)/machine learning (ML) would take a more central role in network operations which is also evident from the presence of a significantly large number of study items centered around ML in current Third Generation Partnership Project (3GPP) discussions. The advent of adversarial ML, however, means that the opacity of these black box solutions renders them questionable from a trustworthiness perspective, which is one of the primary goals of next-generation networks [7].

1.2 The need for security in wireless communication

As mentioned earlier, wireless communication plays a pivotal role in significantly diverse applications such as entertainment, trading, health monitoring, autonomous driving, smart grids, public safety, and the military. One could see that not all those applications can be considered equally critical. For instance, entertainment services like video streaming are nowhere near as critical as collision avoidance signaling between autonomous vehicles in a high-speed scenario. Not only do these mission-critical applications require reliable, but also, secure connectivity. Considering the previous example of autonomous driving, any delay or interruption in a collision avoidance command can lead to fatal consequences. A more thorough example of such an attack on a power grid is given in [8], where it is shown that the manipulation of power demand for high-wattage appliances such as air conditioners and heaters

*The definition of flexibility given in the text is primarily from the physical (PHY) layer perspective. However, in 5G and beyond networks the concept can be extended to include paradigms such as network slicing, network function virtualization, etc.

can be leveraged by a malicious entity to increase the running cost of the power grid, cause local outages, or even large-scale blackouts. Another recent example of an attack on Internet of things (IoT) networks is reported in [9], where the attackers successfully disabled more than 1/3 of the radioactivity sensors in Spain, illustrating the vulnerability of the system.

Before diving into the security techniques, it is important to first identify the general categories of security threats and the corresponding security requirements for wireless communication. As Figure 1.2 illustrates, there are three main types of attacks considered in wireless systems. In *eavesdropping* [10], the malicious node or attacker tries to intercept the ongoing transmission between the legitimate transmitter (Alice) and receiver (Bob). Here, it should be noted that eavesdropping is not only limited to an attempt to extract the data being transmitted between the legitimate nodes, but can also be expanded to include other information such as user behavior or network usage patterns. The *jammer* [11], on the other hand, tries to disrupt legitimate wireless transmission. This is generally achieved by transmitting noise-like signals on the same time–frequency resources to degrade the perceived link quality at the legitimate transceivers. A *spoofer* [12] represents the most sophisticated attack, where its aim is to deceive the legitimate nodes into believing that it (spoofer) is also a legitimate device. Once the spoofer establishes itself in the network, it can send erroneous messages to mislead the legitimate receiver.

Given a glimpse of the various types of threats, we can now go and list the requirements to ensure secure communication. Usually, the *CIAA* (confidentiality, integrity, authenticity, availability) quartet is used to represent these requirements [13]. Here confidentiality refers to the protection of data/information

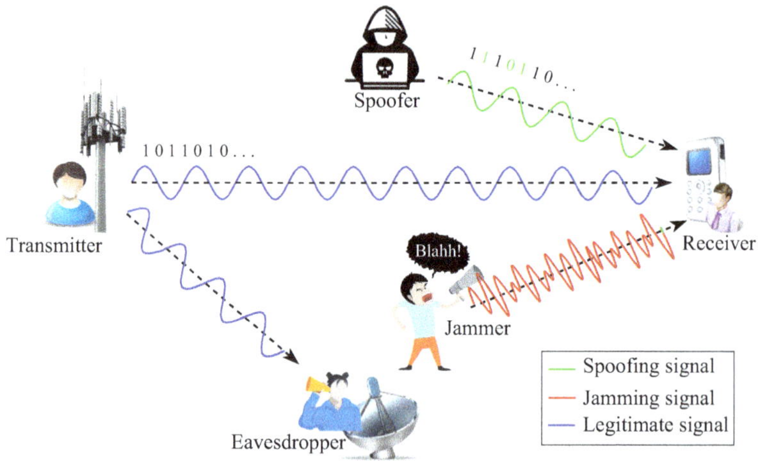

Figure 1.2 Illustration of different security threats at the physical layer

from unauthorized access. These requirements map directly to eavesdropping attacks and are extremely difficult to satisfy owing to the broadcast nature of wireless transmissions. Integrity ensures that the data is free from any manipulation/modification, while authenticity means that the transceiver's identity is secure and no impersonation occurs. Both integrity and authenticity are, therefore, primarily associated with spoofing attacks. Availability refers to the ability of the network to provide connectivity to the devices and allocate resources that are sufficient to fulfill the users' requirements. As such, availability is closely associated with jamming attacks mentioned earlier.

Having looked at the main threats and security requirements in wireless systems, it seems an opportune moment to take a look at the major approaches utilized for securing today's wireless systems.

1.3 Cryptography vs. physical layer security

The popular techniques for securing wireless communication networks include *cryptography* and *PLS*. Cryptography focuses on securing the contents of a communication while PLS tries to secure the physical signal along with the wireless link through which that signal traverses.

1.3.1 Cryptography

Cryptography-based security approaches ensure the privacy of communication by converting the *plaintext* (message) to *ciphertext* using different encryption mechanisms [14]. The commonly used cryptography based approaches include *hashing*, *symmetric*, and *asymmetric* encryption.

The main purpose of hashing is to verify the shared message or data's integrity. To make sure that the transmitted data has not been tampered with, it uses a one-way function, which is simple to calculate in one direction but very challenging to reverse. In this scenario, the legitimate transmitter shares the hash function output with the data, and the receiver then applies the same hash function to the received data. The message is deemed to be authentic if the hash function's output matches that of the original. Making sure the hash function and its outcome are communicated securely between nodes is one of the major challenges of hashing. Some of the most popular hashing algorithms include Secure Hash Algorithm (SHA)-256 and SHA3-256 which convert any input to 256-bit output. Symmetric cryptography, also known as private-key cryptography, encrypts and decrypts messages using the same private key. This technique, like hashing, relies on the safe transmission of keys between authorized nodes. Popular symmetric encryption methods include Data Encryption Standard (DES), Advanced Encryption Standard (AES), and Triple Data Encryption Standard (TDES). In asymmetric (public-key) cryptography, the key is effectively divided into two parts. The receiver broadcasts or shares its public key, and any node wishing to communicate uses this key to encrypt the message. The receiver's private key is used to perform the decryption. The biggest benefit of this method is that no secure key exchange mechanism is required [15]. The use of one-way functions in public-key cryptography makes it challenging to decrypt data without the knowledge of the private key. The Rivest–Shamir–Adleman (RSA) algorithm, which makes use of the difficulty of prime factorization to secure the shared messages, is one of the most notable realizations of asymmetric cryptography while other examples include Digital Signature Standard (DSS) and elliptic-curve cryptography (ECC) [16].

It is clear from the previous discussion that cryptography-based approaches (hashing, symmetric encryption, and asymmetric encryption) can provide security solutions for wireless communication systems. However, key sharing and management (in the case of hashing and symmetric keys) are quite challenging in heterogeneous networks, applications, and devices in upcoming generations due to the requirement of trustable third-party-based key sharing. Moreover, the continuously changing network topology in high-mobility applications such as vehicle-to-everything (V2X) communications, NTNs, and high-speed trains require renewed key management and authentication procedures [17], which is challenging to achieve practically. Although, asymmetric cryptography can provide security without a third party, it requires considerable computations which make it unsuitable for computation-limited and power-constrained IoT and mMTC devices [18]. Additionally, some of the conventional cryptographic methods are too time-consuming to support latency-sensitive applications such as uRLLC/ERLLC applications [19]. Finally, the assumption of cryptographic-based algorithms in terms of the computational complexity of key breaking is rendered a naive assumption with the advent of quantum computing [20].

The above-mentioned challenges faced by conventional cryptography-based algorithms necessitate an alternative solution, which is relatively lightweight in terms of complexity and scalable with the heterogeneity of future networks. In this regard, PLS can be considered as that alternative solution and will be introduced in the next subsection.

1.3.2 Physical layer security

The goal behind PLS is to provide proven and quantifiable secrecy based on information theory by leveraging properties of the wireless channel, propagation environment, and the communicating devices [21]. In practice, confidentiality is attained by ensuring a better quality of the legitimate link compared to that of the eavesdropper's [22–24]. Apart from confidentiality, properties of the wireless links are also useful in ensuring the authenticity of the communicating nodes [25,26]. Similarly, hardware properties of the transceivers [27] and other physical features of the environment such as the angle/distance between devices [28] have also been shown to be effective authentication tools.

Figure 1.3 illustrates certain scenarios which are challenging for cryptography but can be catered to by PLS approaches. For example, unlike cryptography which requires a trustable third party for key sharing, PLS enables communicating nodes to extract keys from the common propagation channel thus eliminating the need for trustable third party-based key sharing [29]. Moreover, PLS also supports

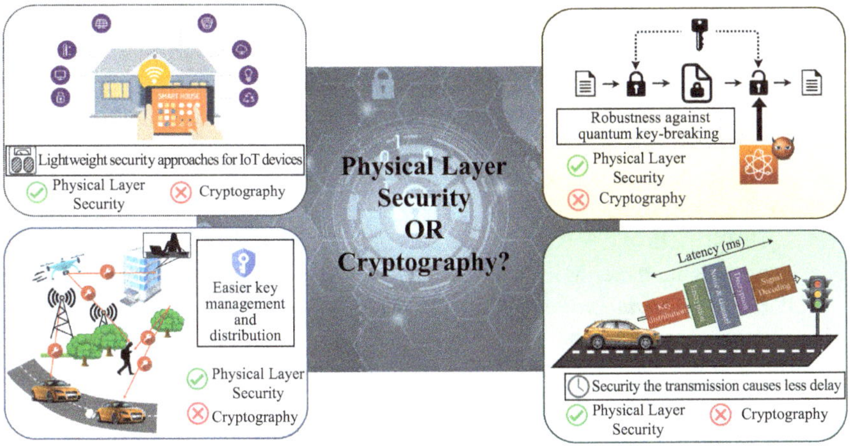

Figure 1.3 Illustration of selected communication scenarios where PLS can complement cryptographic approaches

asymmetrical security mechanisms that allow to shift the complexity towards base station (BS)/access point (AP) side. This makes it suitable for IoT terminals [30]. Also, compared to cryptographic approaches, PLS boasts reduced latency since there is no key exchange. Moreover, depending on the specific technique, encryption and decryption processes may not be necessary [31]. Finally, PLS is immune to being broken by quantum computing due to its dependence on the physical laws of the wireless propagation characteristics instead of relying on the computational complexity of key breaking.

Another advantage of PLS over cryptographic methods is its ability to provide adaptive security. This is particularly relevant for next-generation wireless networks where not only the user/application requirements but also the capabilities of network entities vary significantly. As such, inspired by quality of service (QoS), a similar metric for security, i.e., quality of security (QoSec) is being discussed. QoSec is targeted to enable the identification of the required security level and then provide adaptive, dynamic, and situation-aware security mechanisms [32]. Such a cognitive PLS approach was initially proposed in [33], adapted for V2X communication in [17], with [34] providing a general framework.

In essence, PLS can be said to have two major aspects [35]. First, it involves techniques that rely on the observation of wireless parameters. Chapters 4–6 will look in more detail at the different domains that can provide useful parameters for PLS including channel, hardware of the transceivers, and the radio environment. The second aspect of PLS revolves around the design of wireless transmission specifically aimed at improving security. Chapters 7–9 detail how these transmissions can be adapted in terms of signal design, scheduling, or network coordination to achieve this goal.

References

[1] Jonsson P, Lundvall A, Möller R, *et al.* Ericsson Mobility Report; Accessed July 24, 2022. Available from: https://www.ericsson.com/49d3a0/assets/local/reports-papers/mobility-report/documents/2022/ericsson-mobility-report-june-2022.pdf.

[2] Zhang Z, Xiao Y, Ma Z, *et al.* 6G wireless networks: vision, requirements, architecture, and key technologies. *IEEE Vehicular Technology Magazine*. 2019;14(3):28–41.

[3] Tariq F, Khandaker MR, Wong KK, *et al.* A speculative study on 6G. *IEEE Wireless Communications*. 2020;27(4):118–125.

[4] Yazar A, Dogan-Tusha S, Arslan H. 6G vision: an ultra-flexible perspective. *ITU Journal on Future and Evolving Technologies*. 2020;1(1):121–140.

[5] Türkmen H, Solaija MSJ, Tusha A, *et al.* Wireless sensing – enabler of future wireless technologies. *Turkish Journal of Electrical Engineering & Computer Sciences*. 2021;29(1):1–17.

[6] Zaidi AA, Baldemair R, Tullberg H, *et al.* Waveform and numerology to support 5G services and requirements. *IEEE Communications Magazine*. 2016;54(11):90–98.

[7] Ziegler V, Schneider P, Viswanathan H, *et al.* Security and trust in the 6G era. *IEEE Access*. 2021;9:142314–142327.

[8] Soltan S, Mittal P, Poor HV. BlackIoT: IoT botnet of high wattage devices can disrupt the power grid. In: *27th USENIX Security Symposium*, 2018. p. 15–32.

[9] Bradburry D. Spanish Police Arrest Alleged Radioactive Monitoring Hackers. Accessed August 14, 2022. Available from: https://www.infosecurity-magazine.com/news/spanish-police-arrestradioactive/.

[10] Hamamreh JM, Furqan HM, Arslan H. Classifications and applications of physical layer security techniques for confidentiality: a comprehensive survey. *IEEE Communications Surveys & Tutorials*. 2018;21(2):1773–1828.

[11] Grover K, Lim A, Yang Q. Jamming and anti-jamming techniques in wireless networks: a survey. *International Journal of Ad Hoc and Ubiquitous Computing*. 2014;17(4):197–215.

[12] Yılmaz MH, Arslan H. A survey: Spoofing attacks in physical layer security. In: *40th Local Computer Networks Conference Workshops (LCN Workshops)*. New York, NY: IEEE; 2015. p. 812–817.

[13] Carsten Reffgen. Protection Goals: CIA and CIAA. Accessed July 26, 2022. Available from: https://eosgmbh.com/en/protection-goals-cia-and-ciaa.

[14] Rivest RL. Cryptography. In: Van Leeuwen J, Leeuwen J, editors. *Handbook of Theoretical Computer Science*, vol. A. New York, NY: Elsevier; 1990. p. 718–755.

[15] Hellman ME, Diffie BW, Merkle RC. Cryptographic apparatus and method. Google Patents; 1980. US Patent 4,200,770.

[16] Chandra S, Paira S, Alam SS, *et al.* A comparative survey of symmetric and asymmetric key cryptography. In: *International Conference on Electronics,*

Communication and Computational Engineering (ICECCE). New York, NY: IEEE; 2014. p. 83–93.

[17] Furqan HM, Solaija MSJ, Hamamreh JM, *et al.* Intelligent physical layer security approach for V2X communication. arXiv preprint arXiv:190505075. 2019; p. 1–7.

[18] Qi Q, Chen X, Zhong C, *et al.* Physical layer security for massive access in cellular Internet of Things. *Science China Information Sciences*. 2020;63(2):1–12.

[19] Li C, Li CP, Hosseini K, *et al.* 5G-based systems design for tactile Internet. *Proceedings of the IEEE*. 2018;107(2):307–324.

[20] Mavroeidis V, Vishi K, Zych MD, *et al.* The impact of quantum computing on present cryptography. arXiv preprint arXiv:180400200. 2018;p. 1–10.

[21] Shakiba-Herfeh M, Chorti A, Poor HV. Physical layer security: authentication, integrity, and confidentiality. In: *Physical Layer Security*. New York, NY: Springer; 2021. p. 129–150.

[22] Schaefer RF, Boche H, Poor HV. Secure communication under channel uncertainty and adversarial attacks. *Proceedings of the IEEE*. 2015;103(10): 1796–1813.

[23] Zou Y, Zhu J, Wang X, *et al.* A survey on wireless security: Technical challenges, recent advances, and future trends. *Proceedings of the IEEE*. 2016;104(9):1727–1765.

[24] Wu Y, Khisti A, Xiao C, *et al.* A survey of physical layer security techniques for 5G wireless networks and challenges ahead. *IEEE Journal on Selected Areas in Communications*. 2018;36(4):679–695.

[25] Liu Y, Chen HH, Wang L. Physical layer security for next generation wireless networks: theories, technologies, and challenges. *IEEE Communications Surveys & Tutorials*. 2016;19(1):347–376.

[26] Sun L, Du Q. Physical layer security with its applications in 5G networks: a review. *China Communications*. 2017;14(12):1–14.

[27] Wang W, Sun Z, Piao S, *et al.* Wireless physical-layer identification: modeling and validation. *IEEE Transactions on Information Forensics and Security*. 2016;11(9):2091–2106.

[28] Gungor O, Chen F, Koksal CE. Secret key generation via localization and mobility. *IEEE Transactions on Vehicular Technology*. 2014;64(6): 2214–2230.

[29] Zeng K. Physical layer key generation in wireless networks: challenges and opportunities. *IEEE Communications Magazine*. 2015;53(6):33–39.

[30] Wang N, Wang P, Alipour-Fanid A, *et al.* Physical-layer security of 5G wireless networks for IoT: challenges and opportunities. *IEEE Internet of Things Journal*. 2019;6(5):8169–8181.

[31] Chen R, Li C, Yan S, *et al.* Physical layer security for ultra-reliable and low-latency communications. *IEEE Wireless Communications*. 2019;26(5):6–11.

[32] Chorti A, Barreto AN, Köpsell S, *et al.* Context-aware security for 6G wireless: the role of physical layer security. *IEEE Communications Standards Magazine*. 2022;6(1):102–108.

[33] Yılmaz MH, Güvenkaya E, Furqan HM, *et al.* Cognitive security of wireless communication systems in the physical layer. *Wireless Communications and Mobile Computing*. 2017;2017.

[34] Furqan HM, Solaija MSJ, Türkmen H, *et al.* Wireless communication, sensing, and REM: a security perspective. *IEEE Open Journal of the Communications Society*. 2021;2:287–321.

[35] Solaija MSJ, Salman H, Arslan H. Towards a unified framework for physical layer security in 5G and beyond networks. *IEEE Open Journal of Vehicular Technology*. 2022;3:321–343.

Chapter 2

Information theoretic perspective of physical layer security

Talha Yılmaz[1] *and Hüseyin Arslan*[1]

Different perspectives have been formed on the security concept in communication technologies. As a result, academic researchers and experts in the field have developed many different approaches to data security. The dominant approach for data security has been cryptography for many years. However, recently physical layer security (PLS) research gathered enormous attention from the researchers. Even though mentioned approaches look quite different, they share the same fundamentals and philosophy. In this regard, this chapter first introduces readers to the history of security concepts in communication systems. Next, the information theory-based philosophy of secrecy is established. Finally, performance metrics that can be used to evaluate a security system are given to the readers. This chapter aims to explain the overall idea of PLS systems and form a handbook for researchers, engineers, and experts to consult in their research.

2.1 History of information theory

Before getting introduced to the security concept, one needs to understand the basis of information theory as it forms the foundation of data security. In 1948, Claude E. Shannon established the information theory in [1]. In his work, he integrated probability theory, statistics, computer science, and electrical engineering to study digital information's quantification, storage, and transmission. Shannon's key observation was that the amount of information obtained from a particular outcome is directly related to the level of surprise the outcome had. Let X be a discrete random variable with the set of N possible outcomes $\mathscr{X} = \{x_1, \ldots, x_n\}$. Then self-information, $I_X(x)$ for a certain event x is defined by

$$I_X(x) = -\log(Pr(x)). \tag{2.1}$$

[1]Department of Electrical and Electronics Engineering, Istanbul Medipol University, Turkey

Note that this chapter is solely about digital communications. Thus, log refers to the base-2 logarithm, and $Pr()$ is defined as the probability of an event. Self-information quantifies the amount of information a certain event yields. As the likelihood of an event decreases, self-information for the same event increases. Using self-information, Shannon introduced entropy in [1]. Entropy quantifies the expected information content of a measurement from a random variable. Entropy ($H(X)$) is defined as expected value of the self-information for all $x \in \mathscr{X}$

$$H(X) = E[I_X(x)] = -\sum_{x \in \mathscr{X}} p(x) \log(p(x)), \tag{2.2}$$

where $p(x)$ is the probability mass function for the discrete random variable X. Using the probability theory, $H(X)$ can be expanded into joint and conditional versions to analyze complicated scenarios further. Definition for joint entropy is given by

$$H(X, Y) = E_{X,Y}[-\log(p(x, y))] = -\sum_{x \in \mathscr{X}, y \in \mathscr{Y}} p(x, y) \log(x, y), \tag{2.3}$$

where $p(x, y)$ is joint probability mass function for the discrete random variables X and Y, and $\mathscr{Y} = \{y_1, \ldots, y_m\}$. Conditional entropy (equivocation) is the entropy of a random variable X given a random variable Y. Without vigorous proof, conditional entropy can be defined and obtained as

$$H(X|Y) = -\sum_{x \in \mathscr{X}} p(x) H(X|Y = y) \tag{2.4}$$

$$H(X|Y) = -\sum_{x \in \mathscr{X}, y \in \mathscr{Y}} p(x, y) \log\left(\frac{p(x, y)}{p(y)}\right) \tag{2.5}$$

Another fundamental quantity information theory introduced for communication theory is mutual information. Mutual information ($I(X; Y)$) can be defined as

$$I(X; Y) = \sum_{x \in \mathscr{X}, y \in \mathscr{Y}} p(x, y) \log\left(\frac{p(x, y)}{p(x)p(y)}\right) \tag{2.6}$$

$$I(X; Y) = H(X) - H(X|Y) \tag{2.7}$$

In essence, mutual information measures the amount of information a random variable carries about another variable. In other words, $I(X; Y)$ quantifies the amount of information obtained from Y about X. Mutual information is a beneficial quantity for obtaining a limit to a communication system. First let us consider the communication system given in Figure 2.1. In this system, a message (M) is encoded ($\hat{X}$) for transmission through a channel. The decoder receives $\hat{Y}$, which is the deformed version of $\hat{X}$ due to various channel phenomena (noise, fading, interference, etc.).

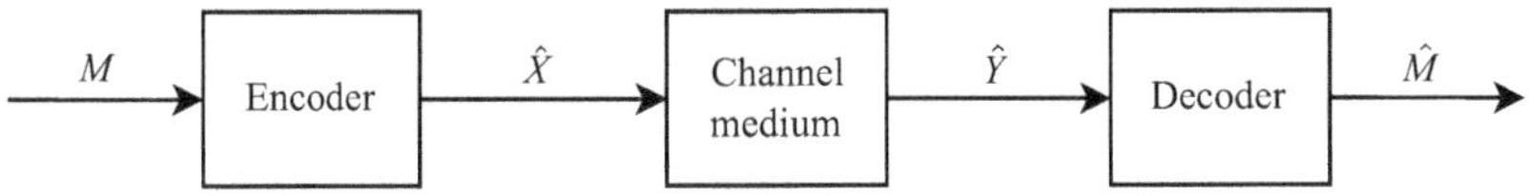

Figure 2.1 A basic communication system

For a given communication system like Figure 2.1, Shannon has concluded that the channel is the fundamental limiting factor for the amount of information that can be reliably transmitted. Hence, channel capacity, which is the maximum amount of information that can be reliably transmitted over a given channel, can be obtained as

$$C = \max_{p(x)} I(\hat{X}; \hat{Y}), \tag{2.8}$$

which is the maximization of mutual information over all possible input distributions $p(x)$ of $\hat{X}$.

2.2 Fundamentals of security and security notions

Shannon again proposed the first theoretic approach for data security. He proposed the first notion of secrecy using the information theory he established. It is more appropriate to understand his objective and system model of Shannon before explaining his work. In [2] Shannon considered a system as shown in Figure 2.2. There are three different nodes in Shannon's system. The legitimate transmitter (Alice) encodes the message (M) and sends it to the legitimate receiver (Bob). The illegitimate receiver (Eve) tries to retrieve the message (M) by intercepting X. However, Alice encoded her message using a secret key (K) that is assumed to be shared with Bob, while Eve had no idea about the key. Shannon tried to answer the question "*Is transmission without information leakage to Eve possible between Alice and Bob?*". Ref. [2] has brought the first security notion, called perfect secrecy. Answer the Shannon found in [2] includes both good and bad news. The good news was that it was possible to develop a ciphering system that guarantees secure transmission between Alice and Bob, given that Eve does not have access to the secret key. Or mathematically, it is possible to develop a system such that

$$I(M; X) = 0 \implies H(M) = H(M|X) \tag{2.9}$$

By utilizing properties of entropy we can show that (2.9) also implies the following inequality

$$H(K) \geq H(M) \tag{2.10}$$

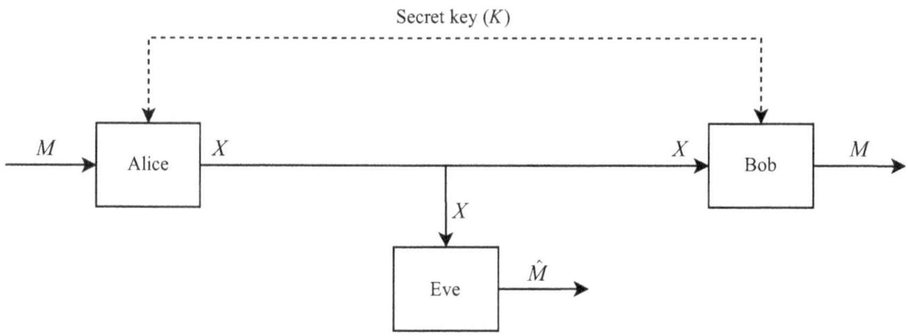

Figure 2.2 Shannon's cipher system

In other words, according to Shannon, by his cipher system, the **perfect secrecy** is obtainable if and only if the entropy of the key ($H(K)$) is at least the source's entropy ($H(M)$). This result suggests that a secret key with as much information as the source message has to be shared between Alice and Bob to obtain perfect secrecy. In this regard, key sharing is not plausible considering the volume of data in today's communication system. However, this constraint does not imply that obtaining perfect secrecy with Shannon's cipher system is impractical. For example, a system may prioritize secrecy over capacity-related performance metrics and would rather have perfect secrecy over high throughput. Applications like banking and healthcare are great examples that prioritize secrecy. Modern cryptography is built on Shannon's cipher system. However, modern encryption systems utilize keys that do not satisfy (2.10). The main idea of modern cryptography is to provide security through computationally difficult problems to solve. However, it is theoretically possible to obtain plain text from cipher-text by solving these problems. Today, with the rise of quantum computing technology, this theoretical drawback has started to oppose a significant challenge to cryptography techniques. On top of that, developing a secure key exchanging and storing mechanism is an inherent challenge for this secrecy approach.

Motivated by the drawbacks mentioned for cryptography-based methods, PLS has attracted many researchers to provide solutions to secrecy by exploiting the physical properties of communication channel medium. Fundamental idea of PLS is first encountered in [3]. In [3], Wyner has introduced his wiretap channel for a different approach to secrecy. To compare with Shannon's approach to secrecy, let us examine Wyner's system model and objective. Figure 2.3 shows Wyner's wiretap channel. Instead of sharing a secret key, Wyner considers message (M) to be encoded (X) by Alice and sent through a channel medium. The philosophy of this modeling is that channel medium inherits randomness. Therefore, the channel medium Bob differs from the Eve's channel. Under these circumstances, Wyner tries to answer the following question "*Is it possible for Alice to encode her message so that Bob can reliably obtain the message while Eve cannot?*" Wyner has proved that it is possible to obtain secrecy given that Eve's received signal is more *degraded* than Bob. In

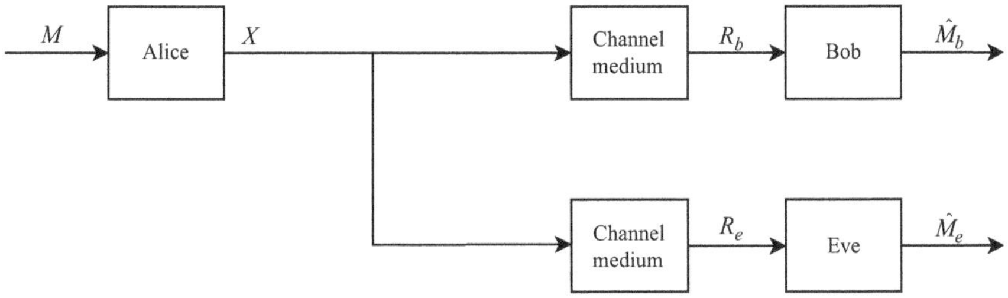

Figure 2.3 Wyner's wiretap channel

other words, as long as Bob has better channel conditions than Eve, Alice can design an encoding method to ensure secrecy. Compared to perfect secrecy definition of Shannon's, Wyner's secrecy notion is much less strict. Wyner-based works consider secrecy to be perfect as long as secrecy capacity has a positive value regardless of the amount of information leaks to Eve. Thus, Wyner's findings result in what is called **weak secrecy**. Weak secrecy is defined as

$$\lim_{n\to\infty} \frac{1}{n} I(M; R_e) = 0; \tag{2.11}$$

which states that as codeword length (n) increases, the information leaked approaches zero on average channel use. Besides weak secrecy, **strong secrecy** is also defined as

$$\lim_{n\to\infty} I(M; R_e) = 0 \tag{2.12}$$

Strong secrecy states that as codeword length (n) increases, the information leakage to Eve approaches zero [4] on each channel use.

Although Wyner's and Shannon's work presents different approaches to secrecy, we can see some similarities. The first similarity is that both perspectives assume Bob has some advantage over Eve. While Shannon-based works assume that Bob has the secret key to decrypt the cipher-text, Wyner-based works consider situations where Bob has a favorable channel condition compared to Eve. Both Shannon and Wyner exploit their advantage to guarantee secrecy. Nonetheless, both assumptions also bring great drawbacks with them. First, if Eve somehow obtains the secret key in Shannon's cipher system, that is the end of the secrecy. As for Wyner's wiretap channel, Eve does not necessarily have to have worse channel conditions than Bob. For example, Eve could be closer to Alice in a wireless network and have a favorable channel than Bob. Even assuming Bob is closer to the transmitter may not be enough since Eve could be close to Bob and have similar channels. Furthermore, the channel is a random phenomenon. Fading conditions may randomly favor Eve, while Bob has degraded channel condition. Similar to Shannon's entropy constraint (2.10), which leads to overhead problems due to large key lengths, PLS gives up the capacity to guarantee secrecy as well. However, the variety of channel conditions makes PLS more versatile. Accordingly, there are various information-theoretic works that investigates secrecy in non-degraded [5], Gaussian [6,7], fading [8–11], multiple access [12–14], interference [15–17], MIMO [18,19], and relay [20,21] channels.

2.3 Physical layer security performance metrics

PLS approaches can be divided into signal-to-interference-plus-noise ratio (SINR)-based (keyless) and complexity-based (key) techniques. Both classes have inherently different performance metrics for different types of attacks. Nonetheless, each security approach has its own merits and demerits. Thus, each technique's feasibility depends on the application, scenario, and communication system. The performance metrics mentioned in the section should help the reader evaluate the secrecy performance of their system design.

2.3.1 SINR-based physical layer security techniques and performance metrics

SINR-based techniques provide secrecy when Eve's SINR is either naturally (channel medium) or artificially (due to some developed technique) lower than Bob's SINR. Through SINR-based techniques, many security measures are developed for eavesdropping, spoofing, and jamming attacks. We can categorize each SINR-based PLS technique for each attack and investigate developed performance metrics respectively.

2.3.1.1 Performance metrics for systems against eavesdropping

Eavesdropping was the first issue that both Wyner and Shannon investigated in their respective frameworks to lay down the foundations of secrecy. By adopting Wyner's wiretap channel, weak and strong secrecy notions have already been established in the previous chapters. However, we can expand on those ideas to further establish information theoretic performance metrics that can be useful to evaluate the security of a communication system against eavesdroppers. Using Figure 2.3 as our reference framework and channel capacity of Bob and Eve as C_b, C_e respectively, we can define the **secrecy capacity** as

$$C_s = \max_{p(x)} \{I(X; R_b) - I(X; R_e)\} \implies C_s = C_b - C_e. \tag{2.13}$$

Secrecy capacity for a secrecy system is similar to the channel capacity for a communication system. Secrecy capacity defines the maximum achievable secrecy rate over all input distributions for the transmitted signal (X). Secrecy capacity gives a deterministic secrecy metric which is plausible for many situations. What if randomness in the environment does not allow for a deterministic metric? In these situations, secrecy capacity itself becomes a random variable. Generally, we adopt statistical measures to deal with these types of situations. Hence, we can define an **outage probability** [22] to measure the performance of our system. First let us define R_T^t as our target **secrecy rate**. Then secrecy outage probability can be defined as

$$\mathbf{Pr}_{\text{out}}(R_T^t) = \mathbf{Pr}(C_i < R_T^t), \tag{2.14}$$

where C_i is the instantaneous secrecy capacity. Chapter 8 demonstrates how to optimize these metrics by resource allocation and scheduling. Readers also referred to Chapter 12 for closed-form secrecy outage probability and average secrecy capacity derivations for non-terrestrial networks (NTNs). Since outage probability is a statistical metric, we may want to obtain a deterministic metric out of the randomness of our system. A classic approach to this is to calculate the average value. Following the average value definition from probability theory and our derived secrecy outage probability, we can derive the measure of the average confidential transmission rate as

$$\text{ST} = R_T^t(1 - \mathbf{Pr}_{\text{out}}(R_T^t)), \tag{2.15}$$

which is called **secrecy throughput**. We can extend secrecy throughput to include both confidential and reliable information transmission. Then for a target transmission rate for Bob R_B^t we can derive our **tight secrecy throughput** metric as

$$\text{TST} = R_T^t(1 - \mathbf{Pr}_{\text{out}}(R_T^t))(1 - \mathbf{Pr}_{\text{out}}(R_B^t)). \tag{2.16}$$

All of the performance metrics we have covered so far are information-theoretic. In various works in the literature, these basic performance metrics are used to evaluate a system's secrecy performance theoretically. Nevertheless, these theoretical metrics are difficult to realize and measure practically since non-Gaussian codes, and finite block lengths are adopted [23]. Fortunately, these metrics can be mapped to parameters that can be easily measured, such as bit error rate (BER). Driven by the previous argument, Refs. [24,25] proposed the **secrecy gap** concept as a practical approach for assessing the secrecy performance. The secrecy gap is a performance metric that quantifies the secrecy level based on the BER performance of Bob and Eve. In Figure 2.4, the concept of security gap is explained in a general manner. Following the BER curve of Eve, we can deduce that $\text{SNR}_{\text{max}}^{(E)}$ is the maximum allowable signal-to-noise ratio (SNR) for the minimum allowable error probability $P_{e,\min}^{(E)}$ that provides enough security against Eve. Similarly, $\text{SNR}_{\min}^{(B)}$ is the minimum SNR that Bob requires for a reliable reception for the maximum allowed error probability $P_{e,\max}^{(B)}$ for Bob. Then we can evaluate the security gap as

$$S_g = \text{SNR}_{min}^{(B)} - \text{SNR}_{max}^{(E)}. \tag{2.17}$$

From Figure 2.4, we can evaluate the security gap as 10 dB for our example. This SNR difference showcases the amount of advantage Bob has over Eve, as mentioned

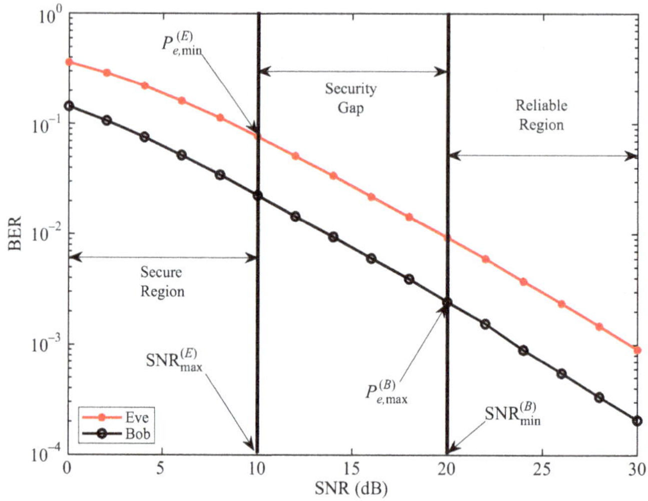

Figure 2.4 Generalized security gap for different BER performances for Bob and Eve

with secrecy capacity. Even though Figure 2.4 is a generalized illustration with different BER curves for Bob and Eve, generally, it is assumed that both Bob and Eve have the same BER curve. However, Eve's BER performance can differ from Bob's due to several reasons:

1. Eve and Bob may have different fading characteristics. For example, Bob may have a line-of-sight (LoS) to the transmitter, yielding Rician distribution for his multipath components. On the other hand, Eve may have non-LoS (NLoS) link with the transmitter, which would result in Rayleigh distributed multipath components.
2. Artificial effects such as artificial noise (AN) could affect Bob and Eve differently to cause different BER performances [26].

Apart from defining security gap, BER can also be used directly to quantify the secrecy performance of a system. Assume that we have implemented a PLS technique to prevent eavesdropping. Through our simulations or real-life measurements, we obtain BER performances for our legitimate receiver Bob and our illegitimate receiver Eve as shown in Figure 2.5. Assuming Eve and Bob would have the same BER performance before the method was applied, we can say that there is some performance improvement from the secrecy aspect. From Figure 2.5 we can observe that there is an *error floor* for Eve. This error floor shows that even if Eve obtained infinite signal strength, Eve would be stuck with a certain error probability. However, the performance of this particular security measure is not enough. With this method minimum error probability Eve gets is still well below 10^{-1}. In practice,

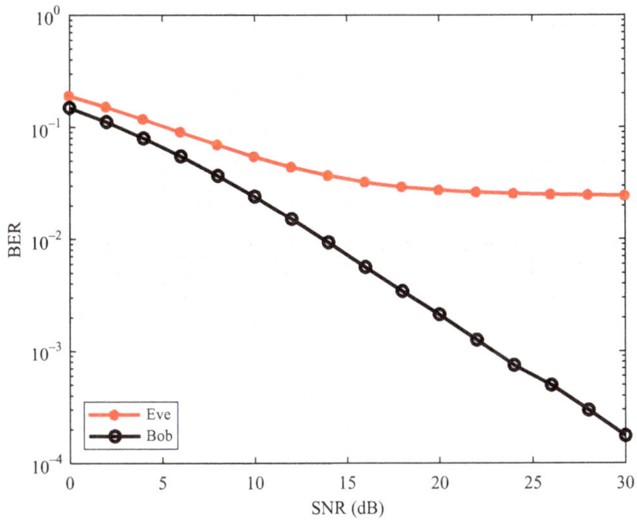

Figure 2.5 Illustration of BER performances for any PLS technique that prevents eavesdropping

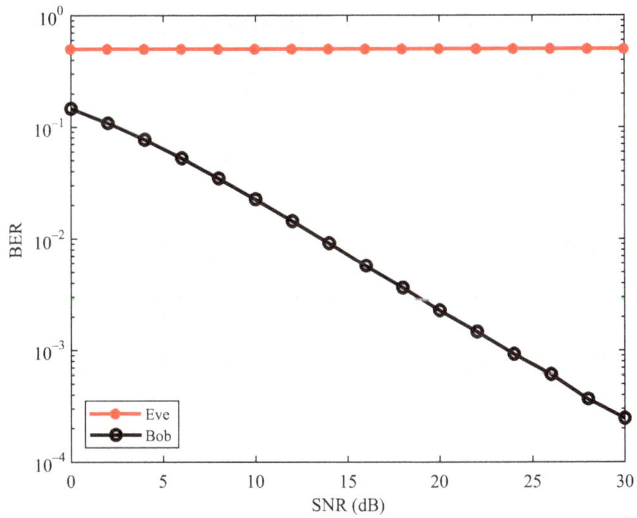

Figure 2.6 Illustration of BER performances for a practical PLS technique that prevents eavesdropping

this probability should be above 0.5 [23]. Then, BER performance of a practical PLS solution against eavesdropping would look like in Figure 2.6.

Various other performance metrics (fractional equivocation-based, packet error rate (PER), etc.) can be used to assess system performance against eavesdropping.

For more detailed information on performance metrics and PLS measures against eavesdropping readers are referred to [4,23].

2.3.1.2 Performance metrics against spoofing

In a spoofing attack, Eve tries to act like Alice to deceive Bob. Eve can do this with any malicious intent. Security measures against spoofing involve detection of the spoofer and rejecting its messages. Extending this idea,`legitimate parties may also be interested in the localization of the spoofer. We can classify some common techniques used for spoofing detection and localization as follows:

- *Received signal strength (RSS)-based techniques*: Through the RSS of the transmitted signal receiver can obtain some information about the transmitter. One critical piece of information is the transmitter's location due to pathloss. While user detection can be done by one receiver, multiple receivers are required for user localization. Readers can refer to the following works for detailed technical information on detection and localization based on RSS measurements [27–29].
- *Angle of arrival (AoA)-based techniques*: AoA-based techniques are used for localization purposes. Compared to RSS-based localization, AoA-based techniques are shown to be more accurate [30]. However, the applicability of this technique depends on the number of antennas in the receiver. Furthermore, its performance degrades indoors due to massive multipath components and NLoS. Ref. [31] showcases the use of AoA for spoofing detection.
- *Time difference of arrival (TDoA)-based techniques*: When a transmitter sends a message, the receiver can use the arrival time difference of two consecutive pulses to obtain the transmitter's location. As localization requires multiple receivers, inherently, this approach requires synchronization between these receivers. Ref. [32] is an excellent example for this approach.
- *Channel-based techniques*: Channel-based methods exploit physical properties of the wireless transmission, such as fingerprints or link signatures. Link signatures are obtained from channel impulse response (CIR) measurements. Since position difference would cause changes in the obtained CIR, a transmitter from a different location (likely to be a spoofer) can be detected by comparing CIR properties. Usually, the receiver starts the transmission with the legitimate transmitter and saves its CIR. The receiver can use this CIR for authentication in the following transmissions. However, this approach assumes that both legitimate receivers and transmitters are static. In the case of mobility, authors in [33] provided a detection algorithm based on channel frequency response (CFR) statistics that consider time variations.
- *Game-theoretical techniques*: Game theory is a mathematical tool to make objective decisions for some problem. Game theory involves players, strategies, and utility functions to formulate a problem mathematically as a game. Readers can refer to [34–36] for the application of game theory on PLS.

Generally, detection relies on building a hypothesis and hypothesis testing on obtained measurements. For our spoofing case, we can form a binary hypothesis in the following form:

$$K_0 : \text{Null hypothesis (legitimate user)} \tag{2.18}$$

$$K_1 : \text{Alternative hypothesis (illegitimate user)} \tag{2.19}$$

A test statistic is formed by using measured quantities for hypothesis testing to detect which one is true. In this regard, two types of errors can be made. **Type-1 error** occurs when K_1 decided while K_0 is true. This error is called a false alarm. Moreover, the probability of false alarm (PFA) P_{FA} is the first performance metric for detection performance. By definition, P_{FA} is given as

$$P_{FA} = Pr(K_1|K_0), \tag{2.20}$$

On the contrary **type-2 error** occurs as hypothesis testing decides on K_0 when K_1 is true. This error is also called missed detection. Similarly, probability of missed detection (PMD) P_{MD} is the second performance metric for detection performance and is defined as

$$P_{MD} = Pr(K_0|K_1), \tag{2.21}$$

Following the framework of [37], we can investigate Figure 2.7. Curves represent power delay profile (PDP) of the users of different distances. As indicated in the figure, the blue curve represents the PDP of a user 15 m away, while

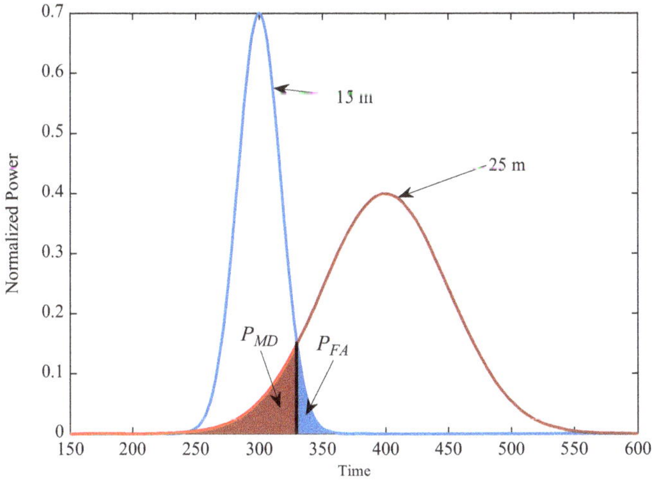

Figure 2.7 Practical example for P_{FA} and P_{MD}

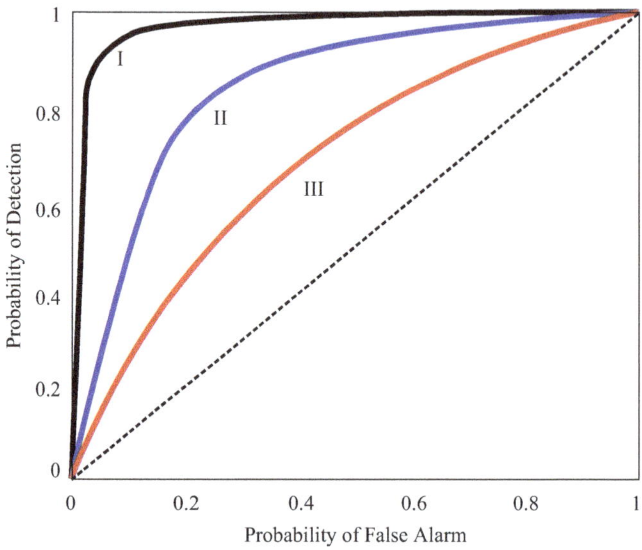

Figure 2.8 Illustration of ROC curves

the red curve represents a user that is 25 m away. P_{FA} can be calculated as the area under the blue curve that overlaps with the red one. While P_{MD} is the area under the red curve that overlaps with the blue curve. In a different method, these curves could be different if a different detection statistic was used. Hence detection performance would be different. Cumulative distribution function (CDF) of the localization errors is usually investigated for localization performance assessment. Finally, spoofing detection performance can also be evaluated by receiver operating characteristic (ROC). ROC curves show the accuracy of signal detection against false alarms. Some example ROC curves are shown in Figure 2.8. As ROC curve bows up, we can deduce that our detection performance is increasing. Another intuitive way to approach ROC curves is to measure the area under each curve. A larger area would imply better detection algorithm performance. Upon further investigation we can see that the area is maximum (1) when $P_D = 1$ and area is minimum (0) when $P_{FA} = 1$. Readers are referred to Chapter 4 for detailed authentication methods against spoofing and ROC analysis for performance evaluation.

2.3.1.3 Performance metrics against jamming

Jamming is a type of attack that cannot be prevented via data security methods (such as cryptography). SINR-based techniques can provide solutions to this type of attack. In general, jammers can use an arbitrary waveform within a limited bandwidth for the attack. The most significant drawbacks of jammers are their power constraints and energy efficiency. However, advanced jammer designs can drastically improve these drawbacks. In line with this, jammer types can be separated into five classes [38,39]:

- *Constant jammers*: Constant jammers are the most simplistic type of jammers. These jammers constantly broadcast a jamming signal to deny legitimate users service within a network. Constant jammers are the most inefficient type of jammers in terms of energy consumption. For detailed explanation of jamming efficiency criteria readers are referred to [39,40].
- *Intermittent jammers*: Contrary to constant jammers, intermittent jammers only employ jamming from time to time. Due to this partially working mechanism, intermittent jammers are more energy efficient than constant jammers.
- *Reactive jammers*: During partial operation time of an intermittent jammer there could be no users communication. In this case, this jammer is just wasting energy trying to block communication that does not exist. Reactive jammers constantly sense the channel and starts jamming when there is a packet transmission.
- *Adaptive jammers*: Adaptive jammers are the ideal version of a jammer as they adjust their jamming power level based on the power level of the transmission. As mentioned in previous sections, wireless communication systems suffer from time-varying fading. For instance, for a while, the channel medium between legitimate transmitter and receiver could favor the transmission, resulting in high RSS at the receiver. On the other hand, a given time channel could be in a deep fade and cause the received signal to be weak. In this case, it would be enough for the jammer to transmit high power only when the transmission link is good and transmit low power when it is bad. However, this jammer is not practical since it requires real-time RSS or receiver channel information, which is hard to obtain practically. Therefore, this jammer is usually used for the worst-case analysis for simulations as the ideal version of a jammer.
- *Intelligent jammers*: Intelligent jammers are different from conventional ones. An intelligent jammer exploits upper-layer functionalities like the MAC layer to deny service. Since the scope of this book is for the physical layer, these jammers will not be covered in detail. However, readers interested in this concept are referred to [41] for further reading.

Securing the wireless link against jamming attacks starts with detecting the attack. Similar to spoofing attacks, a binary hypothesis can be formulated. Then we can form a test statistic to subject against hypothesis testing for detection of the attack. In this regard, jamming detection performance can be evaluated similarly to spoofing. As for jamming prevention, several approaches are particularly effective against jamming. These approaches include but are not limited to:

- *Frequency hopping*: Frequency hopping has been traditionally employed to overcome the presence of a jammer [42–44]. Frequency hopping has been traditionally employed to overcome the presence of a jammer [42–44]. Frequency hopping can be done reactively or proactively. Reactive frequency hopping is changing the transmission frequency channel to a non-jammed channel in the presence of a jamming signal. Proactive frequency hopping corresponds to frequency hopping spread spectrum (FHSS) transmission scheme where the user hops between all available channels in an order that is decided by a generated pseudo-random key.

- *Spread spectrum*: There are two types of spread spectrum techniques. As we already covered FHSS above, direct sequence spread spectrum (DSSS) is the other spread spectrum technique to mention. DSSS provides unique protection against jammers. For jammers that sense the spectrum and attack when it is busy, DSSS can offer excellent security. As mentioned with reactive jammers, the performance of these types of jammers has a great connection to their sensing capabilities. DSSS hides the signal below the noise floor. Thus, it aggravates spectrum sensing for the jammer.
- *Improvement of the wireless link*: A jamming attack tries to deny service via interfering with the legitimate receiver. Thus, methods to improve RSS of the legitimate transmission signal in any method would give some prevention against jamming.
- *Directional antennas and beamforming*: Usage of directional antennas or multiple antennas with beamforming can potentially reduce the jamming effects of an attacker. Depending on the beam pattern and position of the jammer, the link between the legitimate transmitter and receiver can be improved while the jamming signal is rejected.
- *Forward error correction (FEC)*: If the interfering jammer signal is not strong enough, the whole transmission frame may not be corrupted. In these situations, FEC can be used for error correction to ensure reliable transmission.

In order to evaluate the performance of a PLS method against jamming, BER and PER performance would suffice. An example scenario is shown in Figure 2.9. In the figure, there are two different BER curves, one corresponding

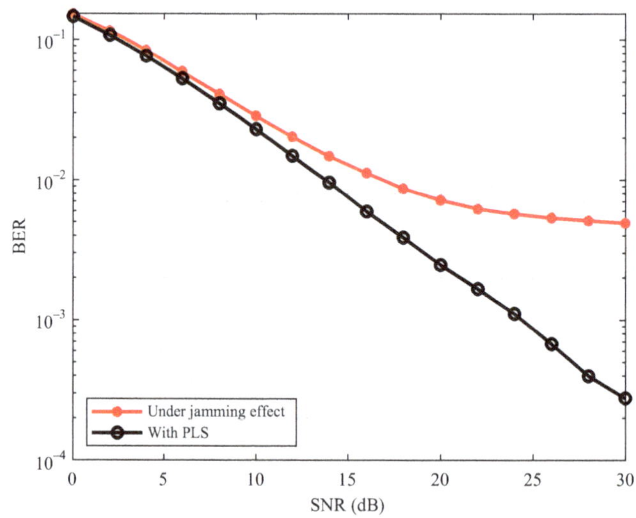

Figure 2.9 BER performances for a sample jamming attack scenario

to the instance of jamming attack and the other after the PLS solution employment. Readers are referred to Chapter 9, which demonstrates the usage of multiple nodes against jamming, and Chapter 11, to read about the effect of jamming on sensing applications.

2.3.2 Complexity-based physical layer security techniques and performance metrics

Complexity-based PLS involves generating a key by exploiting the randomness of the wireless channel and encrypting it with this key to ensure data security. At first glance, it looks exactly like a cryptography technique. However, specific differences make key generation PLS techniques superior to conventional cryptography:

1. Assuming the wireless channel is reciprocal, key sharing is not required. Compared to Shannon's cipher system, this eliminates the burden of crucial sharing and Eve's chances of retrieving the secret key.
2. Assuming the extracted key satisfies the perfect secrecy constraint, securing the message against a quantum computer is possible.

In summary, PLS key generation techniques eliminate modern cryptography's challenges. For the detailed key generation process, readers can refer to Chapter 4. Performance evaluation for key generation involves three metrics:

1. *Randomness*: Randomness is the most important metric for key generation systems since the key generator can be considered a Random Number Generator (RNG). As information theory states, entropy is directly related to randomness. By the notion of perfect secrecy, the entropy of the key impacts the secrecy offered by the secret key, and absolute secrecy is possible if the entropy of the key is at least the entropy of the source message. Nonetheless, not all RNG methods are equal. Even if a generated sequence looks random to a human, there might be an unapparent relationship. Figure 2.10 demonstrates an RNG

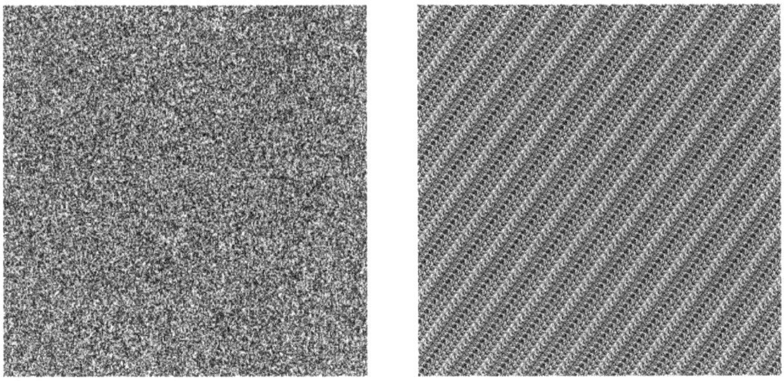

(a) (b)

Figure 2.10 Visualization of randomness generated by two different algorithms. (a) MATLAB® rand() and (b) linear congruential generator

algorithm may not be truly random. On the left, we can see MATLAB's rand() function, and on the right, we see a linear congruential generator with peculiar parameters. It is referred to as peculiar since investigating a random bitmap generated with these parameters shows us generated sequence is not exactly random. There is some deterministic pattern in the generated sequence, even though it is hard to detect at first glance. However, there could also be some hidden pattern with MATLAB's rand() function. Still, we are trying to detect a possibly well-hidden pattern with our perception. Thus, statistical tests that quantify randomness are developed for the randomness evaluation. One statistical randomness test suite is developed by National Institute of Standards and Technology (NIST). This test suite performs a series of statistical analyses to test some chosen features of random sequences in order to classify whether the test sequence is either random or not. For detailed information on this test suite, readers are referred to [45].

2. *Key generation rate (KGR)*: KGR refers to the amount of bit generated for key generation per unit. The unit here could be an instance, second, or measurement. It depends on the evaluation scenario. KGR could be important for real-time applications since particular encryption mechanisms require a key with exact length. Thus, the developed key generation mechanism must generate enough random bits to keep the real-time application continuous.
3. *Key disagreement rate (KDR)*: KDR is the percentage of different bits between the keys of Alice and Bob

$$\text{KDR} = \frac{\sum_{i=1}^{N} \left|K^A(i) - K^B(i)\right|}{N}, \tag{2.22}$$

where K^A and K^B denote keys, Alice and Bob generated, respectively, and N is the length of the key. Intuitively in a symmetric encryption scheme, if keys were different, then proper communication between these users would be impossible. One of the critical steps to key generation is information reconciliation. Information reconciliation involves measuring KDR and, if possible, improving KDR. If KDR is below some rate, then correction of keys would not be possible, and both users would attempt to generate a new key.

2.4 Conclusion

In this chapter, we first introduced the central philosophy of information theory. Then, we explained that the first secrecy analyses were done for eavesdropping by utilizing information theory. Different perspectives of Shannon and Wyner introduced cryptography and PLS to the literature, respectively. Next, the secrecy notions of Shannon and Wyner are explained and compared to lay down the similarities and differences between them. Finally, this chapter discusses the most common performance metrics for PLS system evaluations and standard security measures against various attacks.

References

[1] Shannon CE. A mathematical theory of communication. *The Bell System Technical Journal*. 1948;27(3):379–423.

[2] Shannon CE. Communication theory of secrecy systems. *The Bell System Technical Journal*. 1949;28(4):656–715.

[3] Wyner AD. The wire-tap channel. *Bell System Technical Journal*. 1975; 54(8):1355–1387.

[4] Hamamreh JM, Furqan HM, Arslan H. Classifications and applications of physical layer security techniques for confidentiality: a comprehensive survey. *IEEE Communications Surveys & Tutorials*. 2018;21(2):1773–1828.

[5] Csiszár I, Korner J. Broadcast channels with confidential messages. *IEEE Transactions on Information Theory*. 1978;24(3):339–348.

[6] Yamamoto H. A coding theorem for secret sharing communication systems with two Gaussian wiretap channels. *IEEE Transactions on Information Theory*. 1991;37(3):634–638.

[7] Leung-Yan-Cheong S, Hellman M. The Gaussian wire-tap channel. *IEEE Transactions on Information Theory*. 1978;24(4):451–456.

[8] Gopala PK, Lai L, El Gamal H. On the secrecy capacity of fading channels. *IEEE Transactions on Information Theory*. 2008;54(10):4687–4698.

[9] Liang Y, Poor HV, Shamai S. Secure communication over fading channels. *IEEE Transactions on Information Theory*. 2008;54(6):2470–2492.

[10] Khisti A, Tchamkerten A, Wornell GW. Secure broadcasting over fading channels. *IEEE Transactions on Information Theory*. 2008;54(6): 2453–2469.

[11] Jeon H, Kim N, Choi J, *et al*. Bounds on secrecy capacity over correlated ergodic fading channels at high SNR. *IEEE Transactions on Information Theory*. 2011;57(4):1975–1983.

[12] Liang Y, Poor HV. Multiple-access channels with confidential messages. *IEEE Transactions on Information Theory*. 2008;54(3):976–1002.

[13] Tang X, Liu R, Spasojevic P, *et al*. Multiple access channels with generalized feedback and confidential messages. In: *2007 IEEE Information Theory Workshop*. New York, NY: IEEE; 2007. p. 608–613.

[14] Tekin E, Yener A. The Gaussian multiple access wire-tap channel. *IEEE Transactions on Information Theory*. 2008;54(12):5747–5755.

[15] Koyluoglu OO, El Gamal H, Lai L, *et al*. Interference alignment for secrecy. *IEEE Transactions on Information Theory*. 2011;57(6):3323–3332.

[16] Liu R, Maric I, Spasojevic P, *et al*. Discrete memoryless interference and broadcast channels with confidential messages: Secrecy rate regions. *IEEE Transactions on Information Theory*. 2008;54(6):2493–2507.

[17] Liang Y, Somekh-Baruch A, Poor HV, *et al*. Capacity of cognitive interference channels with and without secrecy. *IEEE Transactions on Information Theory*. 2009;55(2):604–619.

[18] Fakoorian SAA, Swindlehurst AL. Full rank solutions for the MIMO Gaussian wiretap channel with an average power constraint. *IEEE Transactions on Signal Processing*. 2013;61(10):2620–2631.
[19] Cumanan K, Ding Z, Sharif B, *et al*. Secrecy rate optimizations for a MIMO secrecy channel with a multiple-antenna eavesdropper. *IEEE Transactions on Vehicular Technology*. 2013;63(4):1678–1690.
[20] Richter J, Scheunert C, Engelmann S, *et al*. Weak secrecy in the multiway untrusted relay channel with compute-and-forward. *IEEE Transactions on Information Forensics and Security*. 2015;10(6):1262–1273.
[21] Suraweera HA, Garg HK, Nallanathan A. Performance analysis of two hop amplify-and-forward systems with interference at the relay. *IEEE Communications Letters*. 2010;14(8):692–694.
[22] Barros J, Rodrigues MR. Secrecy capacity of wireless channels. In: *2006 IEEE International Symposium on Information Theory*. New York, NY: IEEE; 2006. p. 356–360.
[23] Güvenkaya E, Hamamreh JM, Arslan H. On physical-layer concepts and metrics in secure signal transmission. *Physical Communication*. 2017;25:14–25.
[24] Klinc D, Ha J, McLaughlin SW, *et al*. LDPC for physical layer security. In: *GLOBECOM 2009-2009 IEEE Global Telecommunications Conference*. New York, NY: IEEE; 2009. p. 1–6.
[25] Klinc D, Ha J, McLaughlin SW, *et al*. LDPC codes for the Gaussian wiretap channel. *IEEE Transactions on Information Forensics and Security*. 2011;6(3):532–540.
[26] Goel S, Negi R. Guaranteeing secrecy using artificial noise. *IEEE Transactions on Wireless Communications*. 2008;7(6):2180–2189.
[27] Chen Y, Trappe W, Martin RP. Detecting and localizing wireless spoofing attacks. *In: 2007 4th Annual IEEE Communications Society Conference on Sensor, Mesh and Ad Hoc Communications and Networks*. New York, NY: IEEE; 2007. p. 193–202.
[28] Yang J, Chen Y, Trappe W, *et al*. Detection and localization of multiple spoofing attackers in wireless networks. *IEEE Transactions on Parallel and Distributed Systems*. 2012;24(1):44–58.
[29] Wang T, Yang Y. Analysis on perfect location spoofing attacks using beamforming. In: *2013 Proceedings IEEE INFOCOM*. New York, NY: IEEE; 2013. p. 2778–2786.
[30] Chen HC, Lin TH, Kung H, *et al*. Determining RF angle of arrival using COTS antenna arrays: a field evaluation. *In: MILCOM 2012-2012 IEEE Military Communications Conference*. New York, NY: IEEE; 2012. p. 1–6.
[31] Xiong J, Jamieson K. SecureAngle: improving wireless security using angle-of-arrival information. In: *Proceedings of the 9th ACM SIGCOMM Workshop on Hot Topics in Networks*; 2010. p. 1–6.
[32] Wen M, Li H, Zheng YF, *et al*. TDOA-based Sybil attack detection scheme for wireless sensor networks. *Journal of Shanghai University* (English Edition). 2008;12(1):66–70.

[33] Xiao L, Greenstein LJ, Mandayam NB, *et al.* Channel-based spoofing detection in frequency-selective Rayleigh channels. *IEEE Transactions on Wireless Communications*. 2009;8(12):5948–5956.
[34] Thomas RW, Komali RS, Borghetti BJ, *et al.* A Bayesian game analysis of emulation attacks in dynamic spectrum access networks. In: *2010 IEEE Symposium on New Frontiers in Dynamic Spectrum (DySPAN)*. New York, NY: IEEE; 2010. p. 1–11.
[35] Thomas RW, Borghetti BJ, Komali RS, *et al.* Understanding conditions that lead to emulation attacks in dynamic spectrum access. *IEEE Communications Magazine*. 2011;49(3):32–37.
[36] Tan Y, Sengupta S, Subbalakshmi KP. Primary user emulation attack in dynamic spectrum access networks: a game-theoretic approach. *IET Communications*. 2012;6(8):964–973.
[37] Yılmaz MH, Arslan H. Impersonation attack identification for secure communication. In: *2013 IEEE Globecom Workshops (GC Wkshps)*. New York, NY: IEEE; 2013. p. 1275–1279.
[38] Zou Y, Zhu J, Wang X, *et al.* A survey on wireless security: Technical challenges, recent advances, and future trends. *Proceedings of the IEEE*. 2016;104(9):1727–1765.
[39] Pelechrinis K, Iliofotou M, Krishnamurthy SV. Denial of service attacks in wireless networks: the case of jammers. *IEEE Communications Surveys & Tutorials*. 2010;13(2):245–257.
[40] Acharya M, Thuente D. Intelligent jamming attacks, counterattacks and (counter) 2 attacks in 802.11 b wireless networks. In: *Proceedings of the OPNETWORK—2005 Conference*, Washington DC, USA. Citeseer; 2005.
[41] Liu X, Noubir G, Sundaram R, *et al.* Spread: foiling smart jammers using multi-layer agility. In: *IEEE INFOCOM 2007—26th IEEE International Conference on Computer Communications*. New York, NY: IEEE; 2007. p. 2536–2540.
[42] Xu W, Ma K, Trappe W, *et al.* Jamming sensor networks: attack and defense strategies. *IEEE Network*. 2006;20(3):41–47.
[43] Xu W, Wood T, Trappe W, *et al.* Channel surfing and spatial retreats: defenses against wireless denial of service. In: *Proceedings of the 3rd ACM Workshop on Wireless Security*, 2004. p. 80–89.
[44] Navda V, Bohra A, Ganguly S, *et al.* Using channel hopping to increase 802.11 resilience to jamming attacks. In: *IEEE INFOCOM 2007—26th IEEE International Conference on Computer Communications*. New York, NY: IEEE; 2007. p. 2526–2530.
[45] Rukhin A, Soto J, Nechvatal J, *et al.* *A Statistical Test Suite for Random and Pseudorandom Number Generators for Cryptographic Applications*. Mclean, VA: Booz-allen and Hamilton Inc; 2001.

Chapter 3
Physical layer security definition and domains

Hanadi Salman[1] *and Hüseyin Arslan*[1]

Today's fast-paced world is becoming increasingly dependent on wireless communications. Whether it is for businesses or individuals, mobility, portability, and instant access (via the internet) to unlimited information are the mantras of our day. In the past few decades, wireless communication has grown exponentially, with a bright future ahead of it. It is anticipated that this technology will have a profound impact on our lives and allow us to accomplish things we had never imagined. A key element of the success of these technologies will be the ability to communicate securely and reliably. Although cryptographic security mechanisms are widely adopted, they cannot cope with increasingly decentralized and heterogeneous networks. As a complementary solution, physical layer security (PLS) exploits the dynamic features of wireless environments to ensure that legitimate transmissions are subject to CIAA* requirements. A generalized PLS framework is provided in this chapter, which not only incorporates the existing work but also allows for the development of next generation PLS methods. To reflect this, PLS fabric is divided into observation and modification planes. Furthermore, different domains of PLS are discussed under these plane concepts, allowing a vision of future PLS mechanisms.

3.1 Physical layer security definition

Based on information theory, the core concept behind PLS is to provide unbreakable, verifiable, and quantified secrecy [1]. This is commonly assumed to be accomplished by inherent wireless channel characteristics such as fading, interference, and multipath propagation [2–4]. In either case, these approaches are used to verify the identity of the user or to secure transmissions by ensuring superior signal reception at the legitimate receiver versus an illegitimate/malicious attacker [5,6]. In other attempts, hardware/radio frequency (RF) features of transceivers have been used as well as the environment (such as distance/angle between devices) to authenticate devices [7].

There have been several articles in the current literature that provide a detailed review of existing PLS approaches, focusing either on certain attack types or their

[1]Department of Electrical and Electronics Engineering, Istanbul Medipol University, Turkey
*CIAA is referred to confidentiality, integrity, authenticity, and availability.

countermeasures- for example [8,9] investigating confidentiality and [10,11] looking at anti-jamming PLS mechanisms. By examining these approaches, there are three main types of approaches that PLS defines:

- A secret key is extracted to encrypt/decrypt data bits.
- By using securely shared keys, adjust physical signals/transmissions.
- Physical signals/transmissions are modified based on keys that have been extracted.

However, a single definition and framework that integrates existing approaches and also allows for next-generation PLS methods is still lacking. Due to this, the objective of this chapter is to fill the aforementioned gap in the literature by presenting a generalized PLS framework discussed in [12]. Using PLS to secure wireless transmissions requires observing and then utilizing the dynamic characteristics of wireless signals, the transmission medium, and the radio propagation environment, as well as the RF front-end of the devices.

3.2 Generalized physical layer security framework

Given how PLS has been previously defined, we believe its scope needs to be expanded from protecting data and communication to safeguarding the whole radio environment map [13]. The PLS framework is depicted in high-level illustration in Figure 3.1. An observable plane can be used to provide randomness/uniqueness that can be used to secure or authenticate wireless communication. In turn, the parameters derived from the observation plane are used to modify the transmissions at the bit, signal, or network level. It should be noted that no matter what approach is adopted, either the observation or modification parameters must have a physical basis. To illustrate this point, PLS is not responsible for the use of shared sequences in order to modify/secure data bits (rather, it belongs to the realm of cryptography). More information on the observation/modification planes and the connection between them is provided in the following sections.

3.2.1 Observation plane

In order to secure the wireless link, the *observation* plane contains a number of wireless propagation environment parameters that can be used as sources of entropy and randomness. In any case, the observations must meet certain criteria, whether they come from the channel, the RF front-end, or the radio environment. Therefore, the following characteristics should be fulfilled to be selected as observable parameter:

- *Measurability*: This phrase denotes how obvious, visible, and detectable the observable parameter is. In particular, it must be quantifiable and can be measured scientifically.
- *Reciprocity*: As wireless transmission occurs between locations A and B, the observable parameter's response measured at A and B should be theoretically equal.

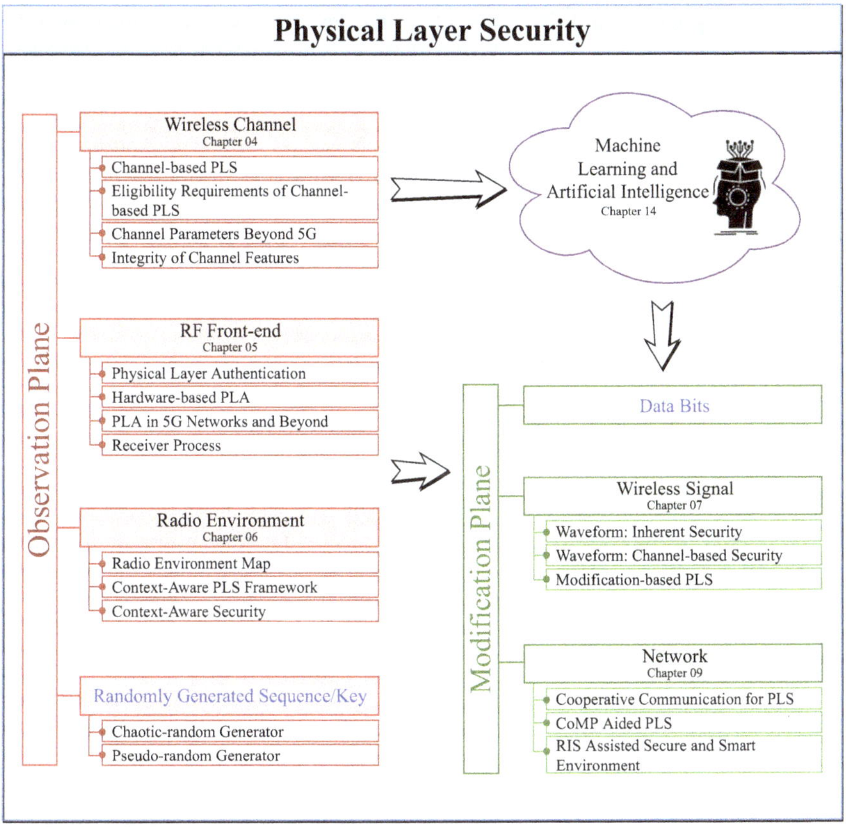

Figure 3.1 *Illustration of PLS framework. Transitions between observation and modification planes can be direct or automated using machine learning and artificial intelligence. Additionally, the combination between the blue headings is not considered to be within PLS definition, rather it is cryptography*

- *Uniqueness*: For a given transmission, the observable parameter should be distinct and solitary. In this case, a third party located separately from the legitimate transceivers should receive another parameter unrelated to what is being sent between the legitimate parties.
- *Randomness*: Statistically, the observable parameter should be random and unpredictably distributed in terms of its individual values. This involves that random models can be used to model the values of the observable parameter.

3.2.2 *Modification plane*

The second plane is the *modification* plane which includes the processes that affect the data either at the bit, the signal, or the network domain such that an attacker can not recover the transmitted data correctly. Modulation and coding are well-known physical processes where PLS falls out. Husain *et al.* in [14] propose a PLS approach by alerting the modulation scheme based on a shared sequence between legitimate parties. Any attacker who receives the signal cannot correctly decode the received symbols as long as the modulation scheme is unknown. Another related approach has been presented in [15] where the parameters of the reconfigurable convolutional coder are changed based on a shared sequence. A packet with changed coding parameters will indicate an error to an eavesdropper. However, a packet with the same coding parameters can be successfully decoded by the intended/legitimate receiver. Artificial noise, beamforming, multiple transmission, and RF impairments are another examples of physical modification processes. Sections 3.3.4, 3.3.5, and 3.3.6 provide more details of existing PLS approaches under this plane concept.

3.2.3 From observation plane to modification plane

It has already been mentioned that the PLS framework is made up of two essential parts. The observation planes provide unique randomness that can be used to protect wireless communications through modification planes. There is, however, a question as to how the decision-making mechanism links these two planes. In other words, what parameters should be observed in a given scenario, how long should they be observed, and how should they be processed and analyzed, and so on. In light of the fact that user security requirements could differ based on the application being used, it is critical that the system can adapt in real-time.

Thus, artificial intelligence (AI) and self-organizing networks (SONs) play an increasingly important role in this process. A cognitive PLS framework was initially presented in [16], was further justified in terms of vehicle-to-everything (V2X) communication in [17], and finally offered in [13]. Like many next-generation systems, on-the-fly learning is a necessary component of optimized PLS. A major role for AI in observation plane extends far beyond the analysis of signals to modeling the behavior of the users. As a result, the information of the environment can be utilized to identify any anomalies that might point to the presence of malicious entities [18]. Thereafter, it is necessary to select an appropriate PLS mechanism, resource allocation, signal processing method, and node selection (in the network domain) in order to implement the design. Nonetheless, AI itself is susceptible to various potential threats, including adversarial machine learning, which must be mitigated [19].

3.3 Physical layer security domains

The different *domains* of PLS are discussed in this section, where each domain is essentially a subset of PLS methods based on whatever aspect of the wireless channel, RF front-end and/or radio environment is utilized.

3.3.1 Wireless channel

Through the propagation of wireless signal, it interacts with objects in the environment, resulting in absorption, reflection, refraction, and diffraction. Because of the mobility in the environment, these phenomena are time-variant and random. Especially in rich scattering environments, the random nature of the channel poses a communication challenge due to its limited coherence distance, bandwidth, and time. However, from PLS perspective, it is beneficial for legitimate and illegitimate nodes to observe their channels independently (as long as they are half-wavelengths apart). More information on channel-based PLS techniques can be found in Chapter 4.

In order to understand the wireless channel, one needs to consider observable plane parameters in terms of their ideal properties. Despite the complexity of the interaction between wireless signals and the environment, several mathematical models have been developed to demonstrate the impact of the environment on wireless signals. A variety of quantities are used to measure a channel, including received signal strength indicator (RSSI), channel state information (CSI), channel impulse response (CIR) and channel frequency response (CFR). For modeling and analysis purposes, the channel is typically represented as a finite impulse response (FIR) filter. In a reciprocal channel, all other parameters (especially frequency) remain constant, so that both uplink and downlink responses are the same. In fact, a major motivation for using time division duplex (TDD) systems is reciprocity. In addition, there are a considerable number of objects in the propagation environment whose reflection/absorption abilities differ, resulting in multipath propagation, which occurs when different replicas of the signal arrive at the receiver in different phases and power levels. Each of these multipath components is treated as random and is expected to add up in either a constructive or destructive manner. Based on this randomness, each wireless link exhibits unique and distinct channel characteristics, see Figure 3.2.

As part of wireless communication, channel estimation is one of the primary advantages of utilizing wireless channels for PLS. For the nodes communicating through the wireless channel to remove the effect of the environment on the signal

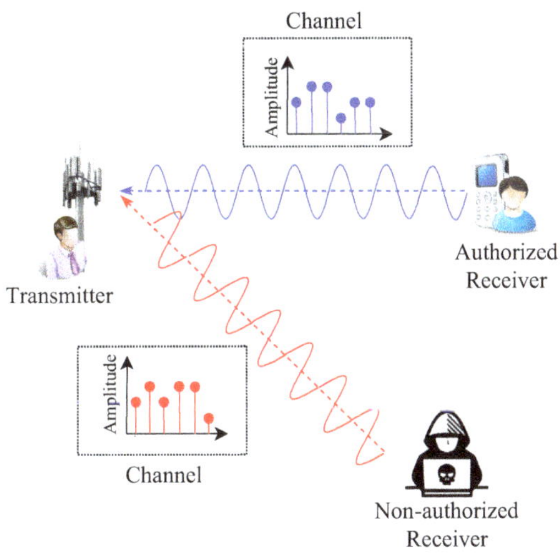

Figure 3.2 Link/device authentication based on the CIR observation

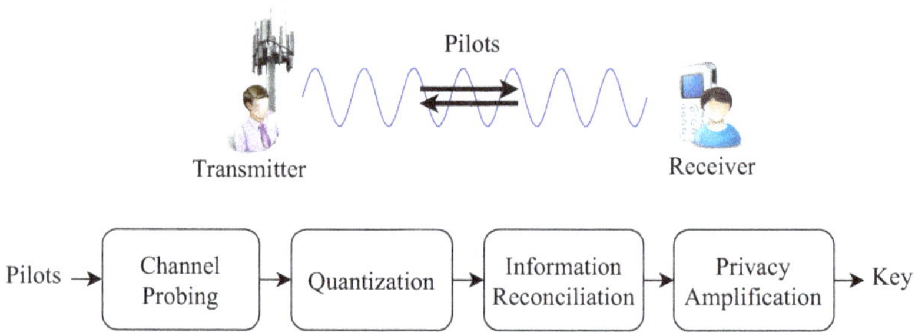

Figure 3.3 Illustration of the steps involved in channel-based key generation

and restore a clean signal at the receiver, they need to know the effect of the environment on the signal. In this way, the PLS approaches do not impose an excessive overhead in channel estimation. Due to this, we have seen the widespread use of wireless channels in PLS for generating channel-based keys [20] (the main steps are illustrated in Figure 3.3), adapting links [21], and injecting interference signals into the communication channel [22].

It should be noted at this stage that the channel observations depend on transmission parameters. As an example, path loss can be affected by carrier frequency, where higher frequencies, like millimeter wave (mmWave) and terahertz (THz), experience increased attenuation than conventional systems. Similarly, more bandwidth can give better resolution in time (delay) domain and vice versa. It is, therefore, advisable to select the appropriate transmission parameters for any given scenario such that the necessary criteria of the observable parameters are met.

3.3.2 RF front-end

Aside from the wireless medium itself, the RF front-end is also subject to imperfections resulting in impairments such as clock jitter, phase noise, carrier frequency offset (CFO), in-phase/quadrature imbalance (IQI), power amplifier non-linearity, and antenna imperfections [23]. These impairments vary based on the device, so they can be used as device "fingerprints" for distinguishing between them [24]. More details are provided in Chapter 5. In this regard, RF fingerprinting is a physical layer authentication mechanism (see Figure 3.4), designed to prevent (or at least detect) attacks on node identity or message integrity. A further benefit of using RF fingerprints is that they complement channel-based authentication. As illustrated in [25], devices are authenticated by fingerprints, while secure communication is allowed by channel-based key generation. In general, RF-based approaches can be leveraged in mobile environments due to their stability. While channel-based techniques are more ideal for indoor and relatively stationary environments (so authentication is not needed too frequently).

In real network conditions, the fingerprint's reliability poses a problem for RF-based PLS. For example, one impairment may not be able to differentiate between

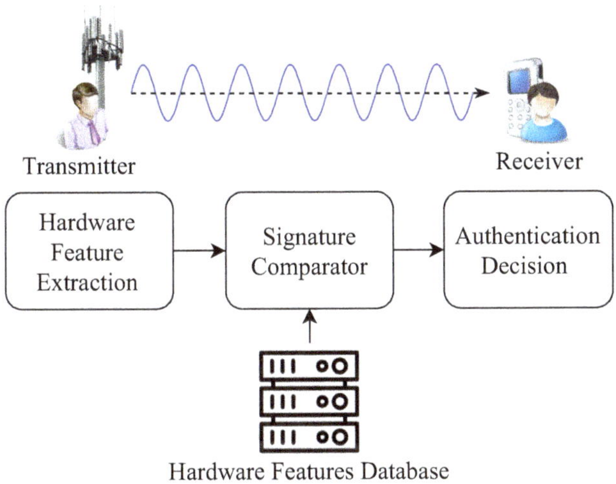

Figure 3.4 Illustration of RF fingerprinting for physical layer authentication

devices due to a small dynamic range. To address this, Hao and Wang [26] suggested combining multiple device characteristics in a weighted manner, while [27] discussed a collaborative approach where observations from multiple nodes are combined and used. In RF impairment-based security solutions, as with channel-based PLS, reliability is ensured by identifying and measuring all impairments to determine their significance. However, the main issue arises when certain channel-related phenomena influence the signal in the same way as hardware impairments. As an example, mobility in the environment causes Doppler spread/shift, resulting in a change in frequencies seen by the receiver. This is similar to a local oscillator imperfection causing CFO and unsynchronized frequency. It might be possible to separate the channel effects from device impairments in such scenarios by considering the fact that the former vary at a much smaller rate and the latter are relatively stable [18]. In alternative to this, the channel impairment and the RF impairment can be combined into a time-varying device fingerprint, as illustrated in [28], which combines the CFO induced by oscillator mismatch with channel-induced Doppler to form a time-varying CFO for authentication of the device. Ref. [29] uses a similar approach. The former uses imperfect or chaotic antenna geometries and activation sequences as a means of authentication, while the latter uses beamspace representations of the mutual coupling between multiple antennas in a multiple-input multiple-output (MIMO) system.

3.3.3 Radio environment/sensing

Due to the objects and their properties in the surrounding environment, wireless signals are subjected to different phenomena during their transmission through the air (such as absorption, reflection, refraction, diffraction, and so on). In the same way as the channel might be independently determined in a rich scattering environment, the surrounding objects and their properties could also be independently determined for a variety of links, giving out an indication of the environment in which they are located. Distance, speed, angle, and size of objects, or their constituent materials, are examples of different observable parameters used to authenticate or secure wireless links [30]. An illustration of how different physical parameters can be used to generate keys is shown in Figure 3.5. In light of this, Chapter 6 provides additional information regarding radio environment.

The most popular environmental measurements for PLS is the distance or angle between communicating nodes. Ref. [31], for example, uses angle of arrival (AoA)-based key generation to generate the secret key, using azimuth, elevation, or both angles. The authors argue that low signal-to-noise ratio (SNR) scenarios can benefit from this approach due to its lower mismatch rate than channel-based key generation. A further proposal in [7] is to generate keys based on nodes' relative locations. This approach eliminates the need to share the entropy source between the devices since relative location is a reciprocal quantity. Different approaches to distance calculation exist, such as time difference of arrival (TDoA)-based approaches and received signal strength (RSS)-based approaches [32]. As in the RF-based approaches discussed earlier, parameters gathered from the radio environment or sensing can be incorporated with channel knowledge, as illustrated in [33]. It should be noted that sensing is not exclusively restricted to RF domain in our generic PLS framework. Also, external sensors may be integrated into the system, including cameras, LiDar, humidity sensors, and temperature sensors.

3.3.4 Data bits

A bit-level security mechanism has conventionally been used for wireless systems. As a rule of thumb, data is secured by encrypting *plaintext* or messages

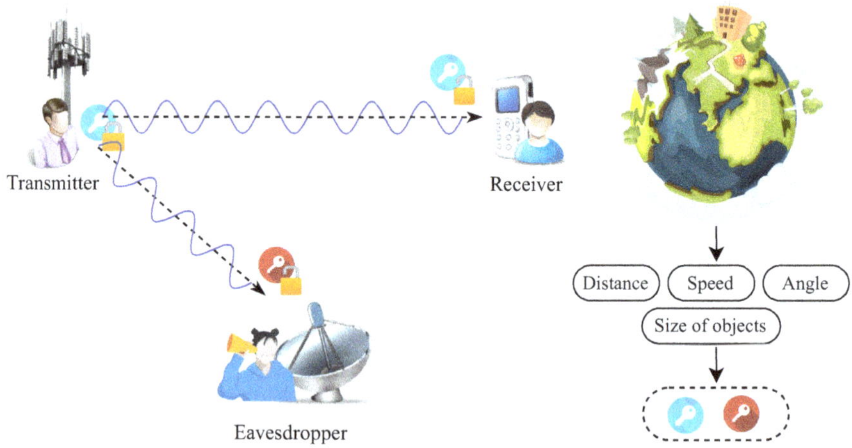

Figure 3.5 Illustration of how different parameters extracted from the environment can be used to generate secure keys

into *ciphertext* using some cryptographic algorithm [34]. During this discussion, it is imperative to distinguish between cryptography and key-based PLS. Both cases involve bit-level transformations. As for the former case, both symmetric and asymmetric encryption use a known or shared sequence of keys, and the key is shared or not between the communicating nodes depending on the encryption method. Since this process, including key sharing/management, is typically performed at a higher layer, the PLS paradigm does not cover it. As discussed in the previous section, PLS develops its key-generation mechanisms according to the observable parameters associated with the wireless channel and radio environment around the communicating nodes. Thus, key exchange is not required with PLS mechanisms because both transceivers observe the same channel and environment [35]. Besides key-based PLS, channel coding is also used to provide bit-level security. In spite of the fact that coding-based mechanisms exist in a variety of forms (including polar and low-density parity-check (LDPC)), one of the major limitations of them is that the eavesdropper's (wiretap) channel must be compromised compared with the legitimate link [36]. Note that bit-level modifications are merely intended to prevent eavesdropping. Moreover, this paradigm does not address jamming and spoofing, which is a significant limitation of standard cryptographic solutions.

3.3.5 Wireless signal

There is no doubt that the majority of PLS work falls under the wireless signal domain. Note that in the modification plane, wireless signal (discussed in Chapter 7) is supposed to cover all blocks between the transceivers' antenna and the coded bit stream. The goal of these security solutions is essentially to improve the data decoding capability at legitimate receivers compared to malicious attackers. In order to accomplish this, either the eavesdropper's performance can be intentionally degraded or the legitimate receiver can benefit from improved quality of service (QoS). It should be noted that in order to counter eavesdropping, a host of PLS signal modification solutions have been developed. Additionally, modifying signal parameters can mitigate jamming more effectively than anything else. In contrast, signal modification is often not the best (or most popular) means of protecting against spoofing.

3.3.5.1 Beamforming

It is possible to increase the security level of legitimate nodes by utilizing methods such as adaptive resource allocation [37] or beamforming [38] (shown in Figure 3.6), both aimed at improving the quality of their links. In spite of the fact that beamforming can be designed according to several criteria (linear [39] or nonlinear [40]), a common goal is to direct the legitimate signal toward the legitimate receiver, while using spatial degrees of freedom to reduce signal strength in the direction of the eavesdropper. However, it is difficult to guarantee PLS if the eavesdropper is located closer to the transmitter than the legitimate receiver. In this scenario, spatial beamforming may not provide adequate secrecy.

3.3.5.2 Interfering signal

Various realizations exist where the interfering signal may be generated at the eavesdropper, such as lying in the null space of the legitimate receiver. This means there will be no interference with the legitimate receiver. In order to accomplish this, certain works assume they know where eavesdroppers are or how to decode them, and manipulate signals so they are impaired for decoding at that particular node. Refs. [41,42], for instance, add artificial noise to the transmitted signal on the assumption that the eavesdropper's CSI is known (see Figure 3.7). As with passive eavesdroppers, this assumption cannot be relied upon. An alternative would be to design interfering signals solely based on the information provided by the legitimate

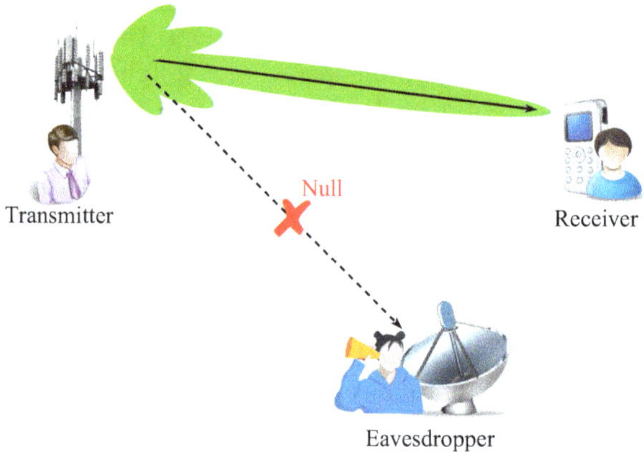

Figure 3.6 Beamforming-based PLS to degrade eavesdropper's performance

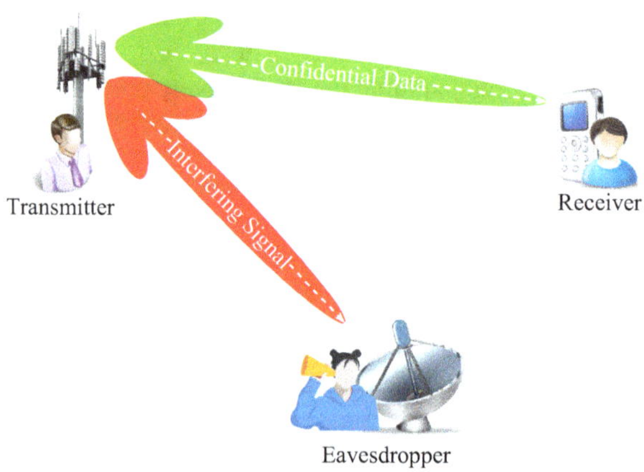

Figure 3.7 Intentional interference at eavesdropper to degrade its interception capability

link. Ref. [43] demonstrates that as the legitimate transmitter has fewer antennas and only knows the legitimate channel. Similarly, noise-loop modulation is also proposed in [9] as a means of ensuring secure and reliable communications. This approach involves jamming the transmission by deliberately introducing noise in the channel, resulting in the illegitimate node being unable to receive the information, irrespective of its computational power.

3.3.5.3 Signal design

It is also possible to maintain reliable data transmission to eavesdroppers through another PLS approach referred to as signal design [44]. In order to achieve this, altering the signal structure (such as modulation scheme, constellation structure, extra processes, etc.) is necessary so that a potential eavesdropper will be unable to decode the signal. It was proposed in [45] to adapt constellation order (and mapping) according to the legitimate CSI. Thus, as shown in Figure 3.8, the eavesdropper is not aware of what modulation scheme or order the transmission block is using. Also, shared sequences beyond channel-based sequences have been utilized for constellation rotation [46,47]. As a result of all these approaches, the eavesdropper observes a seemingly chaotic signal [48], which is associated with a cloudy and distorted constellation. The channel-based shortening in [21] is based on the development of shortening filters in order to reduce the effective delay spread at the receiver. This results in a reduction of the cyclic prefix (CP) at the eavesdropper resulting in inter-symbol interference (ISI). Ref. [49] presents adaptive and flexible PLS algorithms that jointly secure data and pilots. Particularly, an all-pass minimum-phase channel decomposition approach is employed in the proposed algorithms, in which the data

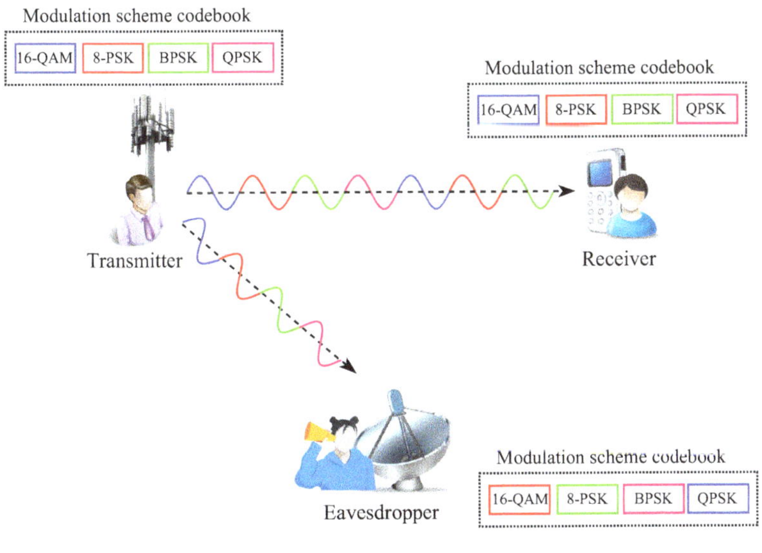

Figure 3.8 Modulation order/scheme-based PLS, the sequence can be pre-shared or extracted

and pilots are precoded using the all-pass component of the channel. This component is random enough to ensure security without affecting the peak-to-average power ratio (PAPR), thus ensuring that legitimate users are not adversely affected. Additionally, it has also been possible to secure communication through RF impairments. As an example, Ref. [50] utilizes the channel to produce secure communication by pre-equalizing the CFO. Due to the fact that the CFO of the legitimate link is independent of, and unknown to the eavesdropper, there is a degradation in eavesdropping quality.

3.3.5.4 Spread spectrum

To disrupt jamming, the most commonly utilized spread spectrum approach is to dynamically change the frequency at which legitimate transmissions take place. Frequency hopping (shown in Figure 3.9) can be implemented using a shared sequence [51] or a channel-dependent sequence [52]. It may be sufficient to use these approaches for simple jamming attacks where the attacker is not capable of monitoring and adapting to frequency hopping. However, more sophisticated and intelligent jammers that can monitor transmissions can still pose significant problems. In [53], a novel approach is used to deal with such attacks: the legitimate transmitter employs deep reinforcement learning to understand the jammer's strategy and then derive the most effective countermeasures. Aside from adapting its own transmission parameters, it also harvests energy from the jamming signal. By using ambient backscatter communication, the legitimate node is able to augment its own transmissions while wasting the attacker's power resources.

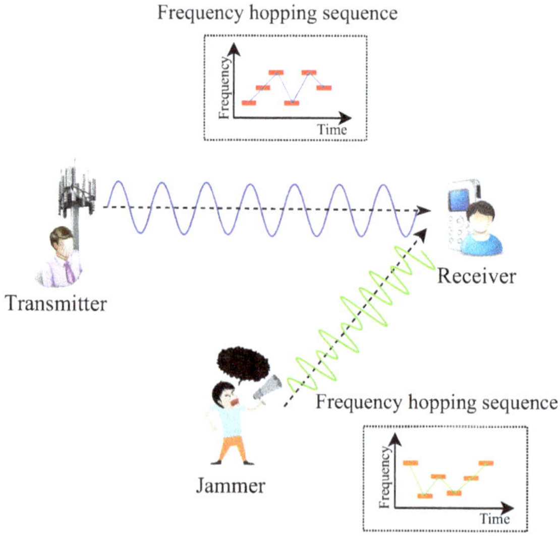

Figure 3.9 Frequency hopping approach to disrupt jamming attacks

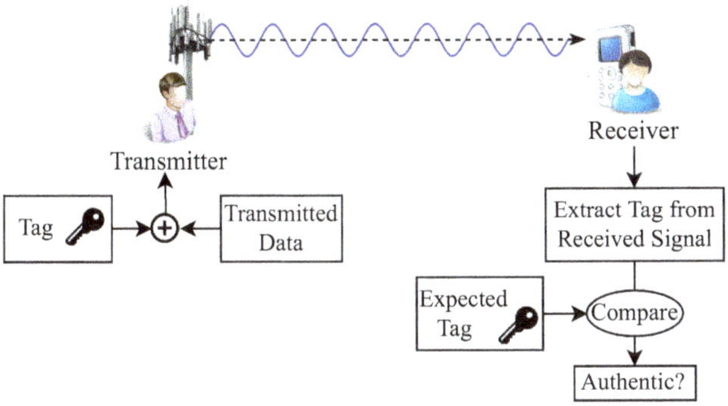

Figure 3.10 Basic steps of tag-based authentication

3.3.5.5 Authentication tag

In terms of spoofing, most PLS solutions use either the wireless channel or RF impairments to verify authentication of the communicating node (and thus the message itself). No modification of the transmitted signal is required for either of these approaches. Despite this, a few works propose adding an authentication tag to wireless transmissions. The tag is embedded into the transmitted signal, independent of the message, to allow the legitimate device to be determined from an illegitimate one [54]. A summary of the basic steps can be found in Figure 3.10.

3.3.6 Network

In the context of the modification plane, the network refers to the various nodes that exist in the environment. As examples of this, relays may be used in cooperative communication, base station (BS) as a part of a coordinated multipoint architecture, and reconfigurable intelligent surfaces (RISs) as part of smart radio systems. An in-depth exploration of the effects of network changes is provided in Chapter 9. In network-based PLS, macro-diversity is used to increase the reliability of the user's communication and to degrade the effectiveness of the attacker at the same time. With limited effort put into implementing network-based authentication, these techniques are primarily used for eavesdropping and jamming attacks.

3.3.6.1 Cooperative communication

It has become increasingly popular to use cooperative communication paradigms as they enable otherwise resource-constrained devices to benefit from the diversity of MIMO technology with the help of relays and helper nodes [55] (Figure 3.11 illustrates this). As a general rule, the cooperative communication process typically consists of two phases: the first entails broadcast transmission from the source (to both the relay and destination) and the second involves the relay retransmitting the signal [56]. A common deployment of these systems is in ad hoc or wireless sensor

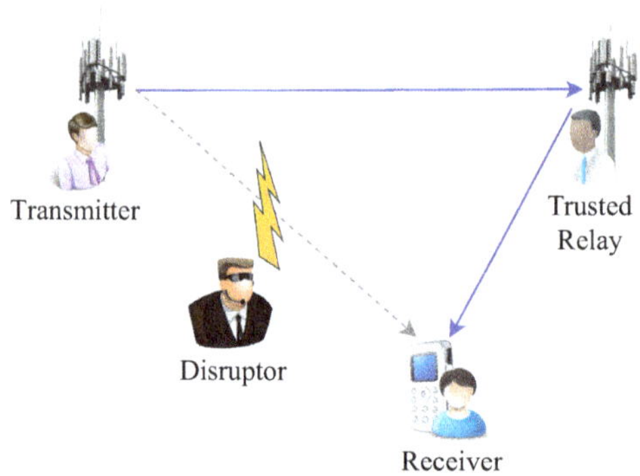

Figure 3.11 Illustration of the use of cooperative communication in PLS

networks, which are decentralized and have device limitations, making authentication slightly more difficult compared to conventional cellular and WiFi systems. As a result of the lack of authentication, malicious attackers can use relays to disrupt communication.

Consequently, several approaches have been developed that rely on cooperative relaying and jamming to overcome the problem of untrustable relays [57]. When cooperative relaying is used, trustworthy relays are chosen to prevent message interception in the case of an eavesdropping possibility. As a result, the diversity benefit that is the primary driving force behind cooperation might be reduced. Alternatively, cooperative jamming uses a known jamming signal transmitted from either a source, a destination, or a helper node to prevent eavesdropping. Although destination-based jamming [58] eliminates the need for a helper, the system cannot exploit diversity unless the destination supports full duplexing. Additionally, jamming is also an energy-intensive approach. In [59], an interesting workaround is proposed by using fast Fourier transform (FFT) operation for orthogonal frequency division multiplexing (OFDM) transmissions, in which the destination node transmits a jamming signal only during the CP period of OFDM symbols. As a result of the FFT operation, this jamming signal spreads throughout all subcarriers at the relay, resulting in inter-carrier interference and reduced interception. This solution requires less power since the signal is only transmitted for a limited (CP) duration, and fully duplexing is not necessary.

3.3.6.2 Coordinated multipoint

Coordinated multipoint (CoMP) is a cellular concept intended to alleviate the inter-cell interference (ICI), particularly for small cells and heterogeneous network deployments. CoMP was first introduced for Long Term Evolution (LTE) in Third Generation Partnership Project (3GPP) Rel-11 [60], with numerous upgrades in the succeeding releases. While security is not the major driver behind CoMP, few works have examined CoMP from the perspective of PLS. In [61], transmissions from multiple BSs in the underwater scenario are scheduled along with controlling their

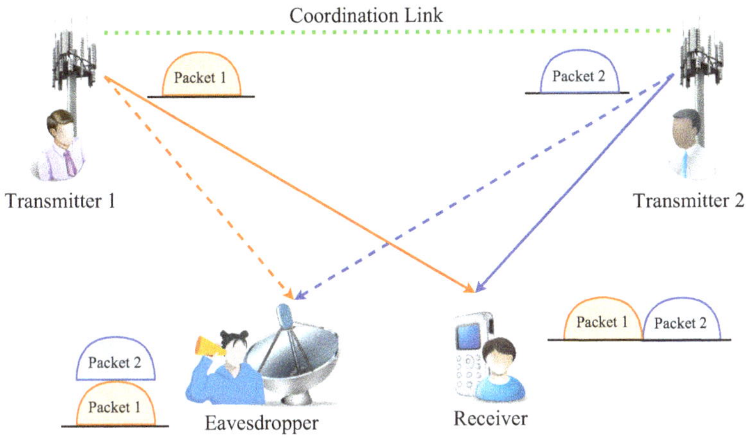

Figure 3.12 CoMP-assisted PLS by creating intentional misalignment between the received packets at the eavesdropper

transmitted power. By doing this, the received signal at the legitimate receiver is non-overlapped and can be decoded correctly. However, the received signal at eavesdropper overlapped leading to degrading its performance (as shown in Figure 3.12). Additionally, the distributed BSs are used to circumvent the constraint of directional modulation, where the eavesdropper is in the same direction as the legitimate receiver [62]. In [63], employing the coordination leads to assuring that the broadcast message is recovered only at the intersection of the transmission area from coordinating BSs. The multipoint (or multi-landmark) approach is also used in authentication when RSSI observations are acquired at several physical locations to authenticate the user's identification [64]. The inclusion of multiple antennas at each landmark also offers improved spatial resolution, which improves authentication accuracy.

3.3.6.3 Reconfigurable intelligent surface

Conventional wireless systems rely on the radio environment to determine the propagation channel. This results in uncontrollable channel properties, and transceivers can only compensate or mitigate them. In conventional wireless systems, the propagation channel is determined by the radio environment. As a result, it is assumed to be uncontrollable and the transceivers can only attempt to compensate/mitigate this impact. However, the smart radio environment paradigm enabled by RISs considers the wireless channel as a controlled entity [65], opening up a slew of new possibilities for PLS employing RISs [66]. As previously stated, a rich scattering channel is more preferable from the PLS perspective since it gives more unpredictability. Controlling the variation in channel over time, which may subsequently be utilized for other purposes such as physical layer key generation, is one of the possible benefits of employing RISs [67]. Furthermore, considering the huge number of antenna elements at the RIS, RIS enables cooperative active/passive beamforming at

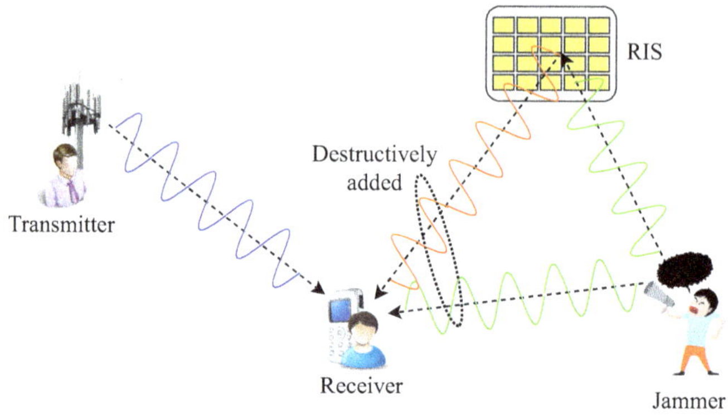

Figure 3.13 Exploit RIS as anti-jamming solution

the transmitter and RIS. This has been utilized to safeguard communication against eavesdropping in [68], even in the presence of a stronger eavesdropper channel than the legitimate one in a line-of-sight (LoS) propagation scenario. RISs were also used in [69] to defend communications from jamming attacks (as shown in Figure 3.13). Specifically, to guarantee that users' QoS needs are satisfied in the presence of a multi-antenna jammer, joint optimization of beamforming and power allocation is investigated.

3.4 Conclusion

Despite the fact that PLS is a well-studied topic, there has yet to be a single definition or unified framework covering the main components (and their realizations). We defined the generic framework of PLS, which was missing in the literature. In this Chapter, we explained the unified PLS framework, which consists of observation and modification planes, where the observation plane is basically the source of entropy while the modification plane shows how the transmissions may be adjusted to guarantee the security gap between legitimate and illegitimate nodes. This framework is not only valid for the classification of existing works but also moves aside for the development of next-generation PLS methods.

References

[1] Shakiba-Herfeh M, Chorti A, Poor HV. Physical layer security: authentication, integrity, and confidentiality. In: *Physical Layer Security*. New York, NY: Springer; 2021. p. 129–150.

[2] Schaefer RF, Boche H, Poor HV. Secure communication under channel uncertainty and adversarial attacks. *Proceedings of the IEEE*. 2015;103(10):1796–1813.

[3] Zou Y, Zhu J, Wang X, *et al.* A survey on wireless security: technical challenges, recent advances, and future trends. *Proceedings of the IEEE.* 2016;104(9):1727–1765.

[4] Wu Y, Khisti A, Xiao C, *et al.* A survey of physical layer security techniques for 5G wireless networks and challenges ahead. *IEEE Journal on Selected Areas in Communications*. 2018;36(4):679–695.

[5] Liu Y, Chen HH, Wang L. Physical layer security for next generation wireless networks: theories, technologies, and challenges. *IEEE Communications Surveys & Tutorials*. 2016;19(1):347–376.

[6] Sun L, Du Q. Physical layer security with its applications in 5G networks: a review. *China Communications*. 2017;14(12):1–14.

[7] Gungor O, Chen F, Koksal CE. Secret key generation via localization and mobility. *IEEE Transactions on Vehicular Technology*. 2014;64(6):2214–2230.

[8] Hamamreh JM, Furqan HM, Arslan H. Classifications and applications of physical layer security techniques for confidentiality: a comprehensive survey. *IEEE Communications Surveys & Tutorials*. 2018;21(2):1773–1828.

[9] Mucchi L, Caputo S, Marcocci P, *et al.* Security and reliability performance of noise-loop modulation: theoretical analysis and experimentation. *IEEE Transactions on Vehicular Technology*. 2022;71(6):6335–6350.

[10] Grover K, Lim A, Yang Q. Jamming and anti-jamming techniques in wireless networks: a survey. *International Journal of Ad Hoc and Ubiquitous Computing*. 2014;17(4):197–215.

[11] Pirayesh H, Zeng H. Jamming attacks and anti-jamming strategies in wireless networks: a comprehensive survey. *IEEE Communications Surveys & Tutorials*. 2022;24(2):767–809.

[12] Solaija MSJ, Salman H, Arslan H. Towards a unified framework for physical layer security in 5G and beyond networks. *IEEE Open Journal of Vehicular Technology*. 2022;3:321–343.

[13] Furqan HM, Solaija MSJ, Türkmen H, *et al.* Wireless communication, sensing, and REM: a security perspective. *IEEE Open Journal of the Communications Society*. 2021;2:287–321.

[14] Husain MI, Mahant S, Sridhar R. CD-PHY: physical layer security in wireless networks through constellation diversity. In: *MILCOM 2012—2012 IEEE Military Communications Conference*. New York, NY: IEEE; 2012. p. 1–9.

[15] Tang L, Ambrose JA, Parameswaran S, *et al.* Reconfigurable convolutional codec for physical layer communication security application. In: *2014 IEEE Military Communications Conference*. New York, NY: IEEE; 2014. p. 82–87.

[16] Yılmaz MH, Güvenkaya E, Furqan HM, *et al.* Cognitive security of wireless communication systems in the physical layer. *Wireless Communications and Mobile Computing*. 2017;2017:1–9.

[17] Furqan HM, Solaija MSJ, Hamamreh JM, *et al.* Intelligent physical layer security approach for V2X communication. arXiv preprint arXiv:190505075. 2019; p. 1–7.

[18] Zhao L, Zhang X, Chen J, *et al.* Physical layer security in the age of artificial intelligence and edge computing. *IEEE Wireless Communications*. 2020;27(5):174–180.

[19] Wang B, Gong NZ. Stealing hyperparameters in machine learning. In: *2018 IEEE Symposium on Security and Privacy (SP)*; 2018. p. 36–52.

[20] Furqan HM, Hamamreh JM, Arslan H. New physical layer key generation dimensions: subcarrier indices/positions-based key generation. *IEEE Communications Letters*. 2020;25(1):59–63.

[21] Furqan HM, Hamamreh JM, Arslan H. Enhancing physical layer security of OFDM systems using channel shortening. In: *Proceedings of the IEEE 28th Annual International Symposium on Personal, Indoor, and Mobile Radio Communications (PIMRC)*. Montreal, Canada; 2017. p. 1–5.

[22] An Y, Zhang S, Ji Z. A tag-based PHY-layer authentication scheme without key distribution. *IEEE Access*. 2021;9:85947–85955.

[23] Arslan H. RF impairments. In: *Wireless Communication Signals: A Laboratory-based Approach. Wiley Online Library*; 2021. p. 99–120.

[24] Wang W, Sun Z, Piao S, *et al.* Wireless physical-layer identification: modeling and validation. *IEEE Transactions on Information Forensics and Security*. 2016;11(9):2091–2106.

[25] Zhang J, Rajendran S, Sun Z, *et al.* Physical layer security for the Internet of Things: authentication and key generation. *IEEE Wireless Communications*. 2019;26(5):92–98.

[26] Hao P, Wang X. Performance enhanced wireless device authentication using multiple weighted device-specific characteristics. In: *IEEE China Summit and International Conference on Signal and Information Processing* (ChinaSIP). Chengdu, China; 2015. p. 438–442.

[27] Wang X, Hao P, Hanzo L. Physical-layer authentication for wireless security enhancement: current challenges and future developments. *IEEE Communications Magazine*. 2016;54(6):152–158.

[28] Hou W, Wang X, Chouinard JY, *et al.* Physical layer authentication for mobile systems with time-varying carrier frequency offsets. *IEEE Transactions on Communications*. 2014;62(5):1658–1667.

[29] Karabacak M, Peköz B, Mumcu G, *et al.* Arraymetrics: authentication through chaotic antenna array geometries. *IEEE Communications Letters*. 2021;25(6):1801–1804.

[30] Badawy A, Elfouly T, Khattab T, *et al.* Unleashing the secure potential of the wireless physical layer: secret key generation methods. *Physical Communication*. 2016;19:1–10.

[31] Badawy A, Khattab T, El-Fouly T, *et al.* Secret key generation based on AoA estimation for low SNR conditions. In: *81st Vehicular Technology Conference* (VTC Spring). New York, NY: IEEE; 2015. p. 1–7.

[32] Perkins C, Lei L, Kuhlman M, *et al.* Distance sensing for mini-robots: RSSI vs. TDOA. In: *International Symposium of Circuits and Systems* (ISCAS). New York, NY: IEEE; 2011. p. 1984–1987.

[33] Badawy A, Khattab T, ElFouly T, *et al.* Secret key generation based on channel and distance measurements. In: *6th International Congress on Ultra*

Modern Telecommunications and Control Systems and Workshops (ICUMT). New York, NY: IEEE; 2014. p. 136–142.

[34] Rivest RL. Cryptography. In: Van Leeuwen J, Leeuwen J, editors. *Handbook of Theoretical Computer Science*, Volume A. New York, NY: Elsevier; 1990. p. 718–755.

[35] Jiao L, Wang N, Wang P, *et al.* Physical layer key generation in 5G wireless networks. *IEEE Wireless Communications Magazine*. 2019;26(5):48–54.

[36] Furqan HM, Hamamreh JM, Arslan H. Physical layer security designs for 5G and beyond. In: Arslan H, Basar E, editors. *Flexible and Cognitive Radio Access Technologies for 5G and Beyond*. IET; 2020. p. 545–587.

[37] Khodakarami H, Lahouti F. Link adaptation for physical layer security over wireless fading channels. *IET Communications*. 2012;6(3):353–362.

[38] Chen X, Ng DWK, Gerstacker WH, *et al.* A survey on multiple-antenna techniques for physical layer security. *IEEE Communications Surveys & Tutorials*. 2016;19(2):1027–1053.

[39] Pei M, Wang L, Ma D. Linear MMSE transceiver optimization for general MIMO wiretap channels with QoS constraints. In: *International Conference on Communications in China* (ICCC). New York, NY: IEEE; 2013. p. 259–263.

[40] Zhang L, Cai Y, Champagne B, *et al.* Tomlinson-Harashima precoding design in MIMO wiretap channels based on the MMSE criterion. In: *International Conference on Communication Workshop* (ICCW). New York, NY: IEEE; 2015. p. 470–474.

[41] Li Z, Trappe W, Yates R. Secret communication via multi-antenna transmission. In: *2007 41st Annual Conference on Information Sciences and Systems*. New York, NY: IEEE; 2007. p. 905–910.

[42] Khisti A, Wornell G, Wiesel A, *et al.* On the Gaussian MIMO wiretap channel. In: *International Symposium on Information Theory*. New York, NY: IEEE; 2007. p. 2471–2475.

[43] Liu S, Hong Y, Viterbo E. Practical secrecy using artificial noise. *IEEE Communications Letters*. 2013;17(7):1483–1486.

[44] Xiong T, Lou W, Zhang J, *et al.* MIO: enhancing wireless communications security through physical layer multiple inter-symbol obfuscation. *IEEE Transactions on Information Forensics and Security*. 2015;10(8): 1678–1691.

[45] Althunibat S, Sucasas V, Rodriguez J. A physical-layer security scheme by phase-based adaptive modulation. *IEEE Transactions on Vehicular Technology*. 2017;66(11):9931–9942.

[46] Li W, Mclernon D, Lei J, *et al.* Cryptographic primitives and design frameworks of physical layer encryption for wireless communications. *IEEE Access*. 2019;7:63660–63673.

[47] Ma R, Dai L, Wang Z, *et al.* Secure communication in TDS-OFDM system using constellation rotation and noise insertion. *IEEE Transactions on Consumer Electronics*. 2010;56(3):1328–1332.

[48] Park D, Ahn J, Choe C, *et al.* A noise-shaped signaling method for vehicle-to-everything security. *IEEE Access*. 2021;9:75385–75397.

[49] Zegrar SE, Furqan HM, Arslan H. Flexible physical layer security for joint data and pilots in future wireless networks. *IEEE Transactions on Communications*. 2022;70(4):2635–2647.

[50] Yusuf M, Arslan H. Controlled inter-carrier interference for physical layer security in OFDM systems. In: *84th Vehicular Technology Conference* (VTC-Fall). New York, NY: IEEE; 2016. p. 1–5.

[51] Hao L, Li T, Ling Q. A highly efficient secure communication interface: collision-free frequency hopping (CFFH). In: *Workshop on Signal Processing Applications for Public Security and Forensics*. New York, NY: IEEE; 2007. p. 1–4.

[52] Wang Q, Zhang H, Lyu Q, *et al.* A novel physical channel characteristics-based channel hopping scheme for jamming-resistant in wireless communication. *International Journal of Network Security*. 2018;20(3):439–446.

[53] Van Huynh N, Nguyen DN, Hoang DT, *et al.* "Jam Me If You Can:" defeating jammer with deep dueling neural network architecture and ambient backscattering augmented communications. *IEEE Journal on Selected Areas in Communications*. 2019;37(11):2603–2620.

[54] Zhang P, Liu J, Shen Y, *et al.* Lightweight tag-based PHY-layer authentication for IoT devices in smart cities. *IEEE Internet of Things Journal*. 2019;7(5):3977–3990.

[55] Nosratinia A, Hunter TE, Hedayat A. Cooperative communication in wireless networks. *IEEE Communications Magazine*. 2004;42(10):74–80.

[56] Liu KR, Sadek AK, Su W, *et al. Cooperative Communications and Networking*. Cambridg: Cambridgee University Press; 2009.

[57] Jameel F, Wyne S, Kaddoum G, *et al.* A comprehensive survey on cooperative relaying and jamming strategies for physical layer security. *IEEE Communications Surveys & Tutorials*. 2018;21(3):2734–2771.

[58] Zhao R, Tan X, Chen DH, *et al.* Secrecy performance of untrusted relay systems with a full-duplex jamming destination. *IEEE Transactions on Vehicular Technology*. 2018;67(12):11511–11524.

[59] Solaija MSJ, Furqan HM, Ankaralı ZE, *et al.* Cyclic prefix (CP) jamming against eavesdropping relays in OFDM systems. In: *Wireless Communications and Networking Conference* (WCNC). New York, NY: IEEE; 2022. p. 1976–1980.

[60] 3rd Generation Partnership Project (3GPP). Coordinated Multi-point Operation for LTE Physical Layer Aspects (Rel-11); Sept. 2013. 36.819, ver 11.2.0.

[61] Wang C, Wang Z. Signal alignment for secure underwater coordinated multipoint transmissions. *IEEE Transactions on Signal Processing*. 2016;64(23):6360–6374.

[62] Yusuf M, Arslan H. Secure multi-user transmission using CoMP directional modulation. In: *82nd IEEE Vehicular Technology Conference* (VTC2015-Fall); 2015. p. 1–2.

[63] Hafez M, Yusuf M, Khattab T, *et al.* Secure spatial multiple access using directional modulation. *IEEE Transactions on Wireless Communications*. 2017;17(1):563–573.

[64] Xiao L, Wan X, Han Z. PHY-layer authentication with multiple landmarks with reduced overhead. *IEEE Transactions on Wireless Communications*. 2017;17(3):1676–1687.

[65] Di Renzo M, Zappone A, Debbah M, *et al.* Smart radio environments empowered by reconfigurable intelligent surfaces: how it works, state of research, and the road ahead. *IEEE Journal on Selected Areas in Communications*. 2020;38(11):2450–2525.

[66] Almohamad A, Tahir AM, Al-Kababji A, *et al.* Smart and secure wireless communications via reflecting intelligent surfaces: a short survey. *IEEE Open Journal of the Communications Society*. 2020;1:1442–1456.

[67] Li G, Hu L, Staat P, *et al.* Reconfigurable intelligent surface for physical layer key generation: constructive or destructive? *IEEE Wireless Communications*. 2022;p. 1–12.

[68] Cui M, Zhang G, Zhang R. Secure wireless communication via intelligent reflecting surface. *IEEE Wireless Communications Letters*. 2019;8(5):1410–1414.

[69] Yang H, Xiong Z, Zhao J, *et al.* Intelligent reflecting surface assisted anti-jamming communications: a fast reinforcement learning approach. *IEEE Transactions on Wireless Communications*. 2020;20(3):1963–1974.

Chapter 4

Wireless channel from physical layer security perspective

Abuu B. Kihero[1], Haji M. Furqan[1], and Hüseyin Arslan[1]

One of the critical requirements of the future wireless networks is the provision of secure communication against wide range of attacks. Contrary to the conventional cryptography-based security, physical layer security (PLS), in the broadest sense, intends to exploit different features of the wireless channel to secure not only the information being communicated but the whole communication process from myriad types of attacks. Future generations of the wireless networks will be built on various new technologies such as massive multiple input multiple output (mMIMO), reconfigurable intelligent surfaces (RIS), and sensing, and are expected to accommodate some newly emerging use-cases that never existed in the legacy networks. Both, the new technologies and the use-cases are accompanied with some new and unique channel characteristics which can be leveraged by the channel-based PLS concept. This chapter explores these newly revealed channel features to highlight their potential advantages and challenges if exploited for PLS. Specifically, the chapter provides a ground work on the critical qualities that define the eligibility of a certain channel feature for the PLS application. Afterward, channel features unique to different scenarios (such as high mobility in vehicle-to-vehicle and high speed train communication) or technologies (such as mMIMO and communication at higher frequency bands) in 5G and beyond networks are discussed form the PLS perspective. The chapter also touches upon the evolving class of security attack that intend to disrupt the channel-based PLS concept by attacking the channel characteristics/features rather than the communication itself. The chapter is then concluded by summarizing some possible research directions for the channel-based PLS concept.

4.1 Introduction

Numerous use cases and scenarios with various sets of performance requirements have emerged as a result of the expansion of societal needs. Along with the conventional pursuit of a higher data rate, specific requirements such as spectral and energy

[1]Department of Electrical and Electronics Engineering, Istanbul Medipol University, Turkey

efficiency, reliability, latency, seamless connectivity, and support for high mobility have emerged as critical. Above all, given that wireless signals are broadcast, the security of the communication process has become extremely important for the majority, if not all, of potential use cases and scenarios.

In contrast to the traditional security concern, which primarily takes into account the privacy of the information being communicated, recent networks are vulnerable to numerous types of attacks like eavesdropping, manipulation, spoofing, and jamming that target the physical signal (its waveform, beam, energy, path, etc.), radio resources (spectrum, power, etc.), user information (location, context, health, etc.), propagation channel, and the radio environment as a whole. As a result, the conventional cryptographic methods fall short of offering the required level of security. PLS developed as a potential and ground-breaking idea to address these security concerns as a result [1].

PLS relies on utilizing the dynamic properties of the wireless environment, such as the propagation channel, radio frequency (RF) front-end, and/or the communication signal itself, in order to secure the communication process without the requirement for key sharing. As mentioned above, there are several communication process aspects that can be used for PLS, however this chapter concentrates on the widely used wireless-channel based PLS implementation. Undoubtedly, PLS has made great use of propagation channel features. However, majority of the PLS approaches have only taken into account a small number of channel characteristics, such as received signal strength (RSS), amplitudes, and/or phases. This is because legacy networks were designed in a way that limited access to certain other channel functionalities. As a result of recent advancements in transceiver topologies, signal processing, and channel estimate approaches (as summarized in Figure 4.1), numerous channel features have emerged and been accessible for use, either individually or collectively, Additionally, it is anticipated that next-generation networks will include sensing and channel management methods. These features will not only enhance the quality of the channel state information (CSI) that is obtained but also allow the network to optimize the propagation characteristics in support of ongoing communication. In terms of their capacities for interacting or becoming aware of the propagation situation, legacy and future networks are compared in Figure 4.1.

Despite the fact that a significant effort has been made in utilizing the wireless channel for PLS, no study has been dedicated to giving in-depth information on the individual channel characteristics from a PLS perspective in light of the recently

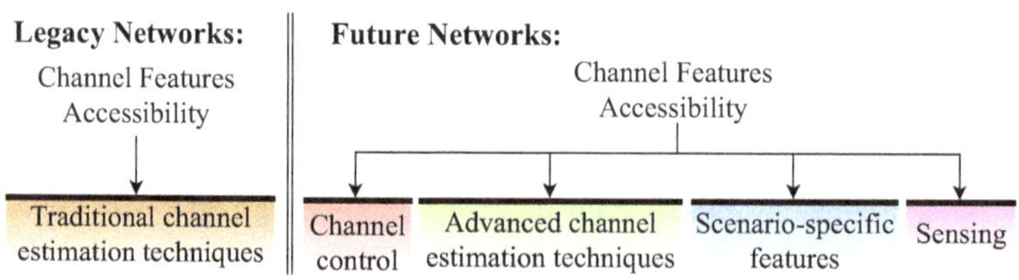

Figure 4.1 Comparison of the legacy and future networks' abilities in interacting with the propagation conditions

obtained network capabilities [1]. Therefore, this chapter aims to investigate the PLS potential of the emerging channel characteristics enabled by enabled by above mentioned latest network capabilities. To this end, this chapter examines the following channel related issues form PLS standpoint:

- *Identifying the eligibility criteria that make a given channel feature suitable for PLS:* Even though it will be possible to observe and access different channel features in the future networks, not every feature is suitable for PLS application. Essentially, the suitability of a given feature strictly depends on the nature of the attack against which a PLS technique is devised. Considering the innumerable types of security threats that have been emerging, wide and proper understanding of which qualities define a good channel characteristic for PLS is indispensable.
- *Exploring the potential of various channel features on PLS:* In relation to new use cases and network features like mMIMO, vehicle-to-everything (V2X), ultra-wide band (UWB), etc., various measurement and numerical investigations have uncovered a number of intriguing channel characteristics that were not visible in the legacy systems. There are already several studies underway on how to exploit these channel characteristics to enhance various aspects of wireless networks, including capacity and channel estimates. The PLS elements of each of these channel properties are specifically explored in this study, highlighting potential advantages and disclosing potential drawbacks.
- *Integrity of the channel features for PLS:* The chapter also outlines attacks that are directly aimed at channel features, rendering them useless for PLS applications. In other words, some attacks focus on disrupting the desired channel characteristics, such as reciprocity and spatial decorrelation, making the channel unsuitable for PLS applications.

4.2 Preliminaries on channel-based PLS approaches

This section reviews some general principles that have been exploited to facilitate PLS through wireless channel. Various channel parameters have been exploited in different ways to facilitate PLS. We categorize the popular channel-based PLS techniques into *key generation-based* and *adaptation-based techniques*, and elaborate them in the subsequent subsections.

4.2.1 Channel-based key generation

The channel-based security key generation (SKG) avoids the key-exchange process by using the wireless channel characteristics as a common source of randomness between legitimate nodes to generate the inherently shared keys. This is in contrast to the traditional Diffie–Hellman key-exchange mechanism, in which the level of communication security depends on the computational complexity of the cryptographic keys. In other words, in the channel-based SKG, the secrecy of the generated key does not rely on the computational difficulty, but rather on the physical rules governing the characteristics of wireless propagation. In particular, reciprocity and randomness are two crucial characteristics that make a given channel feature appropriate for PLS using this method. The major purposes of the channel-based SKG technique are to support confidentiality against eavesdropping attacks or as a method of authentication against spoofing attacks.

The key generation process involves five basic steps, including channel probing, randomness extraction, quantization, information reconciliation, and privacy amplification, illustrated in Figure 4.2. While all of these steps are crucial, some of them can be ignored depending on the specific implementation and environment. For example, information reconciliation, and privacy amplification stages may not be applied when the system achieves perfect key bit agreement after quantization. In the following discussion, these steps are briefly explained:

1. ***Channel probing:*** As the first step, the legitimate nodes (typically referred to as Alice and Bob) exchange probing signals with each other to measure the channel. At time $t_{i,A}$, Alice transmits ith probing signal to Bob who measures the channel parameters of interest from the received signal. Likewise, at time $t_{i,B}$, Bob transmits his ith probing signal to Alice who measures the same channel parameters. Note that the sampling time difference at Alice and Bob, i.e., $\Delta t_i = |t_{i,A} - t_{i,B}|$ must be smaller than the channel's coherence time to ensure that the measurement at Alice and Bob are performed under the same conditions. If channel reciprocity applies on the channel parameters of interest, the measurement at Alice and Bob are ought to be the same. This channel measurement process is repeated until sufficient channel statistics are collected.

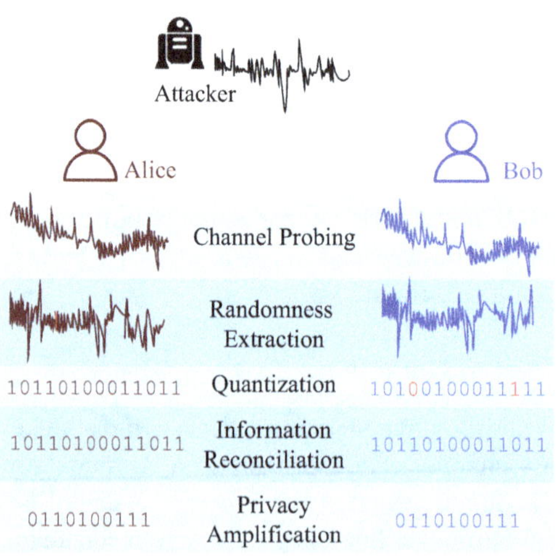

Figure 4.2 Illustration of the channel-based key generation steps between legitimate parts

2. ***Signal pre-processing and randomness extraction:*** The received signal measurements at the legitimate nodes in the channel probing step are generally asymmetric. This is mainly due to the non-simultaneous nature of the conducted measurement, i.e., $\Delta t_i \neq 0$, as well as the independent noise introduced by the two separate hardware platforms at Alice and Bob. Therefore, a signal pre-processing step is normally employed to compensate for this asymmetricity of the measurement [2,3] and get rid of the noise issue [4–6], thereby improving the cross-correlation between the measurement.

   Another critical point regarding the first step is the redundancy between the measurement. Therefore, a resampling process using a probing rate T_{pr} can be considered as well. In this case, T_{pr} has to be set larger than coherence time of the channel. If the channel is assumed to vary with the same rate, T_{pr} is generally determined based on the signal's autocorrelation function model. However, since randomness of the channel may come from the unpredictable movement of the scatterers in the environment, the rate of change of the channel usually vary with time. Therefore, keeping T_{pr} fixed can potentially lead to inefficient channel probing when the channel varies faster or redundancy between samples when the channel varies slower than the selected rate.

   Similarly, the measurement at Alice and Bob may contain some deterministic parts that may be inferred by an attacker. Therefore, a randomness extraction step is also taken in order to get rid of such deterministic aspects of the measured parameters.

3. ***Quantization:*** The legitimate nodes need to convert their channel estimates into strings of identical bits that are suitable to serve as the cryptographic key. To this end, the quantization step is adopted by Alice and Bob to map their measure channel samples into binary values which are then processed further to obtain the inherently shared secrecy key bits. Generally, a good quantization technique has to

   - facilitate a generation of long enough secrecy key bits as the longer the key the higher the security. Normally, the keys with 128–512 bits length are deemed suitable;
   - ensure that the generated key bits at the legitimate nodes have low probability of mismatch or disagreement ratio;
   - guarantee that the generated key bits are statistically random. That is, the bits should not suffer from statistical defects that can be leveraged by the attacker.

   Different approaches of performing quantization have been used in the literature. The main difference between these approaches lies on the number of the thresholds used as well as the methodology of choosing the threshold values. The popular quantization approach in the literature of SKG is the *Censoring Scheme*. In this scheme, two threshold values are defined, let us say $[\eta-, \eta+]$, which in turn define a threshold region. The indices of the measured channel samples that fall within the threshold region are discarded during secrecy bit generation. The rationale behind this practice is that the channel samples close to zero are not used to generate the key bits as they are highly probable to have opposite signs at Alice and Bob which may significantly impact the key agreement rate. Implementation-wise, Alice forms a set $\mathscr{T}_A = \{i : \eta- \leq x_A(i) \leq \eta+\}$ and transmit it to Bob. Note that $x_{\mathscr{C}}$, for $\mathscr{C} \in \{A \text{ (=Alice)}, B \text{ (=Bob)}\}$, denotes the channel measurement at Alice or Bob. Similarly, Bob forms a set $\mathscr{T}_B = \{i : \eta- \leq x_B(i) \leq \eta+\}$ and transmits $\mathscr{T}_B \setminus \mathscr{T}_A$ to Alice. At each of the legitimate nodes, the set $\overline{\mathscr{T}} = \mathscr{T}_A \cup \mathscr{T}_B$ is not used to generate the key bits. Instead, the channel samples whose indices are elements of the set $\mathscr{T} = \{1, 2, \ldots, N\} \setminus \overline{\mathscr{T}}$ are used for the inherently shared secret key generation, where N is length of the measured channel samples. If we let $t_j \in \{1, 2, \ldots, N_{\mathscr{T}}\}$ be the jth element of $\mathscr{T}$ with $N_{\mathscr{T}}$ being the size of $\mathscr{T}$, the secret bit $\mathscr{B}_{\mathscr{C}}$ at a given legitimate node is generated as

$$\mathscr{B}_{\mathscr{C}}(j) = \begin{cases} 1 & , x_{\mathscr{C}}(t_j) > \eta+ \\ 0 & , x_{\mathscr{C}}(t_j) < \eta- \end{cases}. \tag{4.1}$$

It is now pertinent to clarify that although the attacker may get a clue about a sample $x(i)$ as it either lies within or outside the threshold region due to the $\mathscr{T}_A$ and $\mathscr{T}_B$ exchange process between the legitimate nodes, but this process does not reveal any information whether sample is above $\eta+$ or below $\eta+$. Therefore, integrity of the generated key bits is guaranteed. Furthermore, some studies proposed the "excursion" based bit generation approach. An excursion is any sequence of $x_{\mathscr{C}}(t_j)$ samples which are either all above $\eta+$ or below $\eta-$. With this approach, each excursion is encoded into one bit. This allows high reliability in bit encoding while reducing correlation between subsequent bits.

As it can be perceived, the choice of thresholds is very critical in this scheme as well as in quantization process in general. Several techniques of deciding on the threshold values under the censoring scheme are available in the literature. Mean (μ) with/without standard deviation (σ) based thresholding technique is popularly used. In this approach, the thresholds are determined as

$$\begin{aligned} \eta+ &= \mu + \alpha \times \sigma, \\ \eta- &= \mu - \alpha \times \sigma, \end{aligned} \tag{4.2}$$

where $\alpha > 0$.

Despite its simplicity, as we have seen, the censoring quantization scheme involves discarding so many measured samples, which has direct impact on the length of the key bits it can generate. Additionally, this technique allows only two quantization levels "1" and "0," which makes it lack flexibility.

A multi-bit adaptive quantization (MAQ) scheme based on cumulative distribution function (CDF) is also highly used in the literature as an alternative to the above discussed censoring scheme [3]. MAQ is a more generic scheme that allows both single and multi-bit quantization based on the nature of the obtained measurement. If b is the number of bits desired for each codeword, the number of quantization levels is $Q \triangleq 4 \times 2^b$. In order to make these quantization bins equally-likely, distribution of the measured samples must be determined. To this end, the CDF of $x_{\mathscr{C}}$ is calculated as

$$F(x) = P[x_{\mathscr{C}} < x]. \tag{4.3}$$

Thresholds η_q for the equally-likely quantization, bins are then generated by using the above obtained CDF,

$$\eta_q = F^{-1}\left(\frac{q}{Q}\right), \tag{4.4}$$

where $q = 1, 2, \ldots, Q-1$. The qth quantization bin is defined by the interval $(\eta_{q-1}, \eta_q]$. Outer thresholds of the edge bins are set as $\eta_0 = -\inf$ and $\eta_Q = \inf$. Afterward, the gray codewords with b-bits are constructed and assigned to each quantization bin. A Gray codeword list changes only one bit between consecutive codewords in the list. With this coding approach, when a disagreement occur, it is more likely that one 1 bit of the multi-bits codeword will be in error.

4. ***Information reconciliation:*** Once the secret key bit streams are generated at the legitimate nodes, it is crucial to check if they agree with each other. If any disagreement is detected a correction process must be applied. The disagreement may occur due to various reasons such as presence of noise and interference, hardware limitations including manufacturing variations and differences in implementing automatic gain control among the devices, as well as the non-simultaneous sampling during channel probing step due to the half-duplex nature of the communication devices. The main dilemma in the information reconciliation step is how to get rid of this key disagreement at Alice and Bob without exposing information out on the channel that can be leveraged by attackers.

   A number of information reconciliation approaches are used in the literature, including the ones that are based on the elliptic-curve cryptography (ECC) and those that involve interactive information protocols such as Cascade technique [5,7]. With the Cascade approach, one of the legitimate notes (Alice, for example) randomly permutes her key bit stream, chunks it into small blocks and send out the permutation and parity information of each block to Bob. Likewise, Bob permutes his own key bits, chunks into blocks, calculates and checks if parity the blocks agree with those received from Alice. For the blocks whose parities do not agree, Bob performs a binary search to find and accordingly correct the bits such that parity of the block matches. This procedure is iteratively repeated until the desired probability of success is achieved.

   In the ECC-based reconciliation approaches, a code C is adopted and Alice randomly choose a codeword c from C, i.e., $c \in C$. She then computes a sequence s by exclusive OR-ing her key stream $\mathscr{B}_A$ with c as $s = \text{XOR}(\mathscr{B}_A, c)$ which is then transmitted to Bob. Bob computes c'' by exclusive OR-ing his key bit stream $\mathscr{B}_B$ with the sequence received from Alice, i.e., $c'' = \text{XOR}(\mathscr{B}_B, s)$ and decode c' from c''. Using c', Bob calculates $\mathscr{B}'_B$ as $\mathscr{B}'_B = \text{XOR}(c', s)$. Bob agrees on the same key with Alice, i.e., $\mathscr{B}'_B = \mathscr{B}_A$ when the hamming distance between c and c'' are smaller than the error correction capacity of the used code. The key agreement can be confirmed via cyclic redundancy check approach. Generally, the ECC-based techniques are considered to be more efficient than the Cascaded technique discussed before, however, they are also deemed to expose more information to the attackers and are relatively more complex. Different types of codes have been widely used in the literature, including low-density parity-check (LDPC), BCH code, Reed-Solomon code, Golay code, Turbo code, etc. The main criteria of selecting which code to use are their error correcting capacity and they complexity they may cause.

5. ***Privacy amplification:*** This step is applied to deal with two main issues that might have arisen during the whole process of secret key generation. First, to remove portions of the key bit stream that might have been exposed to the attacker during the information reconciliation step. Nevertheless, in practice, it is generally difficult to quantify the amount of the leaked information, or to identify where exactly the leakage occurs. Second, to minimize correlation between key bits and enhance randomness. Correlation between key bits can be contributed by the inability to obtain full independent channel measurement during the channel probing step. This mainly occur due to the fact that, in order to obtain full independent channel measurement we must probe the channel only once during its coherence period. However, in real scenarios, it is usually difficult to estimate the channel's coherence time with adequate accuracy. Erroneous estimate of the coherence time may lead to selection of inadequate channel probing rate that lead to the generation of the correlated secret key bits. Universal hash functions are generally used for privacy amplification step.

4.2.1.1 Channel parameters for SKG in the legacy systems

1. ***Received signal strength (RSS):*** RSS is one of the most popular channel parameters that has been widely considered for PLS application even in the practical networks. It is obtained as average power of the received signal. RSS is the most popular channel parameters exploited for PLS primarily because of its readily availability in the network. Most of the legacy off-the-shelf network interface cards (NICs), without any modification, can measure RSS on a per-frame basis. RSS captures the channel variation over time due to random motion and activities within the propagation environment which induce some randomness required for SKG. However, the mean RSS value is somewhat a predictable function of distance. Such predictable aspects of RSS must be taken care of during key generation as they can be exploited by attackers. Additionally, since RSS is a coarse-grained channel parameter, i.e., only one RSS value can be obtained from the whole frame, it has very low bit generation rate.

2. ***Channel impulse response (CIR):*** CIR is another channel parameter that has been considered for PLS. CIR has two main degrees-of-freedom that can be exploited for SKG. These include the phase and amplitude of the observed channel taps.

Wideband signaling provides adequate resolvability of the channel taps. Phase information of these taps can be easily estimated and used for key generation. Although channel phase can be estimated and exploited in narrowband systems as well, the lack of multipath resolvability in such systems reduces the phase information into a single value parameter per transmission which leads to low key generation rate. Some peculiar characteristics of channel phase makes it appropriate for novel ways of generating security keys. For example, the fact that phases can be accumulated to each other paved a way to the group or cooperative key generation techniques [8,9]. Additionally, phases of the channel paths are distributed uniformly on $[0,2\pi]$, which are not affected by the path power. Despite these advantages, phase information observed at the transceiver devices are highly vulnerable to noise and any type of asynchronicity (carrier frequency offset (CFO) and clock errors) between the devices. This has been limiting the practical realization of the phase based SKG approaches.

Likewise, amplitudes of different channel taps in wideband systems can used for secret key generations. Again, while this is applicable for narrowband systems as well, lack of multipath resolvability in those systems reduces this amplitude information into a single value parameter which reduces the key generation rate. Note that amplitudes of the channel taps usually decrease with tap delays. As such, most of the late-arriving taps have low power and are vulnerable to noise. If such noise infested amplitude information are exploited for key generation may result in high key disagreement rate between Alice and Bob. This problem can be circumvented by employing robust quantization techniques during key generation process. Some studies proposed to exploit the peak/strongest tap only for SKG [10]. However, this may lead to lower key generation rate.

3. ***Channel frequency response (CFR):*** This parameter has been widely studied for SKG in orthogonal frequency division multiplexing (OFDM) systems. Similar to the CIR, channel amplitude and phase information on each subcarrier can be exploited for SKG. In this case too, utilization of the phase is limited by the time and frequency offset problems in the practical implementation.

4.2.1.2 Performance metrics for channel-based SKG

1. ***Randomness and entropy:*** One of the most strict features required for a secret key is the randomness of its constituent bits. This requirement is for preventing an attacker from being able to guess the generated secret key sequence. Provision of a highly random key relies on the channel feature exploited to generate it. Technically, randomness of a generated key is quantified by its entropy. By definition, entropy is a metric that characterizes the uncertainty of a given random variable. For a secret key sequence $\mathscr{B}_{\mathscr{C}}$, its entropy is calculated as

$$H(\mathscr{B}_{\mathscr{C}}) = -\sum_{i}^{N_{\mathrm{Key}}} p\left(\mathscr{B}_{\mathscr{C}}(i)\right) \times \log_2 p\left(\mathscr{B}_{\mathscr{C}}(i)\right), \tag{4.5}$$

where N_{Key} is the length of the key sequence and $p\left(\mathscr{B}_{\mathscr{C}}(i)\right)$ is the probability of occurrence of the ith bit in the sequence.

A statistical randomness test suite provided by National Institute of Standards and Technology (NIST) is widely adopted in SKG studies for PLS for testing the randomness of the generated keys. The test suite consists of about 15 tests that focus on different types of non-random characteristics that may exists in a given sequence. Details about these tests can be found in NIST report [11].

2. ***Key generation rate (KGR):*** KGR describes the amount of secret bits produced in one second/measurement. It mainly depends on environment conditions, which determines the amount of randomness available for extraction. A high KGR is essential for the real time key generation process as the cryptographic schemes require a certain length of keys. For example, advance encryption standard (AES) needs a key sequence with a minimum length of 128 bits.

3. ***Key disagreement rate (KDR):*** KDR is the percentage of the different bits between the keys generated by Alice and Bob, which is defined as

$$KDR = \frac{\sum_i^N |\mathscr{B}_A(i) - \mathscr{B}_B(i)|}{N}, \tag{4.6}$$

where N is the length of keys. The KDR should be smaller than the correction capacity of information reconciliation techniques, otherwise, key generation fails.

4.2.2 Channel-based adaptation PLS techniques

Similar to the SKG, this approach exploit channel characteristics to secure communication. But, instead of generating secrecy keys, legitimate nodes dynamically adapt their transmission parameters based on the instantaneous channel condition which is unique to them. This approach is applicable against wide range of attacks, including eavesdropping, spoofing, jamming, impersonation, etc. Most of the techniques falling under this group involve manipulation of the signal parameters or RF transmission characteristics, which are discussed in details in Chapter 7 under control plane section.

4.3 Eligibility requirements of channel parameters for PLS

Although multitude of channel features are becoming visible and accessible as wireless systems evolve, not all of them are appropriate for PLS applications. As such, deciding which channel aspect or parameter to exploit for this purpose still remains to be a challenge. Generally, a given channel feature must meet a set of minimal requirements in order to be considered exploitable for PLS applications. Figure 4.3 outlines some critical qualities that an ideal channel parameter is ought to exhibit for it to be eligible for PLS. However, it is important to emphasize that the criteria that

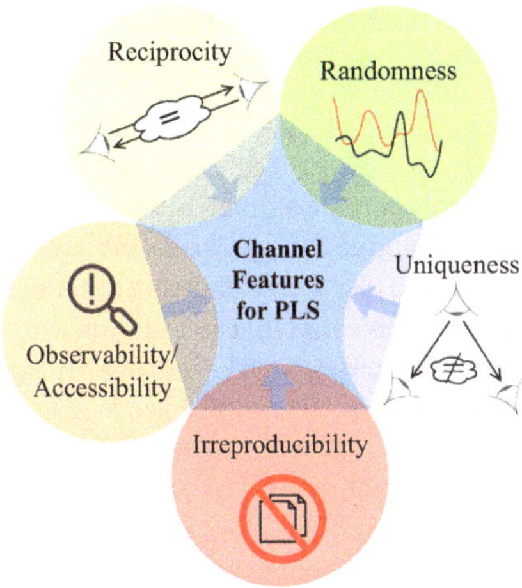

Figure 4.3 Desirable qualities of a channel feature suitable for PLS application

can be used to select one channel parameter over the other strongly depends on the nature of the security attack against which a PLS devised.

4.3.1 Randomness

The random variability of the channel characteristics in different domains is an essential requirement as it increases channel entropy for stronger PLS implementation. Generally, the randomness of the channel response stems from the nature of the propagation environment as well as characteristics of the communicating terminals. Rich scattering environment usually facilitates multipath propagation which allows multiple copies of the transmitted signal with random phases and amplitude to superpose constructively or destructively at the receiver leading to a fluctuating received signal envelope. Note that the phases and amplitudes of the multipath components observed at the receiver depend on the location of the receiver with respect to the randomly distributed environmental scatterers. This guarantees that if an attacker is not co-located with the legitimate user, the variations in his channel response will be different from that at the legitimate user. On the contrary, less scattering environment mostly facilitates line-of-sight (LoS) propagation condition which makes channel response more deterministic, thus can be easily predicted by an attacker. Essentially, propagation environment can naturally be less scattering due to the lack or poor distribution of scattering objects or made appear that way due to the adopted transmission/reception mechanisms. For example, multipath propagation effect can be reduced and LoS propagation enhanced by using directional transmission techniques such as beamforming. With that being said, considering the recent trend toward mMIMO-based networks in which beamforming-based transmission and reception are popular, channel responses are becoming more sparse and deterministic. Although this is advantageous from other aspects of the network such as the reduced channel estimation burden, the loss of randomness quality of the channel imposes great danger on the feasibility of the channel-based PLS concept.

4.3.2 *Uniqueness*

This property ensures that any security parameters taken from the channel response seen by the legitimate node are unique to them exclusively in relation to their immediate context and cannot be copied elsewhere or at any other time instant. In essence, uniqueness is a continuation of the randomness criterion with an emphasis on the dynamicity of the channel characteristics. The channel must be dynamic so that the values of the secrecy parameters generated at one instant differ from those at a later time, in addition to being random in the domain of interest within a single channel usage. This is particularly crucial when using the channel's frequency or spatial unpredictability. For instance, a channel may exhibit strong randomness in the frequency domain in the form of frequency selectivity due to the distant scatterers (longer delay spread), but if the environment is entirely static, the channel response remains the same for all channel usages. This can tempt the enemy to conduct her own test later and calculate the security key that has been utilized in the case of the SKG-based PLS. It is also crucial to note that rapid channel variation may result in the channel aging problem, which can severely affect the feasibility of the channel-based PLS. When channel varies rapidly, it is highly possible that the legitimate nodes may not be able to probe the same channel which may lead to high KDR and complicate the SKG process.

In addition to the temporal uniqueness, channel response also needs to be spatially unique, or what is known as "*spatial decorrelation*." This characteristic stresses that the channel response observed by the legitimate nodes is unique to them only in relation to their immediate context, thus any security parameters taken from it can not be regenerated anywhere else at any other time. In the traditional CIR and CFR-based PLS techniques in rich scattering environment, channel's coherence distance is generally used to quantify the extent with which this requirement can be fullfilled. In this case, uniqueness of the channel response is guaranteed when attacker's node is at least a half-wavelength away from the legitimate nodes. However, this assumption may no longer be valid for the modern wireless systems due to the sparse nature of the observed channels as a result of the use of UWB signaling (temporal sparsity), mMIMO (spatial sparsity), a high-frequency band (spatial and temporal sparsity), etc. It is also difficult to comply with this requirement when the attacker is co-located with one of the legitimate nodes.

4.3.3 *Reciprocity*

Although an appropriate feature of channel is the one that varies randomly and constantly as explained above, measurements of that feature at the legitimate nodes must be highly correlated at any given instant. This is referred to as reciprocity and it is necessary in order to ensure that the channel feature under consideration serves as the common (inherently shared) source of randomness among the legitimate nodes, which is the main driving factor behind the channel-based PLS concept. Imperfect reciprocity might affect communication quality as well as the overall security performance. In order to fulfill this requirement, the legitimate nodes must finish the bidirectional channel probing for feature extraction before the channel decorrelates over time or space (for the mobile terminals). Nevertheless, perfect reciprocity is difficult to achieve in reality. As such, some extra processing are normally employed to make up for the incurred channel mismatch measured by the nodes. For example, in the SKG approach, the information reconciliation step is typically used, as explained earlier. However, if the channel dynamicity is too quick like it could be in the vehicular and high speed train communication scenarios, the channel aging problem may be dominant during the channel probing step, leading to total loss of reciprocity. In this case, considering the non-simultaneous nature of the channel probing process, the channel measure by one node may be totally different from the one measured by the other node. In order to facilitate reciprocity in such scenarios, the choice of the channel probing rate must be done very carefully.

4.3.4 Accessibility/observability

This refers to the amount of processing necessary to extract and use a specific channel feature for PLS. Easy access to the channel features for PLS is very crucial especially in the modern era, considering the exponential rise in the number of low-end, low-energy, and lightweight computing Internet of things (IoT) devices, whose needs for security are in no way less important. For example, the literature have started exploiting the sparse channel responses for PLS, where channel features like individual multipath component (MPC)'s angle of departure (AoD)/angle of arrival (AoA) are being exploited instead of the composite CIR and CFR used in the legacy systems. However, to access and observe these features requires mMIMO transceivers and extra beamspace channel processing capabilities which may not be possible for the IoT networks. In additional to the simplicity, the quality with which the feature is accessed is also important. This also takes into account the feature's robustness to estimation errors, noise, or other RF front-end impairments. As mentioned earlier, some features like channel phase are highly prone to clock and synchronization mismatches. Noise can drastically lower the secrecy bits agreement ratio in SKG.

4.3.5 Irreproducibility

Although the aforementioned uniqueness criterion guarantees that the attacker and the legitimate node perceive different channel responses, it does not restrict the attacker's ability to mimic legitimate nodes' channel properties. This feature is of paramount importance from the impersonation attack perspective such as Camouflage attack [12]. In this case, the attacker can trick the legitimate nodes by injecting fake wireless features between himself and one of them and mimic the legitimate link. In the light of such types of security attacks, it can be concluded that the channel features that are not only unique but also challenging to imitate are more favorable for PLS.

4.4 Channel parameters beyond 5G: PLS perspective

Until now, only a few number of the channel's properties, such as RSS, channel gains, envelope, and phase, have been exploited for PLS. This is mainly because of the fact that the legacy networks did not give much accessibility to other channel features/parameters as illustrated in Figure 4.1. With the recent advancement in different aspects of the modern networks such as transceiver architectures, signaling techniques, signal processing, and channel estimate approaches, numerous channel properties have emerged and become available for exploitation, either individually or collectively. Some of these features are illustrated in Figure 4.4. This section explores these new channel features from PLS perspective. However, it is crucial to emphasize that some of these aspects are still being studied, and thus the discussion herein only tries to highlight their potential from PLS point of view and encourage further research on that direction.

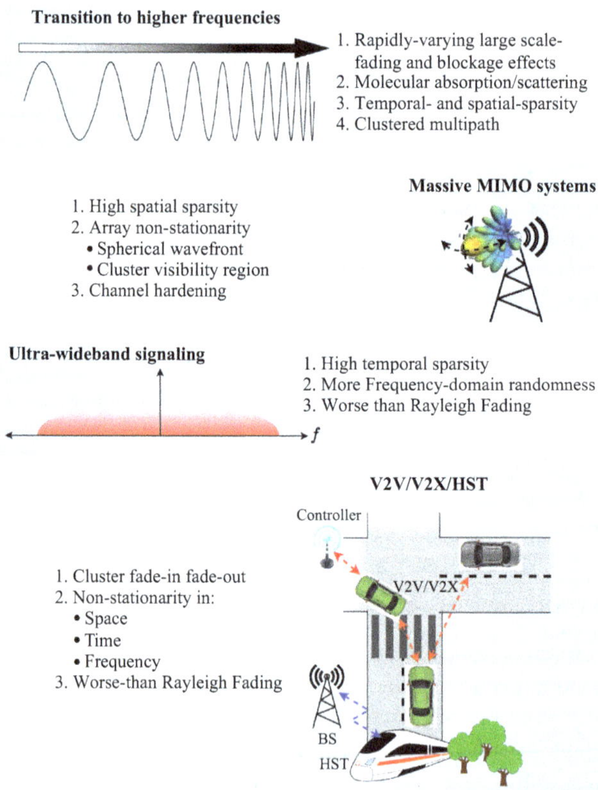

Figure 4.4 Unprecedented channel features due to the use cases and technologies featured in 5G and beyond wireless networks

4.4.1 Large-scale fading

The large-scale fading characteristics of the channel have not been significantly considered PLS because of their deterministic nature. Large-scale fading features such as the distance-dependent PL and the slow-varying shadowing phenomenon observed in the legacy networks are highly deterministic in nature. However, the recent transition to higher frequency bands like millimeter wave (mmWave) in 5G and the prospected THz in 6G and beyond has disrupted this paradigm. Due to the narrow wavelengths at these higher frequencies, even little objects in the surroundings can cause significant momentary shadowing or even total blockage of the signal. This causes the signal strength to fluctuate erratically and quickly across time and/or space. This induces more randomness into the channel that can be beneficial for the key-based and adaptation-based PLS techniques. The study in [13] has revealed that, at higher frequencies, PL coefficients are greatly affected by the orientations of streets and buildings. This results in a significant variation in the signal strength between nearby streets at the same distance from the base station (BS). These nascent characteristics of pathloss (PL) and shadowing introduce an extra layer of randomness into the channel which signifies the potential of large-scale fading in facilitating PLS. Additionally, high PL effect observed at higher frequency bands bestows a high degree of spatial confinement of the communication signal that provides inherent security. Studies like [14] have already demonstrated that large-scale fading characteristics may be individually exploited to provide security against eavesdropping attacks with the right power allocation techniques and transmission methodologies. Essentially, considering the fact that small and large-scale fading phenomena occur concurrently, the traditional PLS approaches that mainly focused on the small-scale fading characteristics of the channel can be revisited to leverage the rapidly varying PL and shadowing phenomena to enhance their achievable secrecy level.

4.4.2 Molecular absorption and scattering

At higher frequencies, the channel exhibits not only peculiar characteristics related to PL and shadowing but also a severe type of frequency-selective absorption referred to as molecular absorption (MoA). At higher frequency bands, some of the frequencies align with natural resonance frequencies of the of the atmospheric contents such as water vapor and oxygen molecules. These molecules absorb a large amount of energy from the signal when activated at their respective resonance frequencies, which raises PL at those particular frequencies. Shortly after, some of the absorbed energy is re-radiated at the same frequency into the surrounding environment. Despite the fact that some studies have treated the re-radiated energy as a source of noise in addition to the usual thermal noise, some investigations have revealed that this energy is highly correlated with the original signal and should be treated as the scattered copy of the same signal, a phenomenon termed as molecular scattering (MoS). Some further studies have shown that MoS raises the multipath richness of the channel (while lowering the channel's K-factor) and raises the randomness, which can be advantageous for PLS. This scattering effect, nevertheless, might also have negative effects. Eavesdroppers can exploit the scattered signal to listen the ongoing conversation. Furthermore, MoA and MoS can be powerful enough to entirely destroy the beam and interfere with communication between the authorized nodes in cases of atmospheric turbulence, heavy rain or snow.

4.4.3 Small-scale fading

Arguably, multipath fading is the channel feature most exploited for PLS. This is simply due to the fact that this feature readily meets most of the eligibility conditions discussed in the previous section. Multipath fading is inherently a random phenomenon as it is constituted by the superposition of the random replicas of the transmitted signal. Multipath characteristics are generally function of the propagation environment in terms of the number, distribution, and mobility of the scatterers with respect to the positions of the transmitting and receiving terminals. This makes the multipath fading features like signal envelope, phases, and spatial distribution not only random but also unique in space domain, which are desirable qualities for PLS. Traditionally, remote scatterers result in significant delay spread, which is the main source of frequency selectivity. In the presence of mobility, however, scatterers local to the transceiver terminals are the main source of time selectivity. The spatial selectivity of the channel is similarly determined by the scatter distribution relative to the communication terminals. In comparison to a static environment with low scattering, a rich scattering environment with some mobile scatterers/terminals imparts more randomness. Additionally, the channel becomes more random in the absence of a strong LoS component.

Reciprocity is another defining feature of the multipath fading phenomenon. Essentially, multipath propagation process is made up of some frequency-dependent and frequency-invariant instances. While frequency-independent instances are generally reciprocal even in the frequency division duplex (FDD) systems, the frequency-dependent instances exhibit reciprocity only when the uplink and downlink communications are performed on the same frequency. Owing to the fact that the legacy networks operates at lower frequency bands in which channel is often rich scattering, it has been difficult to distinguish and exploit these instances individually. As such, PLS techniques in the previous generations are built on the assumption of the validity of reciprocity for time division duplex (TDD) systems only as the same frequency band is used in both uplink and downlink. However, some studies such as [15] have set forth on eliminating the TDD-based reciprocity validity. For example, the study just mentioned above have shown that, by exploiting some basic propagation principles, channel response at one frequency band can be inferred from other bands. It is shown that, for nearby frequency bands, although the overall channel response changes the physical paths traversed by a signal remain the same. Based on this, the authors developed an inter-band *physical-channel* mapping mechanism that estimates the channel coefficient at one band from a known channel at a different band. Evolution of such techniques will facilitate the implementation of the channel-based PLS techniques even in the absence of the traditional TDD-based reciprocity assumption.

4.4.4 *Sparsity*

As mentioned in the Introduction section, the future generations of wireless networks, starting with 5G, are expected to feature some technologies such as mMIMO architectures, UWB signaling, and migration to higher frequency bands which make the observed channel highly sparse. Different from the rich scattering propagation observed in the legacy networks, sparse channel responses put individual channel features, such as AoA, AoD, and delay of the MPCs at the disposal of PLS. The features can now be easily accessed and exploited for PLS individually or jointly rather than relying only on the CSI-based effective channel fading coefficients. For instance, the work in [16] exploits AoA and AoD in mMIMO as a shared source of randomness between the legitimate nodes for key generation. These individual features have been shown to be resilient against noise and demonstrates strong reciprocity even in the low signal-to-noise ratio (SNR) regimes, in addition to being easily accessible or simple to estimate via virtual channel estimating techniques. Additionally, the sparse channel response observed in the modern wireless systems has made it possible to take advantage of the individual multipath channel's frequency-invariant characteristics, including AoA, AoD, and path delays which exhibit reciprocity even under FDD.

4.4.5 *Array non-stationarity*

Large-dimensional arrays, in particular mMIMO, have been discovered to experience other distinctive channel features in addition to the high spatial resolvability of the MPC. These peculiar channel features can be advantageous in different aspects of wireless communication, including PLS. In essence, it has been discovered that there are significant differences across array elements in many channel parameters such as gain, K-factor, angular power distribution, and correlation level. In addition to increasing channel entropy for the production of stronger keys, such considerable fluctuations in different channel parameters also decorrelate the channel coefficients across several domains, hence strengthening uniqueness of the channel coefficient observed by a given user. The peculiar channel characteristics of mMIMO are thought to be mainly caused by two phenomena, spherical wavefront (SW) and cluster visibility region (VR):

Spherical wavefront (SW): With large arrays, majority of the important scatters and user equipment (UE) are found in the array's near-field region, where the conventional planar wavefront assumption is no longer true. The signal in this instance enters the array through SW, which has been proven to add an additional phase shift that significantly improves the spatial channel decorrelation. Early studies have shown that SW propagation can enhance spatial uniqueness by decorrelating the channels of closely located UEs even in LoS conditions. This improves the system's ability to recognize authentic UE from fraudulent ones which improve security.

Cluster VR: Since most significant scatterers are located within the array's near-field region, not all scatterers are "visible" to all array elements. As a result, the concept of cluster VR is introduced in the literature of the mMIMO systems, in which different array components experience different channel responses depending on the cluster of scatterers that are visible to them. The channel becomes extremely non-stationary across the array, resulting in more spatial channel decorrelation an both transmitter and receivers, which can be leveraged for PLS. Different channel characteristics, including AoA/AoD, gain, and fading statistics across the array, can exhibit non-stationarity that can be individually or jointly exploited for PLS.

4.4.6 Temporal, Doppler and spatial non-stationarity

The non-stationarity concept also spans to the temporal and Doppler domains of the channel. The non-stationarity in these domains is observed, particularly in scenarios that involve mobility as in high speed train (HST) and vehicle-to-vehicle (V2V)/V2X communications. These scenarios, which differ from the traditional fixed-to-mobile cellular system, are characterized by an excessive mobility from both, the terminals and the nearby scatterers, which causes the propagation characteristics to vary rapidly. Additionally, the set of scatterers interacting with the signal varies constantly with time, leading to the *cluster fade-in fade-out* effect. As a result, the observed MPCs's spatial, temporal, and Doppler characteristics vary rapidly and continuously. Designing cross-domain PLS approaches can leverage such multi-domain fast variation in the propagation characteristics. For instance, a strong secret key can be generated by exploiting channel randomness from multiple domains as discussed in [17]. However, as mentioned earlier, the rapidly varying channels is often accompanied with channel aging process that destroy reciprocity.

4.5 Integrity of channel features

Numerous types of attacks have been launched as channel-based PLS gains popularity. These are launched specifically to make different channel features unsuitable for PLS application. Additionally, some attackers exploit some basic propagation principles to evade the implemented channel-based PLS. These attacks can be roughly divided into *direct* and *indirect* attacks, which are describe below.

4.5.1 Indirect attacks on the integrity

Under this type of attacks, the attacker implements some signal processing techniques to manipulate certain channel characteristics and deceive the legitimate nodes during the channel probing stage or authentication process. Furthermore, the attacker can also employ machine learning, blind signal analysis, or reverse-engineering processing of the received signal to derive the channel coefficients extracted and used by legitimate nodes for PLS. Specific examples are elaborated below.

As mentioned before, majority of the conventional channel-based PLS techniques rely on CSI, specifically CIR, under the assumption of spatial decorrelation. However, it has been shown that, in practical scenarios, an attacker can make use of the common knowledge about the nature and geometry of the communication scenario and some signal processing techniques to obtain the channel response of the legitimate nodes even under spatial decorrelation effect. For instance, in the CSI-snoop attack [18], the attacker makes use of the filtering mechanisms and knowledge of the standard/known frame structure in multi-user multiple-input multiple-output (MIMO) networks to compute the CSI. Similar to this, in the predictable channel attacks, the CIR between legitimate nodes is pre-computed using knowledge of the propagation environment geometry. In other cases, the attacker may be able to learn the CSI between the legitimate nodes by taking advantage of the actuality of the propagation condition, such as sparse scattering, presence of a significant LoS component or the waveguide propagation effect which makes the channel more deterministic. Such types of attacks urge a reconsideration of PLS strategies that rely on effective CSI, such as CIR. Utilizing the novel, distinctive channel properties discussed above, stronger strategies can be developed. Similar to the cross-layer security concept in which the physical (PHY) and medium access control (MAC) layer characteristics are jointly utilized for PLS, these distinct channel features in different domains can also be utilized simultaneously to gain resilience against these indirect attacks on integrity.

4.5.2 Direct attacks

This type of attacks involves a direct manipulation of the channel characteristics observed by the legitimate nodes. As such, the legitimate nodes may measure and extract wrong/manipulated channel features (i.e., artificially generated by the attacker) for implementing a PLS algorithm or perform authentication process. A typical example of the direct attack on the channel integrity is the *Camouflage attack* [12]. Under Camouflage attack, an attacker may use some of the channel control mechanisms such as those discussed in Chapter 7 to mimic the legitimate link and thus deceive the legitimate transmitter/receiver. The attacker can also manipulate his own channel to disguise his location or identity. An RIS controlled by attacker can also be used to against the legitimate nodes. For instance, the attacker can employ an RIS to manipulate multipath channel characteristics such as reciprocity (exemplified in Figure 4.5). Leveraging the MoS characteristics in terahertz (THz) frequencies, the attacker can create an artificial gaseous cloud to deliberately scatter the legitimate users' signal toward its own direction and boost its listening capability, or just degrade reliability at the legitimate nodes as in the case of the jamming attack. The direct attacks on the environment can also be dedicated to the PLS techniques exploiting propagation features obtained from sensing. For instance, the attacker can re-transmit a modified/manipulated sensing signal or employ an RIS to manipulate the speed, range, or trajectory of the target environmental objects or even cause total misdetection of these objects, leading to the acquisition of the incorrect surrounding information.

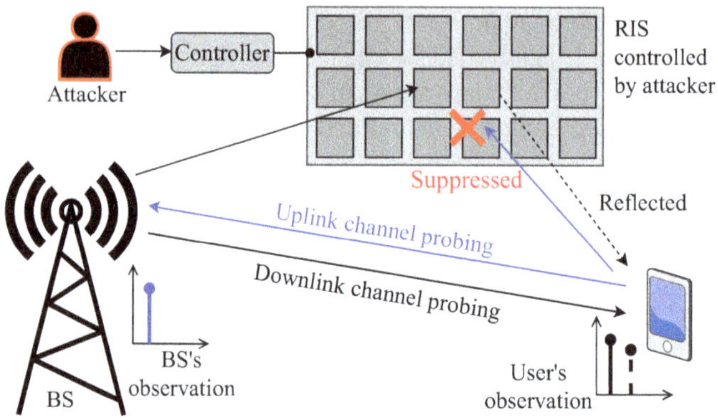

Figure 4.5 An attacker using RIS to manipulate MPCs during channel probing stage and destroy reciprocity

4.6 Future direction and recommendation

4.6.1 Securing integrity of the channel features

As discussed in the previous section, a number of attacks dedicated directly to the propagation channel characteristics have been launched, which put the credibility of the channel-based PLS concept in jeopardy. This urges for the widening of the PLS scope from securing the communication process only in the current paradigm to securing the propagation environment as well. Essentially, securing the propagation characteristics is very important not for PLS alone but for other emerging technologies such as sensing and radio environment map (REM) as well. This opens up a new research frontier for the PLS concept.

4.6.2 High mobility and non-stationarity issues

The presence of mobility in the communication scenario can be a double-edged sword. High mobility as experienced in V2V communication, for example, makes the channel highly random in all domain, which is advantageous for PLS. However, such rapidly varying channels usually comes with the channel aging problem, as discussed before, which may destroy channel reciprocity property. Furthermore, serving mobile users under the recent beam-based communication requires the beam tracking which generally complicates the communication process as whole. Therefore, more efficient PLS algorithms that take into account the consequences of various mobility-related issues need to be developed.

4.6.3 Beam squint issue in (mMIMO)

The received signal at different array elements in antenna array systems is a slightly delayed version of the original signal. With higher array dimension as in mMIMO, this delay across the elements becomes highly significant and leads to the so-called delay-squint phenomenon especially when UWB signaling (short symbol duration) is utilized. The delay-squint problem translates into the beam-squint problem in spatial domain [19]. The beam squint problem has a significant impact on the channel characteristic. For example, under beam-squint, AoD/AoA of the MPC lose their reciprocity property, making them unsuitable for PLS. Impact of beam squint effect on the PLS approaches utilizing AoDs and AoAs in mMIMO systems is an open problem.

4.6.4 Intelligent security frameworks

The necessity for adaptive, intelligent, and adaptable security solutions is driven by the expanding variety of wireless networks, the wide range of device capabilities, and the various security needs of users and applications. By taking use of cross-layer coordinated security architecture, the intelligent and adaptable design can be created depending on information that comes from the application, network, MAC, and physical layers together. Designing effective cross-layer intelligent security frameworks will benefit from understanding the impact of specific channel factors on PLS [20].

4.7 Conclusion

Considering the recently obtained network capabilities in accessing and modifying various channel properties, this chapter revisits the channel-based PLS paradigm. The chapter emphasizes the importance of exploiting the newly emerged unique channel properties in facilitating more effective PLS approaches. To this end, the chapter highlighted and discussed various opportunities and challenges in order to ignite further research. It is however important to emphasize that studies on these new channel features are still ongoing and only their tentative models are available in the literature. Therefore, the discussion presented in this chapter only aims at highlighting their potential from the PLS perspective and encourage further research in that direction.

References

[1] Hamamreh JM, Furqan HM, Arslan H. Classifications and applications of physical layer security techniques for confidentiality: A comprehensive survey. *IEEE Communications Surveys & Tutorials*. 2018;21(2): 1773–1828.

[2] Zhang J, Duong TQ, Marshall A, *et al.* Key generation from wireless channels: A review. *IEEE Access*. 2016;4:614–626.

[3] Patwari N, Croft J, Jana S, *et al.* High-rate uncorrelated bit extraction for shared secret key generation from channel measurements. *IEEE Transactions on Mobile Computing*. 2009;9(1):17–30.

[4] Liu H, Wang Y, Yang J, *et al.* Fast and practical secret key extraction by exploiting channel response. In: *Proc. IEEE Int. Conf. Comput. Commun.* (INFOCOM). Turin, Italy; 2013. p. 3048–3056.

[5] Zhu X, Xu F, Novak E, *et al.* Extracting secret key from wireless link dynamics in vehicular environments. In: *Proc. IEEE Int. Conf. Comput. Commun.* (INFOCOM). Turin, Italy; 2013. p. 2283–2291.

[6] Azimi-Sadjadi B, Kiayias A, Mercado A, *et al.* Robust key generation from signal envelopes in wireless networks. In: *Proc. ACM Conf. Comput. Commun. Security*. Alexandria, Egypt; 2007. p. 401–410.

[7] Wei Y, Zeng K, Mohapatra P. Adaptive wireless channel probing for shared key generation based on PID controller. *IEEE Transactions on Mobile Computing*. 2012;12(9):1842–1852.

[8] Wang Q, Xu K, Ren K. Cooperative secret key generation from phase estimation in narrowband fading channels. *IEEE Journal on Selected Areas in Communications*. 2012;30(9):1666–1674.

[9] Wang Q, Su H, Ren K, *et al.* Fast and scalable secret key generation exploiting channel phase randomness in wireless networks. In: *Proc. IEEE Int. Conf. Comput. Commun.* (INFOCOM). Shanghai, China; 2011. p. 1422–1430.

[10] Mathur S, Trappe W, Mandayam N, *et al.* Radio-telepathy: extracting a secret key from an unauthenticated wireless channel. In: *Proc. ACM Int. Conf. Mobile Comput. Netw.* (MOBICOM). San Francisco, CA; 2008. p. 128–139.

[11] Rukhin A, Soto J, Nechvatal J, *et al. A Statistical Test Suite for Random and Pseudorandom Number gGenerators for Cryptographic Applications*. Mclean, VA: Booz-allen and Hamilton Inc.; 2001.

[12] Fang S, Liu Y, Shen W, *et al.* Where are you from? Confusing location distinction using virtual multipath camouflage. In: *Proc. IEEE Annu. Int. Conf. Mobile Comput. Netw.*; 2014. p. 225–236.

[13] Karttunen A, Molisch AF, Hur S, *et al.* Spatially consistent street-by-street path loss model for 28-GHz channels in micro cell urban environments. *IEEE Transactions on Wireless Communications*. 2017;16(11):7538–7550.

[14] Sarkar MZI, Ratnarajah T. Secrecy capacity over correlated log-normal fading channel. In: *Proc. IEEE Int. Conf. Commun.* (ICC). Ottawa, Canada; 2012. p. 883–887.

[15] Vasisht D, Kumar S, Rahul H, Katabi D. Eliminating channel feedback in next-generation cellular networks. In: *Proc. ACM SIGCOMM Conference*. Florianopolis, Brazil; 2016. p. 398–411.

[16] Jiao L, Tang J, Zeng K. Physical layer key generation using virtual AoA and AoD of mmWave massive MIMO channel. In: *Proc. IEEE Conf. Commun. Netw. Security* (CNS). Beijing, China; 2018. p. 1–9.

[17] Mazloum T, Sibille A. Analysis of secret key randomness exploiting the radio channel variability. *Hindawi International Journal of Antennas and Propagation*. 2015;2015:13.

[18] Zhang X, Knightly EW. *CSIsnoop*: Inferring channel state information in multi-user MIMO WLANs. *IEEE/ACM Transactions on Networking*. 2018;27(1):231–244.

[19] Wang B, Gao F, Jin S, *et al.* Spatial-wideband effect in massive MIMO with application in mmWave systems. *IEEE Communications Magazine*. 2018;56(12):134–141.

[20] Yılmaz MH, Güvenkaya E, Furqan HM, *et al.* Cognitive security of wireless communication systems in the physical layer. *Wireless Communications and Mobile Computing*. 2017;2017:1–10.

Chapter 5
Physical layer authentication in wireless communication systems

Liza Afeef[1] *and Hüseyin Arslan*[1]

This chapter presents the physical-layer security mechanisms in the analog domain under the name of physical layer authentication (PLA). PLA has attracted extensive research interest in recent years for the new generation of communication networks since it provides information-theory security with low complexity. In this chapter, we discuss the PLA system process and its metrics. After that, the PLA techniques that are based on the hardware devices available in the transceiver are presented according to the radio-frequency (RF) components such as an oscillator, power amplifier, device clock, and modulator/demodulator. Furthermore, due to the use of multiple-input multiple-output (MIMO) technology in 5G networks and beyond, new PLA techniques are introduced based on the beam pattern and sparsity of the MIMO channel, which is discussed deeply in this chapter. Finally, we provide some research challenges in PLA systems and future discussion.

5.1 Introduction

Conventional upper-layer security mechanisms address security concerns in wireless communication systems with encryption techniques. However, these techniques involve additional infrastructure overhead and power consumption to enable authentication. Therefore, implementing communication security in the physical layer has recently become a topic of interest. Physical-layer security mechanisms traverse two domains: the digital domain and the analog domain. Providing security in the digital domain is based on various techniques such as coding (i.e., source coding, channel coding, precoding in multiple-input multiple-output (MIMO) systems), key generation, and artificial noise schemes [1]. In general, these methods have inherently high computational complexity. For instance, providing security using key generation methods requires an extensive process of generating, distributing, refreshing, and revoking digital security keys, which results in excessive latency in large-scale

[1]Department of Electrical and Electronics Engineering, Istanbul Medipol University, Turkey

networks [2]. Hence, they are not suitable for delay-sensitive communications, decentralized networks, and low-end devices.

Physical layer authentication (PLA) attributes are introduced in the analog domain as an alternative approach for authentication to support the other security techniques and compensate for the aforementioned challenges. Such attributes rely on hardware imperfections (i.e., radio frequency (RF)/hardware-based attributes) and environment (i.e., channel-based attributes) that are unique, unpredictable, and provide multidimensional protection.

In this chapter, Section 5.2 discusses the advantages of PLA, Section 5.3 introduces the PLA metrics, and Section 5.4 presents the RF/hardware attributes-based PLA. Section 5.5 discusses in details the PLA schemes in 5G networks and beyond where beam pattern-based attributes and channel sparsity-based attributes are explained. Next, Section 5.6 presents the receiver process to detect and extract the PLA attributes while Section 5.7 introduces some PLA challenges and future discussion. Finally, Section 5.8 concludes the chapter.

5.2 Physical-layer authentication

PLA is a process that uses physical-layer attributes to build secure wireless communication networks in the analog domain, where the legitimate users are identified and verified from the illegitimate ones. Comparing to the other conventional security mechanisms, PLA provides the following advantages:

- PLA protects the physical layer attributes using only the noisy observation, where the receiver can measure the attributes instantaneously. This allows for low computational requirement, low processing delay, low network overhead, and moderate energy consumption.
- PLA is an analog-based mechanisms that depends on either device imperfections or its environment, or both. This leads to faster security process compared to the digital-based techniques. Furthermore, the device hardware imperfections and its environments offer unpredictable and unique attributes due to diverse combination available which provide natural refresh mechanisms and multi-dimensional protections.
- Unlike traditional methods, which require an additional trusted third party in the system, PLA techniques use only available knowledge of physical attribute statistics to provide security.

It is useful to note that the PLA mechanisms do not replace the upper-layer authentication methods, but is considered as a support for them, which provides a higher security level in the system.

PLA techniques are categorized into two parts: RF/hardware-based and channel-based attributes. The channel-based PLA techniques employ the channel features between the legitimate transmitter and receiver such as received signal strength (RSS), channel impulse response (CIR), and channel frequency response (CFR) to authenticate the sender by observing the unique instantaneous/average characteristics' of the estimated channel state information (CSI) then comparing the current CSI to the authentic CSI that previously reserved during the past transmission. Further information of exploiting channel features to perform security can be found in Chapter 4. Although channel-based attributes do not require high-end signal analyzers to extract the features, they rely on the assumption that the channel features of different transmitted devices have strong spatial decorrelation. In general, channel-based attributes from different transmitters to the same receiver can be considered totally uncorrelated if the distance between the transmitters is more than half-wavelength spacing. If a transmitter is located closer, the channel-based attributes may fail due to attributes similarity between these near transmitters' channels.

Different from the existing security schemes based channel attributes discussed in Chapter 4, the new channel representation of MIMO in 5G communication networks present unique features including uniqueness of beam patterns, distinguishability of sector level sweep (SLS) signal-to-noise ratio (SNR) trances, and sparsity of angle/virtual channel representation. These new features will be discussed in Section 5.4 of this chapter.

5.3 PLA metrics

The PLA scheme uses some natural/intrinsic physical-layer attributes to authenticate the transmitter based only on the received signals. This can be done by having robust framework that mitigates the noise interference while preventing the authentication process from being invaded by the attackers.

The PLA system operates in a two-stage process: the learning stage and identification stage, as shown in Figure 5.1. In the learning stage, only Alice (i.e., the

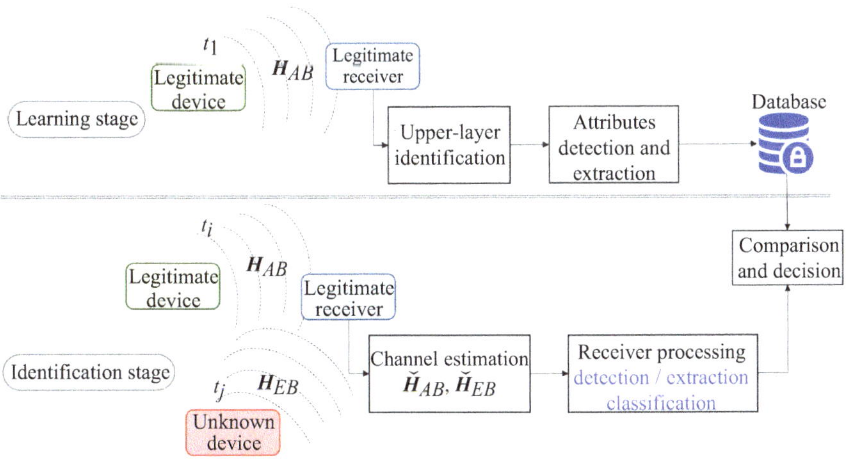

Figure 5.1 General PLA system operation

legitimate transmitter) and Bob (i.e., the legitimate receiver) exist in the system. Alice sends a request to Bob for message transmissions over a wireless channel while Bob checks whether the request is legitimate or not through upper-layers security mechanisms so that the attributes extracted in this stage and then stored are assumed to be legitimate. In the identification stage, Bob receives a signal from unknown transmitter, then it extracts the physical-layer attributes and compare it with the legitimate attributes stored in the database to decide whether the signal is transmitted from a legitimate node or not.

In general, a PLA process starts by generating an authentication message at Alice side and sending it to Bob. Bob receives the authentication message and extracts the identity information, then performs a verification comparison; if the extracted identity information equal to the transmitted one, Alice is verified, otherwise, the authentication fails. However, due to imperfect channel estimation and additive noise effects, a threshold-based hypothesis test needs to be introduced to decide whether the received message is from the legitimate transmitter or not, as follows:

$$\begin{aligned} K_0 &: \eta_L = ||\mathscr{D}(\mathbf{H}) - \mathscr{D}(\mathbf{H}_{\text{str}})|| \leq \eta_{\text{thr}}, \\ K_1 &: \eta_L = ||\mathscr{D}(\mathbf{H}) - \mathscr{D}(\mathbf{H}_{\text{str}})|| > \eta_{\text{thr}}, \end{aligned} \tag{5.1}$$

where K_0 is the null hypothesis representing that Alice sent the received signal while K_1 is the alternative hypothesis representing that the received signal was not sent by Alice. η_{thr} is a threshold used to separate the distance between K_0 and K_1, $\mathscr{D}(\mathbf{H})$ is the extracted identity information from the estimated channel $\mathbf{H}$, and $\mathscr{D}(\mathbf{H}_{\text{str}})$ is the stored identity information.

In order to evaluate the performance of a given authentication scheme, a receiver operating characteristic (ROC) is introduced. ROC concept is a graphical means of separating the process of discrimination and decision. The ROC graph plots two specific stimulus/metrics: A and B. These metrics have a valence such that the y-axis is the proportion of true-positive responses while the x-axis is the proportion of false-positive responses. These stimulus/metrics can be:

- Probability of detection (PD): it is known as the probability of correctly detect a forged signal when Alice does not transmit,

  $$P_D = Pr(\eta_L > \eta_{\text{thr}} | K_1). \tag{5.2}$$

  Probability of missed detection (PMD) can be also considered here which reflects the probability of wrongly detect a legitimate signal while Alice is not transmitting

  $$P_{\text{MD}} = Pr(\eta_L \leq \eta_{\text{thr}} | K_1). \tag{5.3}$$

  PMD can be also determined from the PD as $P_{\text{MD}} = 1 - P_D$.
- Probability of false alarm (PFA): it is the probability of wrongly detect a forged signal while Alice is sending,

  $$P_{\text{FA}} = Pr(\eta_L > \eta_{\text{thr}} | K_0). \tag{5.4}$$

5.4 RF/hardware-based PLA

Apart from exploiting the CSI, hardware attributes in the wireless transceiver devices can be utilized to perform security. These attributes are considered unique even if the same manufacturer produces the transmitting devices. RF/hardware-based PLA techniques reflect the hardware imperfection of the devices in the analog and modulation domains. The transient behaviour of the RF signal emission pattern that is observed during the transmission can be utilized for device identity from the analog domain including its power, timing, amplitude, phase, and frequency, while the signal at the basic level in terms of in-phase (I)/quadrature (Q) samples are considered as modulation domain techniques. Note that an effective RF attribute must satisfy two properties: (1) it should be difficult or impossible to forge and (2) the features should be stable in the presence of environment changes and node mobility.

In this chapter, we divide the hardware-based PLA techniques according to the RF components such as local oscillator, power amplifier, device sampling clock, and RF modulator/demodulator. Figure 5.2 illustrates the RF components for a transceiver. Since the attributes that are extracted from these components are usually stable with time, they can be used in many applications such as Bluetooth network [3], ZigBee network [4], WiMax systems [5], orthogonal frequency division multiplexing (OFDM) systems [6], and highly dynamic communication systems.

5.4.1 Local oscillator

The oscillator, one of the RF components, is an electronic circuit that produces a periodic oscillating signal at specific frequency. This device produces two imperfections; phase noise and carrier frequency offset (CFO). The accuracy of the oscillator is measured based on how much clear the generated signal is from any phase noise.

Phase noise is the frequency domain representation of the rapid, short-term, random fluctuations in the phase of a waveform caused by time domain instabilities (i.e., jitter). It can be measured from the signal spectral density graph. In this graph, the frequency is represented in the x-axis and the signal power is represented in the y-axis as shown in Figure 5.3. Ideally, a phase noise free signal has only one pulse at

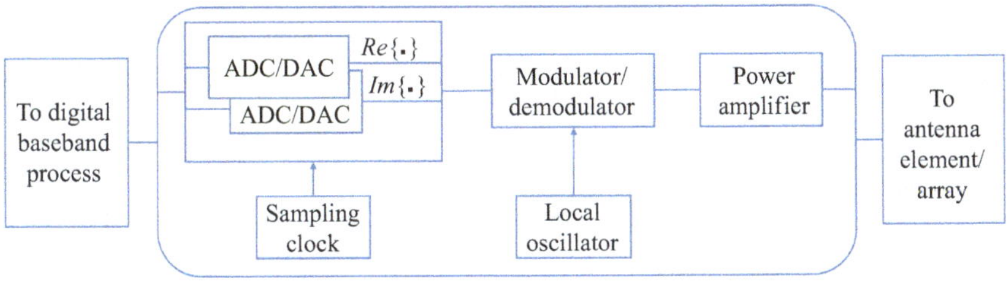

Figure 5.2 RF device equipment

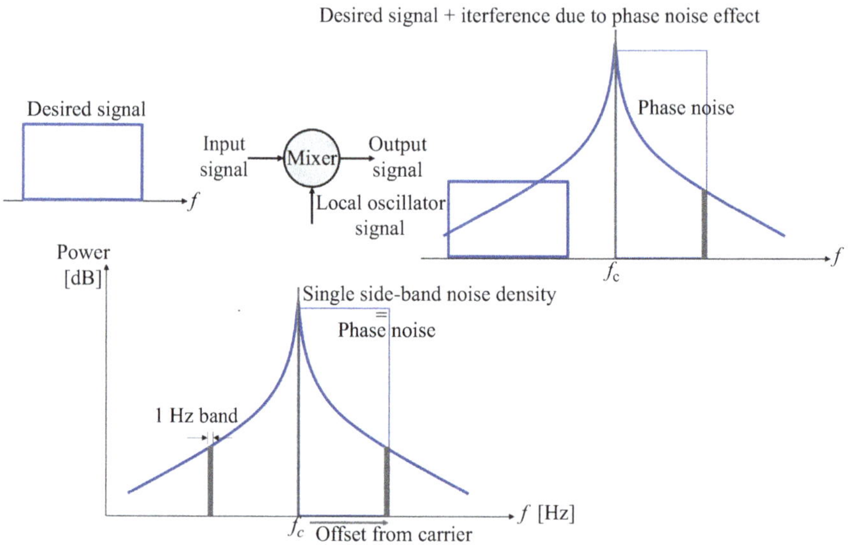

Figure 5.3 Phase noise effect on the signal power spectral density

the frequency f_c. However, the extra unwanted area on the either sides (left and right to the f_c) of the signal represents the phase noise which is called signal side-bands. Thus, a single side-band power density beyond the signal frequency is the phase noise signal. In general, thermal noise, shot noise, Flicker noise, and oscillator aging are the common sources of generating phase noise in an oscillator. This oscillator imperfection is exploited for physical layer authentication (PLA) in various studies such as [7,8].

There are three ways in practice to measure the phase noise:

1. Spectrum analyzer [9]: it is the oldest and simplest method where it directly measures the power spectral density of the oscillator. However, this method has many limitations such as:
   - The noise floor of spectrum analyzer should be below that of the device under test (DUT).
   - Since the carrier signal is not suppressed, the dynamic range of the analyzer is limited.
   - This method has no ability to differentiate between the amplitude noise effect and the phase noise effect.
2. I/Q demodulation [10]: this method is done by down-converting the signal into baseband and measure it through analog-to-digital converter (ADC). This method considers as the only method that works in digital domain to measure the phase noise. However, multiple ADCs are needed within the device for simultaneous measurement which is cost consuming.

3. Phase detector technique [11]: this method is the most popular technique where the device signal is mixed with the reference signal at the same frequency using a phase detector mixer. In order to use this technique, some conditions need to be satisfied such as:
   - The phase noise generated from the reference signal must be below the DUT one.
   - The phase-locked loop (PLL) should have a loop bandwidth less than the lowest required noise frequency.

   One disadvantage that can be found in this method is that only a limited fluctuation range can be compensated in the PLL since it passes a high amount of the phase noise while trying to recompense for phase fluctuation. In order to relax the phase-detector circuit conditions, a cross correlation process is added to this technique resulting in the best accurate phase noise measurement. This is done by having two independent reference sources and phase detector circuit. The output of those two mixers are cross-correlated and averaged which cancel out the noise reference sources. As a result, the noise floor is improved with almost 20 dB over a single-channel phase detector technique.

CFO occurs when the local oscillator signal for the down-conversion at the receiver does not synchronize with the carrier signal contained in the received signal. It causes a phase shift in time domain which reflects as a frequency shift in frequency domain. This can happen due to the carrier frequency mismatch between transmitter's and receiver's local oscillator or Doppler shift from the environment. The CFO has higher impact on the system when a multi-carrier signal is used such as in the case of OFDM signal. The impact of CFO in the system can be seen as scaling, inter-carrier interference, error in channel estimation, and distortion in the constellation. In OFDM signal, the addition of cyclic prefix (CP) can be also used to detect and extract the CFO. Several works use the effect of CFO to perform/improve a PLA scheme with suitable authentication accuracy level [6,12].

5.4.2 Power amplifier

A power amplifier is the last stage of RF transmission that is connected to the antenna element in the system which amplifies the analog high-frequency signals used in radio communications. It can also tune over the desired range of input frequencies. Key characteristics of the power amplifier are frequency response, amplitude response, and distortion. Some design factors need to be considered in designing the power amplifier such as power gain, output power, bandwidth, power efficiency, linearity, input and output impedance matching, and heat dissipation. It can work in three power regions as illustrated in Figure 5.4, where the figure shows the general transfer characteristic between the input power and output power for the power amplifier. When it works in the linear amplifier area, the efficiency is low. However, when the power amplifier enters the saturated working area, serious nonlinear

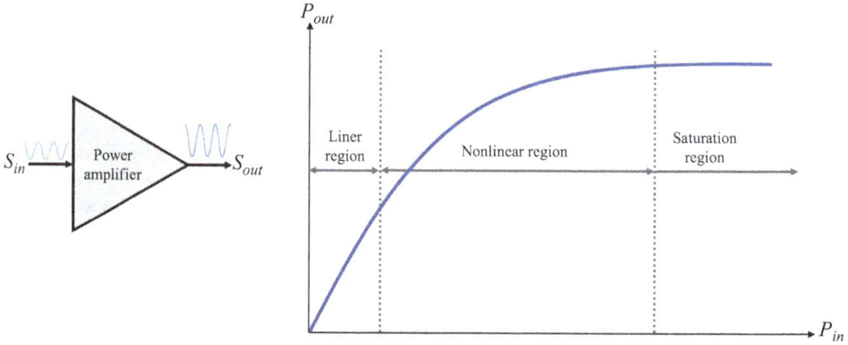

Figure 5.4 Power amplifier transfer characteristic

distortion will occur, which will spread the signal spectrum. Due to the nonlinearity of the device, it is a good choice for extracting the spectrum characteristics of the signal for classification. Since the power amplifier is located at the RF end, it is hard to forge this imperfection via software manipulations which makes it a promising RF attribute to use in building a PLA scheme against the malicious adversaries [13,14].

5.4.3 Device clock

One major part of the RF components is the ADC/digital-to-analog converter (DAC) where the analog signals are allowed to be processed in the digital electronic equipment and vice versa. Device clock represents the sampling clock for the ADC and DAC components. For an imperfect device clock, sampling clock offset and clock skew are introduced to the system. This is due to the mismatch in the sampling frequencies of the ADC at the receiver and DAC at the transmitter. This mismatch is attributed to two main reasons: (1) the transceiver nodes initialized at different moments so there is clock offset between the nodes, and (2) the clock sources have variable clock frequency offset due to different clock manufacturing techniques and/or changing environments.

The sampling clock offset results in scaling, phase rotation, and inter-carrier interference in multi-carrier signals. Many time synchronization algorithms are proposed to detect and correct the time offset caused by device clock imperfections [15–17]. These imperfections are exploited to build efficient PLA schemes [18,19] under the fact that clock offsets between every node pair is unique. This can be done by tracking the legitimate and illegitimate users' clocks using Kalman filters [19] to calculate the time-varying clock offsets. The work in [18] proposed a transmission control protocol (TCP) timestamps-based approach to estimate the clock skew. Note that it is difficult for attackers to reproduce the observed attribute without switching chipset since clock skew offsets are not controllable by the user.

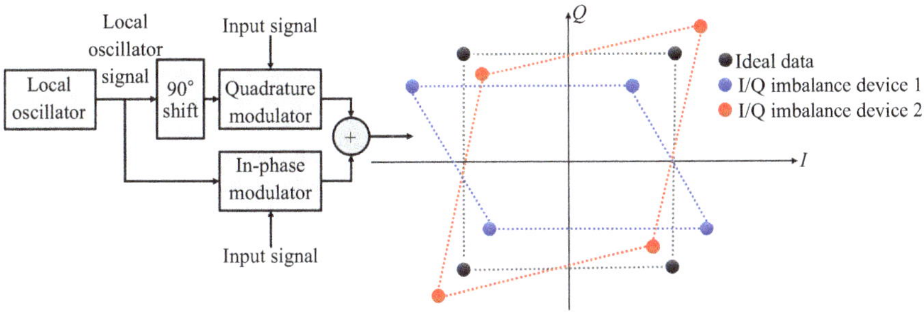

Figure 5.5 I/Q imbalance effects on the constellation diagram

5.4.4 RF modulator/demodulator

The sinusoidal signal generated from the local oscillator, as mentioned in Section 5.4.1, is used to modulate/demodulate the analog RF signal to a quadrature signal at the modulator/demodulator part of the RF chain. This is done by taking the sinusoidal signal and its 90° phase shifted version and individually mixing them with the analog RF signal, producing the I and Q signals. In an ideal case, the shifted version of the sinusoidal signal is exactly the $1/4$ cycle of the original one and the signal can be perfectly modulated/demodulated. However, in practice, neither the phase difference is precisely 90° nor is the amplitude matched. This result in what is known as I/Q imbalance problem which affects the constellation points as shown in Figure 5.5.

It is noteworthy that I/Q imbalances are commonly time-invariant once the hardware of the device is fabricated. Consequently, I/Q imbalance could be utilized as a unique device fingerprint to identify direct conversion architecture-based wireless transceivers. Hence, the work in [20] verified the transceiver I/Q imbalances for the authentication process by involving multiple collaborative receivers. In a relay communication scenario, the work in [21] proposed a novel I/Q imbalance-based PLA scheme to secure the relay communications. Furthermore, in cooperative Internet of things (IoT) non-orthogonal multiple access (NOMA) networks, the work in [22] considered the I/Q imbalance at both transmitter and receiver to build reliable and secure communication systems.

5.4.5 Multiple hardware attributes

Using one RF device attribute may not always be reliable to provide accurate authentication due to the limited dynamic range for each attribute. Therefore, to overcome this, several RF attributes are used to perform authentication in various ways: (1) they can be utilized in a straightforward way to verify the claimed device identity in multiple aspects [13,23,24]; (2) they can be utilized in a weighted combination mechanism to optimize the authentication performance by finding the optimal weight of each selected attribute that maximize the detection probability [25]; and (3) they can be utilized in multi-attribute-based correlation mechanism where the correlated analysis can reduce the effect of estimation errors and enhance the reliability of the authentication process [26].

The conventional PLA schemes based on multiple RF/hardware attributes suffer from high similarity where two or more attributes can have one observation. In addition, the need for high-end signal analyzers or software defined radio (SDR) devices to extract the attributes with high accuracy imposes high-cost overhead and huge computational complexity to the authentication process. For the channel-based PLA schemes, the conventional channel-based methods may not be feasible in practice due to the sparse nature of the channel in fifth generation (5G) networks. Therefore, new characteristics are exploited for 5G communication networks as discussed in the next section.

5.5 PLA in 5G networks and beyond

In order to unlock the full potential of the new spectrum of 5G networks at high frequency bands such as millimeter wave (mmWave), MIMO technology is incorporated into the 5G network deployment. This provides multiple radio and feature options with different characteristics to meet various requirements in the network in a cost-efficient way. With this efficient deployment, new unique attributes can be exploited for better PLA design. For example, the transceivers' positions can be extracted from the 5G MIMO channel due to the high directivity transmission in the design while the conventional channel features such as RSS and CSI that are extracted for the conventional PLA scheme cannot possess the unique attributes of 5G MIMO networks. Hence, the conventional PLA techniques mentioned in the previous sections do not provide high authentication performance when applied to 5G MIMO networks since the unique attributes of 5G MIMO cannot be fully exploited with these traditional methods.

To cope with these changes, new PLA schemes are introduced for the 5G MIMO networks. The main unique attributes that can be extracted from 5G MIMO design are: beam patterns and channel sparsity.

5.5.1 Beam pattern

In 5G MIMO devices, the antenna array has distinct radiation characteristics of its beam pattern that are unique among devices due to the electronic circuit imperfection during the manufacturing processes and the way of packing and placing the antenna elements in the array. Since the beam pattern does not depend on the signal being transmitted, it can be exploited for PLA to provide high dimensionality and resilient to impersonation attacks. Therefore, the beam pattern is exploited for PLA in several ways, such as:

- Detect the rich spatial-temporal information of the beam pattern using multiple access points (APs), as shown in Figure 5.6, for the indoor environment [27]

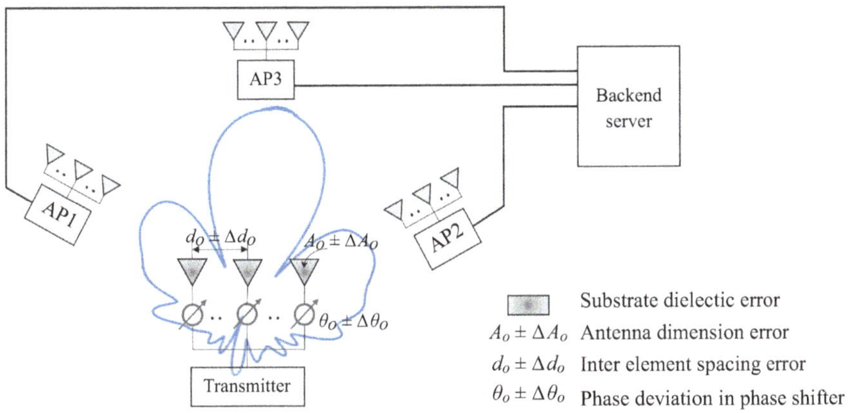

Figure 5.6 PLA technique using rich spatial-temporal information of the beam pattern

by extracting the power spectral density (PSD) of the beam pattern. This is done by initiating a beam search mechanism periodically at the APs side using an SLS introduced in the IEEE 802.11ad standard, extracting the beam pattern attribute vector, and communicating these attributes to the backend server. The backend server builds the beam pattern attribute database for a given user where it is responsible for identifying the user when it intends to join the network again. The beam pattern attributes built here are based on: (1) dielectric properties of the substrate used in the antenna, (2) dimensional tolerance of the antenna fabrication, and (3) phase deviation error in the phase shifters. Moreover, due to energy leakages between the antenna elements during transmission/reception of the electromagnetic waves, a mutual coupling effect is introduced to the system which affects the beam pattern and degrades the performance. However, in [28], the mutual coupling effect is considered as a unique attribute for a given antenna array and thus exploited to build an RF-based PLA scheme. Note that the amount of mutual coupling in the system can be different depending on the self and load impedance connected to each antenna element which permutes the circuit current distribution and distorts the radiation pattern of the other antenna elements as compared to their isolated radiation patterns.

- Determine the variation of the channel gains to estimate the unique beam perturbation, as shown in Figure 5.7(a). This technique is exploited to provide a channel-based PLA to counter the co-located spoofing attacks.
- Perform an SLS to find the antenna pattern pair with the best channel gain based on the SNR trace of each beam pattern [29] to initiate the communication link as in IEEE 802.11ad standard, as shown in Figure 5.7(b). This technique is exploited to provide an RF-based PLA to counter spoofing identity attacks.

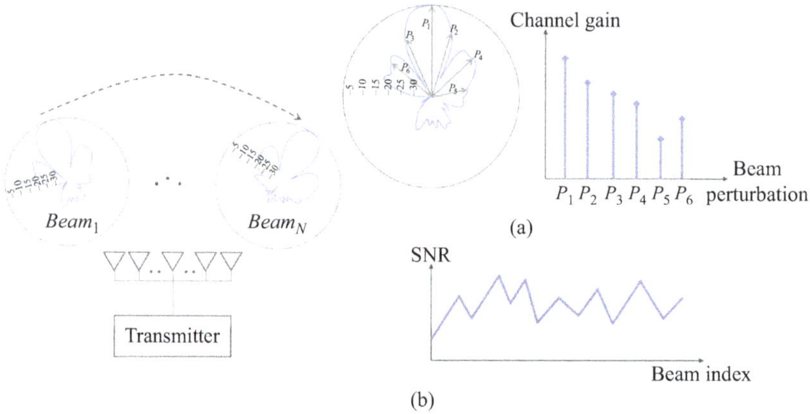

Figure 5.7 Beam pattern based PLA using (a) channel gain at each beam perturbation and (b) beam level sweep in the IEEE 802.11ad standard

Although the aforementioned 5G MIMO PLA techniques that are based on spatial attributes of beam pattern are built on the assumption that the antenna circuit imperfection is intrinsic which creates an intrinsic RF/channel-based PLA scheme, the variation in the beam pattern among different devices may be tiny and can be simply interrupted by additive noise leading to a limitation in system capacity. Accordingly, an intentional manipulation in the antenna circuits (i.e., antenna element shapes, material under/above antenna elements, pre-selected bandpass filters, power dividers, etc.) can be done to improve the authentication performance in the given system setup while ensuring that the communication performance does not have noticeable degradation. One example of an intentional RF-based PLA scheme is the work in [30], where a chaotic antenna array geometry is created along with a particular antenna activation sequence. Specifically, this work displaces the antenna elements independently in two dimensions, in a uniform linear array (ULA) in each dimension, then each antenna element is translated, rotated, scaled, or skewed chaotically from the ULA design. This technique ensures having non-cloneable authentication devices.

5.5.2 Channel sparsity

Due to the sparse nature of the MIMO channel in 5G networks described in Chapter 4, the channel can be transformed into another domain to exploit the unique characteristics of the 5G MIMO channels. This representation is called beamspace domain, introduced for the first time in [31]. Many terminologies have been

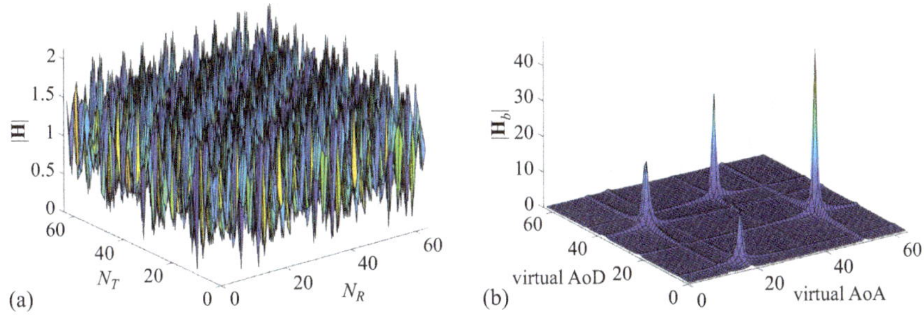

Figure 5.8 (a) Spatial MIMO channel representation, and (b) its beamspace MIMO channel representation

established in the field of beamspace representation including virtual and extended virtual channel representation. 5G MIMO channel can be transformed from its spatial representation (see Figure 5.8(a)) to the beamspace one (see Figure 5.8(b)) using a discrete Fourier transform (DFT) approach. The beamspace domain provides a discretized approximation for the spatial channel by uniformly sampling the angle-delay-Doppler space at the Nyquist rate [32]. Beamspace channel representation is sparse as the number of antenna elements is large enough with fewer paths contribute to each spatial bin in the beamspace channel, where this sparsity property can reflect the number of transmitters [33].

The use of beamspace channel representation can reduce the hardware transceiver complexity, channel matrix dimension, and precoder design [34], leading to a precise channel estimation approach and a faster beam alignment process. Note that the properties of the beamspace channel representation (principal components with the corresponding group of virtual angle of arrival (AoA)/angle of departure (AoD) bins, as shown in Figure 5.8(b)) are robust to noise and can be accurately estimated bins/angles with low overhead [35]. Therefore, several PLA techniques are developed based on the use of beamspace properties. For example, the study in [36–38] uses the beamspace channel $\mathbf{H}_b$ as the identity information $\mathscr{D}$ to define the legitimate transmitter, and [28] which extracts the distance vector $\mathbf{d}$ from the beamspace representation and use it as the $\mathscr{D}$ for the legitimate transmitter. Based on the aforementioned techniques, it is proved that the channel-based PLA scheme in 5G networks does not need any extra feature extracting processing which make them more suitable for delay sensitive applications.

A comparison between all PLA techniques is summarized in Table 5.1.

Table 5.1 A comparison between different PLA techniques in the wireless communication literature

	Channel-based attributes	RF-based attributes	5G MIMO-based attributes	
			Beam Pattern	**Channel sparsity**
Benefit	– Used to counter identity spoofing attacks based on channel feature. – The attributes are readily extractable from the off-the-shelf devices.	– Robust to channel variations.	– Free from the signal analyzer. – No need for high complex attribute extraction processes. – Robust to co-located attacks. – Stable with time.	– Better protection against spoofing and Sybil attacks. – Detect pilot contamination attacks. – Sensitive to the different transmitters and the number of transmitters. – No need for extra attribute extraction.
Challenges	– The attribute should have strong spatial decorrelation between different devices. – Vulnerable to co-located spoofing attacks and pilot contamination. – The channel should be rich scatters. – The attribute detection and extraction should be within the channel coherence time.	– In some cases, a high-end signal analyzer is needed to extract the attributes. – Increase the communication overhead and computational complexity. – Prior knowledge of legitimate devices is required.	– Can be corrupted by an eavesdropper located at the LoS transmission. – Not applicable in dense environments.	– Sensitive to dynamic environment.
Technology	OFDM, Dual-hop, CDMA, cyber-physical systems, and relay networks.	OFDM, Bluetooth, ZigBee, WiFi, and WiMax.	mmWave MIMO and massive MIMO.	5G NOMA-MIMO, massive MIMO.

5.6 Receiver process

For channel-based PLA schemes, since the estimation of channel in time domain (i.e., CIR) or frequency domain (i.e., CFR) can be directly taken as a feature to build the PLA scheme, there is no need for further process to extract the secure attribute. However, two conditions need to be satisfied to validate the use of channel-based PLA schemes:

1. The authentication process should be done within the coherence time of the channel so that the channel is assumed to vary slowly or highly correlated in

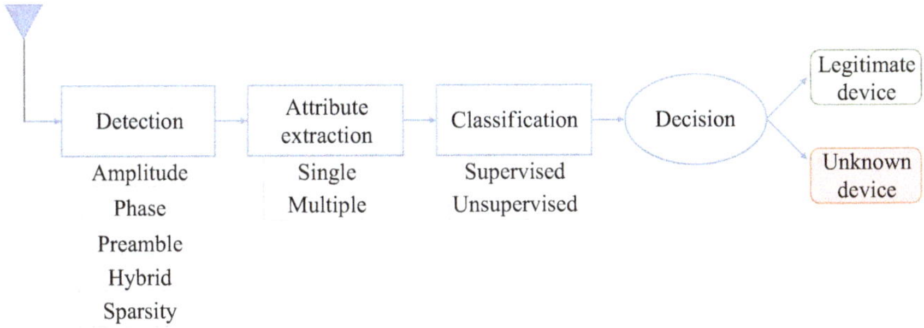

Figure 5.9 Receiver process for the RF-based PLA schemes

time. Hence, the estimation results are almost the same between adjacent time slots.

2. The reciprocity condition for the channel between Alice and Bob should be satisfied.

In RF-based PLA schemes, the legitimate receiver must perform several steps to distinguish between the legitimate transmitter and the forgery one. The receiver starts by detecting the received signal. Then, the target attribute is extracted. After that, the selected attribute is classified to verify the transmitter's identity, either legitimate or forgery [39]. The receiver process in the RF-based PLA scheme is shown in Figure 5.9.

5.6.1 Detection process

This stage is responsible for determining the exact time instant at which the transmitter is turned on. This process deals directly with the received signal. There are several detection techniques:

- Amplitude-based detection: this technique is based on determining the point that demonstrate a sudden change in the received power level which depends on a posteriori probability of a simple step change point detector.
- Phase-based detection: this method is done by measuring the phase variance difference between time slots which relies on the fact that the phase slop associated with the initial transmission is linear. This technique is more susceptible to noise and interference.
- Preamble detection: this method is done by utilizing a fixed preamble to overcome the unstable aligned envelop profiles caused by ramp-on control mode of power amplifier.
- Hybrid detection: this method provides best performance between amplitude-based or phase-based detection technique. For example, this method can be done by a variance trajectory-based approach presented in [40].

- Channel sparsity-based detection: this method is based on estimating the beamspace channel and its properties: basic spatial bins (i.e., principle components) and their AoAs and AoDs.

5.6.2 Attribute extraction

As discussed in Sections 5.4 and 5.5, based on a given scenario, the target attribute is extracted, where a good attribute has low intra-radio variability* and high inter-radio variability† within system requirements. In a static system, extracting the attribute based on specific system imperfection can secure the system and provide a powerful PLA scheme. However, in a time-varying scenario, a continuous PLA scheme is needed by exploiting the variation pattern in the varying attribute. Hence, a prediction-based PLA is required.

In multiple attributes-based PLA schemes, attribute selection algorithms are needed which are mostly based on experimenting and machine learning algorithms such as redundancy maximal relevance minimization algorithm, principle component analysis, Euclidian distance and clustering techniques of multivariate analysis, data-driven density formation and maximum-likelihood classification, and probabilistic neural network.

5.6.3 Classification

The final and main key step of the legitimate receiver process is to develop classification algorithms to identify wireless devices and detect illegitimate ones. Classification algorithms can be divided into two categories based on the need for prior information of legitimate devices or not: supervised and unsupervised learning-based algorithms [41].

- Supervised classification: this method needs to register legitimate devices to set up a database of the attributes of legitimate devices. The supervised classification schemes are: feature distance, Fisher-based machine learning, support vector machines algorithm, K-nearest-neighbor, probabilistic neural network, and bootstrap aggregating [39].
- Unsupervised classification: this method does not require prior knowledge of the legitimate devices since the registration of the device attributes ahead of time is not always feasible in practice. Unsupervised learning is a machine learning-based method (i.e., K-means clustering, Bayesian learning, and principle component analysis) used to report the process of detecting hidden structures in unlabeled data. Due to the lack of legitimate device information, this method does not have the ability to distinguish legitimate devices from forgery ones, however, it can detect the presence of masquerade and Sybil attacks.

*Intra-radio variability attribute is the variation of a given attribute between time samples for a long time period within one radio device.

†Inter-radio variability attribute is the variation of a given attribute between different radio devices.

5.7 Challenges and future discussion

Although the presented channel/RF-based PLA schemes in the literature show a good authentication performance, many challenges still exist for a given scenario. The most important challenges are discussed below.

5.7.1 Optimal attribute selection

For a PLA scheme with a single attribute, the optimal attribute is selected based on the system requirement, where it must be invariant over time. Specifically, one critical criterion that needs to be considered in the system is the sensitivity of the attribute to the environment. For instance, attributes that are robust to location and environmental settings are preferable in dynamic systems. However, they are more vulnerable to adversaries.

For a PLA scheme with multiple attributes, a reliable authentication is made by selecting the best combination of multiple attributes that enhance the system authentication. However, this selection is done at the final authentication decision step since each attribute has its own value range and unit.

In order to achieve the optimal attribute selection in both single and multiple attributes-based PLA schemes, machine learning approaches are promising solutions with the use of large available data stored in the systems currently.

5.7.2 PLA in multiuser communication networks

Compared to the conventional multi-user PLA systems, distributed MIMO systems can use the multiple distributed users at different locations to collect more physical-layer information which enhances the authentication performance, especially for channel-based PLA schemes where the radio channel resolution increases. Note that the current PLA schemes in the distributed MIMO systems are based only on a single authentication entity which keeps the research open to this field, where a collaborative communication between the distributed devices in the network can be also applied. This leads to the need of suitable selection and scheduling methods to achieve higher network authentication accuracy.

In cognitive radio networks, the vacant channel spectrum space is intelligently detected and occupied by secondary users which help in optimizing the use of the available RF spectrum. However, many threats to spectrum sensing such as primary user emulation attack [42] have been introduced to the field. Although many channel/RF-based PLA schemes are proposed to deal with these threats and distinguish the attacker from the legitimate primary users, the type of attacker modes considered in most studies are simple. This keeps several open problems to be investigated in recent technologies with dynamic and more complicated network structures in 4G, 5G networks, and beyond.

Due to the fast and low complexity nature of the PLA processes compared to the standard authentication mechanisms, PLA is considered a promising substitution to guarantee secure communication in IoT networks. However, many problems should be considered before commercially applying PLA schemes on IoT networks such as consideration of the end-to-end PLA, seamless PLA handover mechanisms in dynamic environments, and the integration of PLA and cryptography-based authentication structure.

5.7.3 Mobility or orientation change

In a dynamic time-varying environment, an adaptive PLA framework is needed where single or multiple attributes are to be predicted and tracked. This is due to the non-symmetrical observation at transmitter and receiver caused by the dynamic conditions of the network and/or the mobility/orientation of devices.

For channel-based PLA schemes, if the scheme is built upon observing the channel correlations between adjacent time slots, it faces significant performance degradation in a dynamic environment due to the large mismatch between adjacent channel responses. Hence, it is considered a major challenge in these schemes to design an effective channel tracking method to overcome this degradation due to channel dynamics.

5.8 Conclusion

This chapter revises PLA schemes focusing mainly on the RF/hardware perspective to provide secure communication networks. Furthermore, the beam patterns and channel sparsity were discussed as a major contribution to the 5G networks that can be exploited to perform a reliable PLA scheme without extra system overhead. Afterward, the process of detecting, extracting, and classifying the attributes on the receiver side were discussed in details, which lead to the discussion on the challenges facing PLA schemes, particularly in dynamic environments.

Acknowledgment

This work was supported by the Scientific and Technological Research Council of Turkey (TÜBİTAK) under Grant 5200107, with the cooperation of Turkcell Technology and Istanbul Medipol University.
The authors would like to thank Haji M. Furqan for his valuable comments and guidance in improving this chapter.

References

[1] Yılmaz MH, Güvenkaya E, Furqan HM, *et al.* Cognitive security of wireless communication systems in the physical layer. *Wireless Communications and Mobile Computing*. 2017;2017, https://doi.org/10.1155/2017/3592792.

[2] Ren Y, Chen JC, Chin JC, *et al.* Design and analysis of the key management mechanism in evolved multimedia broadcast/multicast service. *IEEE Transactions on Wireless Communications*. 2016;15(12):8463–8476.

[3] Barbeau M, Hall J, Kranakis E. Detection of rogue devices in bluetooth networks using radio frequency fingerprinting. In: *Proceedings of the 3rd IASTED International Conference on Communications and Computer Networks, CCN*. Citeseer; 2006. p. 4–6.

[4] Dubendorfer CK, Ramsey BW, Temple MA. An RF-DNA verification process for ZigBee networks. In: *MILCOM 2012—2012 IEEE Military Communications Conference*. Orlando, FL; 2012. p. 1–6.

[5] Reising DR, Temple MA. WiMAX mobile subscriber verification using Gabor-based RF-DNA fingerprints. In: *2012 IEEE International Conference on Communications* (ICC). Ottawa, ON, Canada; 2012. p. 1005–1010.

[6] Hou W, Wang X, Chouinard JY, *et al.* Physical layer authentication for mobile systems with time-varying carrier frequency offsets. *IEEE Transactions on Communications*. 2014;62(5):1658–1667.

[7] Zhao C, Huang M, Huang L, *et al.* A robust authentication scheme based on physical-layer phase noise fingerprint for emerging wireless networks. *Computer Networks*. 2017;128:164–171.

[8] Zhang P, Shen Y, Jiang X, *et al.* Physical layer authentication jointly utilizing channel and phase noise in MIMO systems. *IEEE Transactions on Communications*. 2020;68(4):2446–2458.

[9] Owen D. Good Practice Guide to Phase Noise Measurement; 2004.

[10] Shi J, Zhang F, Pan S. Phase noise measurement of RF signals by photonic time delay and digital phase demodulation. *IEEE Transactions on Microwave Theory and Techniques*. 2018;66(9):4306–4315.

[11] Keysight X, Anayzers SS. *Keysight Technologies*. Santa Rosa, CA: Keysight; 2014.

[12] Sankhe K, Belgiovine M, Zhou F, *et al.* ORACLE: Optimized radio classification through convolutional neural networks. In: *IEEE INFOCOM 2019-IEEE Conference on Computer Communications*. Paris, France; 2019. p. 370–378.

[13] Polak AC, Dolatshahi S, Goeckel DL. Identifying wireless users via transmitter imperfections. *IEEE Journal on Selected Areas in Communications*. 2011;29(7):1469–1479.

[14] Li Y, Chen X, Lin Y, *et al.* Wireless transmitter identification based on device imperfections. *IEEE Access*. 2020;8:59305–59314.

[15] Briggs E, Nutter B, McLane D. Sample clock offset detection and correction in the LTE downlink. *Journal of Signal Processing Systems*. 2012;69(1): 31–39.

[16] Jos S, Bynam K, PS CT, *et al.* A low complexity algorithm for sampling clock-offset compensation in low power WPAN receivers. In: *13th IEEE Annual Consumer Communications & Networking Conference* (CCNC). Las Vegas, NV; 2016. p. 833–836.

[17] Harish C, Mohan R, Shashank R. Sampling clock offset estimation and correction in frequency domain for OFDM receivers. In: *TENCON 2017-IEEE Region 10 Conference*. Penang, Malaysia; 2017. p. 1583–1587.

[18] Kohno T, Broido A, Claffy KC. Remote physical device fingerprinting. *IEEE Transactions on Dependable and Secure Computing*. 2005;2(2):93–108.

[19] Rahman MMU, Yasmeen A, Gross J. Physical layer authentication via drifting oscillators. In: *IEEE Global Communications Conference*. Austin, TX; 2014. p. 716–721.
[20] Hao P, Wang X, Behnad A. Performance enhancement of I/Q imbalance based wireless device authentication through collaboration of multiple receivers. In: *IEEE International Conference on Communications* (ICC). Sydney, NSW, Australia; 2014. p. 939–944.
[21] Hao P, Wang X, Behnad A. Relay authentication by exploiting I/Q imbalance in amplify-and-forward system. In: *IEEE Global Communications Conference*. Austin, TX; 2014. p. 613–618.
[22] Li X, Zhao M, Gao XC, *et al.* Physical layer security of cooperative NOMA for IoT networks under I/Q imbalance. *IEEE Access*. 2020;8: 51189–51199.
[23] Brik V, Banerjee S, Gruteser M, *et al.* Wireless device identification with radiometric signatures. In: *Proceedings of the 14th ACM International Conference on Mobile Computing and Networking*; 2008. p. 116–127.
[24] Zhang J, Woods R, Sandell M, *et al.* Radio frequency fingerprint identification for narrowband systems, modelling and classification. *IEEE Transactions on Information Forensics and Security*. 2021;16:3974–3987.
[25] Hao P, Wang X. Performance enhanced wireless device authentication using multiple weighted device-specific characteristics. In: *IEEE China Summit and International Conference on Signal and Information Processing* (ChinaSIP). Chengdu, China; 2015. p. 438–442.
[26] Xia S, Tao X, Li N, *et al.* Multiple correlated attributes based physical layer authentication in wireless networks. *IEEE Transactions on Vehicular Technology*. 2021;70(2):1673–1687.
[27] Balakrishnan S, Gupta S, Bhuyan A, *et al.* Physical layer identification based on spatial-temporal beam features for millimeter-wave wireless networks. *IEEE Transactions on Information Forensics and Security*. 2019;15:1831–1845.
[28] Afeef L, Furqan HM, Arslan H. Physical layer authentication scheme in beamspace MIMO systems. *IEEE Communications Letters*. 2022;26:1484–1488.
[29] Wang N, Jiao L, Wang P, *et al.* Exploiting beam features for spoofing attack detection in mmWave 60-GHz IEEE 802.11 ad networks. *IEEE Transactions on Wireless Communications*. 2021;20(5):3321–3335.
[30] Karabacak M, Peköz B, Mumcu G, *et al.* Arraymetrics: authentication through chaotic antenna array geometries. *IEEE Communications Letters*. 2021;25(6):1801–1804.
[31] Sayeed AM. Deconstructing multiantenna fading channels. *IEEE Transactions on Signal Processing*. 2002;50(10):2563–2579.
[32] Bajwa WU, Haupt J, Sayeed AM, *et al.* Compressed channel sensing: A new approach to estimating sparse multipath channels. *Proceedings of the IEEE*. 2010;98(6):1058–1076.

[33] Wang N, Jiao L, Alipour-Fanid A, *et al.* Pilot contamination attack detection for NOMA in 5G mm-wave massive MIMO networks. *IEEE Transactions on Information Forensics and Security*. 2019;15:1363–1378.
[34] Amadori PV, Masouros C. Low RF-complexity millimeter-wave beamspace-MIMO systems by beam selection. *IEEE Transactions on Communications*. 2015;63(6):2212–2223.
[35] Ma W, Qi C. Beamspace channel estimation for millimeter wave massive MIMO system with hybrid precoding and combining. *IEEE Transactions on Signal Processing*. 2018;66(18):4839–4853.
[36] Tang J, Xu A, Jiang Y, *et al.* MmWave MIMO physical layer authentication by using channel sparsity. In: *2020 IEEE International Conference on Artificial Intelligence and Information Systems* (ICAIIS). Dalian, China; 2020. p. 221–224.
[37] Li W, Wang N, Jiao L, *et al.* Physical layer spoofing attack detection in mmWave massive MIMO 5G networks. *IEEE Access*. 2021;9:60419–60432.
[38] Tang J, Wen H, Song H. Physical layer authentication for 5G/6G millimeter wave communications by using channel sparsity. *IET Communications*. 2022;16(3):206–217.
[39] Bai L, Zhu L, Liu J, *et al.* Physical layer authentication in wireless communication networks: a survey. *Journal of Communications and Information Networks*. 2020;5(3):237–264.
[40] Suski II WC, Temple MA, Mendenhall MJ, *et al.* Using spectral fingerprints to improve wireless network security. In: *IEEE Global Telecommunications Conference GLOBCOM*. New Orleans, LO; 2008. p. 1–5.
[41] Xu Q, Zheng R, Saad W, *et al.* Device fingerprinting in wireless networks: challenges and opportunities. *IEEE Communications Surveys & Tutorials*. 2015;18(1):94–104.
[42] Chen Z, Cooklev T, Chen C, *et al.* Modeling primary user emulation attacks and defenses in cognitive radio networks. In: *IEEE 28th International Performance Computing and Communications Conference*. Scottsdale, AZ; 2009. p. 208–215.

Chapter 6

Context-aware physical layer security for future wireless networks

Halise Turkmen[1] *and Hüseyin Arslan*[1]

The amount information currently generated and contained in wireless networks is paramount. Additionally, wireless sensing through integrated sensing and communication (ISAC) is being incorporated into the next-generation [beyond 5G (B5G), WiFi 7] wireless communication standards, further increasing this information. Faster chips, various network topologies and higher data rates have permitted increased utilization of artificial intelligence (AI) techniques in wireless systems and thus enable accurate predictions of future network and radio environment related phenomenon. Therefore, a capable device can essentially be aware of the user, network, and radio environment activities. These developments have led to the cognitive or context-aware physical layer security (PLS) concept. With this, intelligent wireless devices can anticipate threats and take the necessary precautions dynamically. This chapter will go over radio environment monitoring (REMo), which is the process of collecting, processing and storing the aforementioned information. Then, the context-aware PLS framework will be introduced, with application examples.

6.1 Introduction

Fifth-generation cellular networks were developed with the aim of enabling communication between various devices, thus supporting various industries called *vertical sectors*. These sectors are automotive, manufacturing, media, energy, public safety, eHealth, and smart cities [1]. Enabling communication security in these sectors is critical, as breaches can have catastrophic consequences, such as traffic accidents, power outages, water contamination, and so on, possibly leading to the loss of lives. In addition, an enormous amount of operation-related and personal information is collected from human network users, such as location, usage statistics,

[1]Department of Electrical and Electronics Engineering, Istanbul Medipol University, Turkey

preferences, messages and photographs, and so on. The failure to protect this information is a violation of privacy, and may be punishable by national authorities as well as international human rights communities. As a result, an emphasis of B5G and sixth-generation cellular network standardization research is communication security.

Unfortunately, the diverse capabilities of user equipments (UEs), performance requirements of applications and environmental conditions render a "one-size-fits-all" solution infeasible. For example, an Internet of things (IoT) device on an autonomous vehicle with control or decision-making tasks requires low latency, ultra-reliable and highly secure communication. Lengthy cryptography schemes can provide high security, but at the cost of significant additional delays, given that the IoT device is even capable of carrying out these schemes. Therefore, similar to fifth generation (5G)'s service-based resource allocation, *quality of security (QoSec)* based security techniques should be applied to secure the communication. This will ensure that the minimal security requirements for an application are met at all times. However, this is a resource-greedy approach, where a certain level of security is obtained with the trade-off of resources like power, spectrum, computational abilities, and so on, and is redundant in environments where the probability of attacks is low to none. Additionally, it may be difficult to reduce the plethora of devices with varying capabilities, application security requirements and operation criticality to a few categories. This, and the fact that the majority of attacks mentioned in the technical report [2] can be carried out before performing the 5G security protocols, has motivated research into dynamic security and PLS techniques.

The *cognitive* or *context-aware* PLS concept was incepted [3–5] with adaptability in mind. Unlike the standard security approaches, this concept is preventative, with the aim of fortifying the applied security techniques before an attack actually occurs. To realize this, the device needs to be aware of the activities in the network and environment in addition to the QoSec requirements. This is achievable through the radio environment monitoring (REMo) framework [4,6]. This framework is responsible for collecting, processing and storing information about the network, spectrum, UEs and the physical environment. Using this information and data-driven algorithms, the UE can attain contextual information, such as the presence of illegitimate users, the behavior of the UE and underlying reasons for the changes in the physical environment (and thus the channel), from which it can identify the threat levels, estimate probability of attacks, and determine the optimum security techniques.

That said, this chapter will go over the main features of the REMo architecture and the context-aware PLS framework. Then, selected state-of-the-art works and example scenarios in this area will be discussed. Because the cognitive or context-aware security concept is relatively a novel and unexplored field, some realization obstacles will be touched upon.

6.2 The radio environment map and radio environment monitoring

Radio environment maps (REMs) were first introduced in 2006 [7] as a collection of integrated, dumb databases containing information pertaining to the wireless network, as shown in Figure 6.1. They were envisioned as a flexibility enabler and a prerequisite to awareness, which cognitive radios would utilize in their decision-making processes. Due to the limitations on hardware processing capability and storage, the initial actualizations of REM were limited to spectrum coverage maps. These were used for opportunistic access in the sharing of unlicensed spectrum [8]. Later, the UE mobility information and various mobility models stored in the REMs were utilized for efficient handover management [9]. In 2013, European Telecommunications Standards Institute (ETSI) published a report studying possible architectures for building and utilizing REMs in certain additional use-case scenarios between different operators. These use-cases include coverage/capacity improvement by relays, self-configuration and optimization of femtocells, system optimization, interference mitigation, vertical handover optimization and intra-systems handover optimization [6]. However, there were many obstacles to implementing REM in wireless networks, such as the collection, processing and reliability of REM information.

With the advent of more capable hardware, the cloud, edge computing, and maturing of machine learning (ML) implementations for cellular networks, the full potential of REMs are being realized. Now, rather than just a collection of databases, the REMo framework is an intelligent entity that continually monitors the environment and updates itself. New technologies and study areas, such as wireless sensing and ISAC, are enabling a huge amount of data and information collection through processing the transmitted wireless signals. The radio signals propagating through the channel essentially contain fingerprints of the physical environment, spectrum activities, and transmitting devices. Obtaining these fingerprints through signal processing and AI techniques is known as wireless sensing. Jointly performing communication and wireless sensing utilizing the same resources is known as

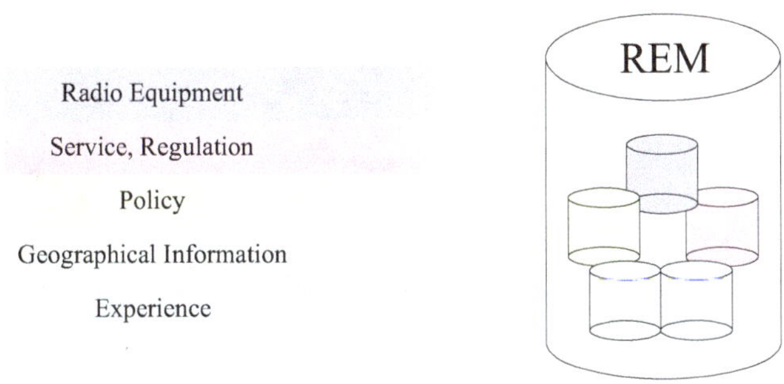

Figure 6.1 Radio environment map containing multiple databases

ISAC. Both wireless sensing and ISAC are being incorporated into the wireless communication standards due to their potential for gathering a large amount of data in a non-invasive manner. The wireless LAN (WLAN) standards have formed a task group, 802.11bf – WLAN Sensing [10], and cellular communication (Third Generation Partnership Project (3GPP)) standards have put ISAC on the list of Release 19 topics [11].

6.2.1 Radio environment monitoring framework

An exemplary REMo framework depiction is given in Figure 6.2. Here, the *framework manager* is responsible for maintaining the REM by initiating data/information collection and processing and removing the outdated data. It is also responsible for responding to the data/information queries made by client devices, such as UE or base station (BS). The contents of the REM database can be acquired from external sources, such as *network operators*, *external sensors*, and *online databases*, or through wireless sensing and data/information processing methods, such as applying ML techniques to the acquired data.

The awareness gained through the REMo framework can undoubtedly be used in risk-of-threat prediction, threat detection, and threat identification. More on this will be discussed in the following sections. The integral REMo components are elaborated below.

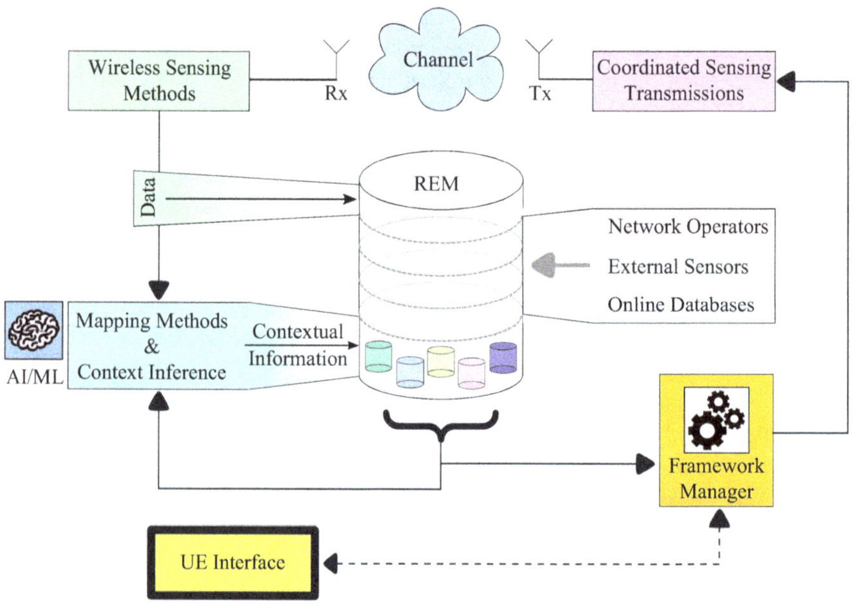

Figure 6.2 Exemplary radio environment monitoring framework

6.2.1.1 Wireless channel

The wireless channel is the result physical phenomenon, such as mobility, obstructions, atmospheric effects, and electromagnetic emissions, such as signals from another source. Therefore, the channel can be viewed essentially as a fingerprint of the physical and spectral activities in that area. Because the transmitted signal propagates through this environment, and thus channel, the received signal also contains this fingerprint. If the appropriate extraction techniques are used or developed, it is possible to extract information about many of the phenomena occurring. Figure 6.3 illustrates some of this information and how/when they are added to the transmitted signal. Here, the radio front-end is also considered part of the channel. The medium access control (MAC) layer of the Open Systems Interconnection model is responsible for selecting the appropriate scheduling/resource allocation schemes and frame designs. Scheduling determines the how the signal will be transmitted, namely, the multiple access or multiplexing schemes to be used, the re-transmission times, the

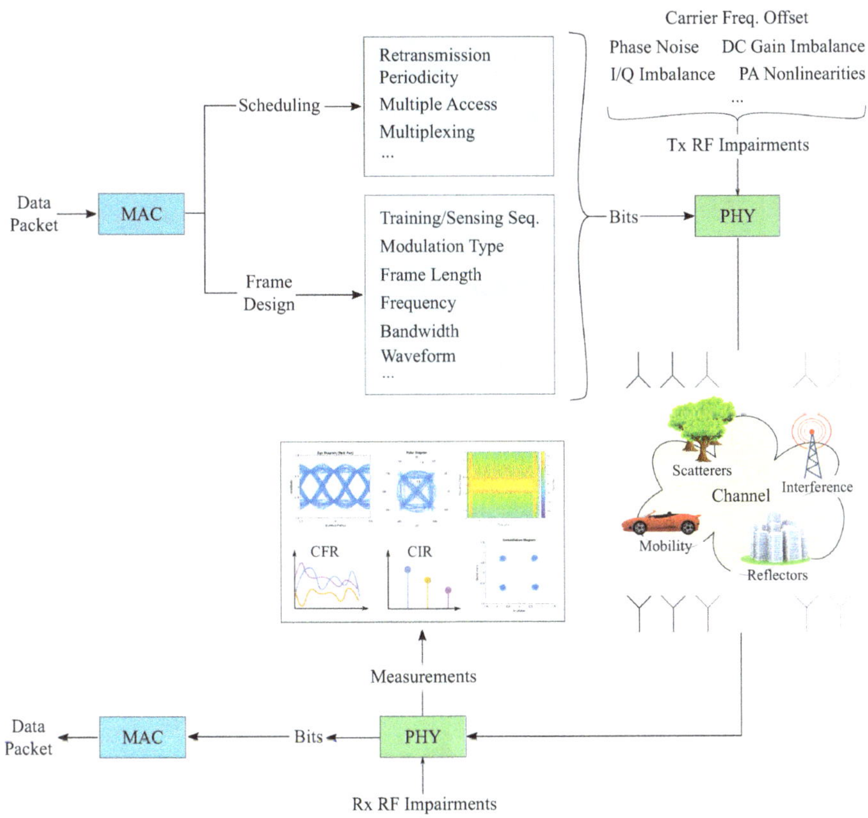

Figure 6.3 Factors affecting and altering the wireless signal

resources to be allocated, and so on. Frame design determines the training/sensing sequences to be used, the modulation schemes, pilot placements and so on. The physical (PHY) layer converts the data frame into a physical signal for transmission. During this process, hardware impairments, such as phase noise and carrier frequency offset, due to non-ideal devices, are added to the signal.

6.2.1.2 Wireless sensing

The *wireless sensing methods* are those which are used to extract the aforementioned fingerprints from the received signal and include radar signal processing, channel state information (CSI) based action/object detection, received signal strength (RSS) based localization, and so on. In this section, these sensing methods will be explained in brief. Readers are welcome to view the references for more in-depth information.

Radar is a well-known and studied wireless sensing technique where a specific signal is transmitted and the echos from objects are received and processed to find the range and velocity through the delays and Doppler shifts of the received echos, respectively [12]. An example of a monostatic radar system is depicted in Figure 6.4. Here, monostatic indicates either a collocated transmitter and receiver or a transceiver. The range, R, and velocity, v, can be found from the delay of the echo, Δt, and Doppler shift, Δf, respectively, using the following formula:

$$R = \frac{c\Delta t}{2}, \tag{6.1}$$

$$v = \frac{c\Delta f}{2f}, \tag{6.2}$$

where c is the speed of light and f is the carrier frequency.

CSI-based wireless sensing, on the other hand, generally involves transmitting known pilots and calculating the CSI, channel frequency response (CFR) or channel impulse response (CIR). There are three main approaches to CSI-based sensing; model-based, pattern-based, and hybrid approaches. Model-based approaches involve first modeling the actions of interest based on a variety of features, such as number of the Fresnel zone crossings and variation in Doppler and delays. Then, these are calculated for known actions and stored. Then, these models are used to detect the action with real-time measurements. While this approach is less computationally less complex once the model is derived, deriving a realistic model which applies to real-life environments is difficult. In pattern-based approaches, the real-time measurements are fed to a ML algorithm and some patterns are deduced. Then,

Figure 6.4 Monostatic radar. The blue signal is the transmitted signal, while the green signal is the echo off of the object

new measurements are compared to these patterns to deduce whether an action has taken place or not. The disadvantage of this approach is the requirements for large data sets and training periods. Also, the ML algorithms must be carefully designed to avoid over-fitting and loss of generality. Hybrid approaches combine both methods. Here, initially, some models are used to reduce the dimensions and size of the data, followed by ML algorithms for the classification [13].

Using RSS measurements for positioning with respect to an anchor or beacon [14], generally a BS or access point (AP), is done in two ways. The first method calculates the distance from one or more beacons using the RSS and the link budget and path loss equations, simplified versions given in (6.3) and (6.4) respectively:

$$RSS = P_{Tx} + G_{Tx} - PL + G_{Rx}, \tag{6.3}$$

$$PL = PL_{ref} + 10\log d^{\eta} + s, \tag{6.4}$$

where P_{Tx} is the transmit power, G_{Tx} and G_{Rx} are the transmitter and receiver antenna gains, PL is the calculated path loss, PL_{ref} is the path loss at a reference distance of 1 m, d is the propagation distance, η is the path loss exponent, and s is the standard deviation in the calculated path loss due to shadowing. The location of the beacons is assumed to be known. Positioning can be done through triangulation if measurements are taken from more than three or more beacons [15]. This is depicted in Figure 6.5. The second method involves surveying the environment and storing the RSS measurements in a look-up table. Then, instantaneous measurements are matched to those in the table and the position corresponding to the closest value is accepted as the device position.

There are numerous wireless sensing methods, and only a few are mentioned here in a simplified and generalized manner. Coordinating and managing these methods appropriately is vital, and is performed by the framework manager. This involves choosing the optimal sensing method and determining the parameters of the sensing signal, such as periodicity, bandwidth, transmission power, beamwidth, and so on,

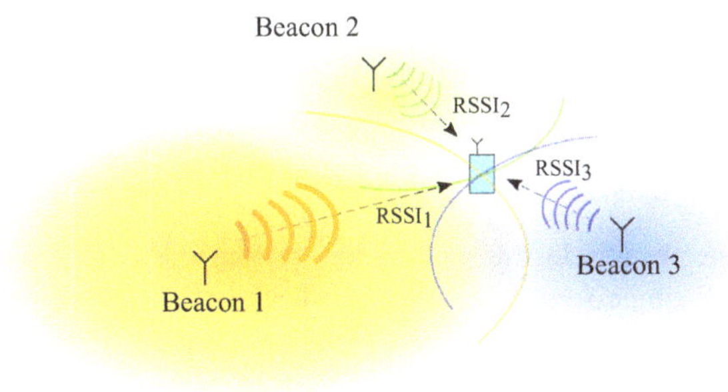

Figure 6.5 Example of positioning with RSSI triangulation. Here, the different colors represent different the coverage areas and signals of different beacons

for the desired sensing performance requirements, like false alarm rate, accuracy, or resolution, while ensuring the minimum acceptable amount of disruption to the communication performance, especially in terms of interference and data rate.

6.2.1.3 Mapping methods and context inference

Mapping methods and context inference block can then process the sensed information along with the information already present in the REM to generate awareness using machine learning (ML) and data fusion techniques. The resulting knowledge acquired can be UE or object mobility information, person detection, tracking, identification, UE identification, interference detection, and gesture identification, among many others. For example, a device processing the mobility patterns of an object/UE can infer the underlying reason for its mobility, and thus accurately *know* its next location. This can be useful in beam forming and management.

6.2.1.4 Other sources of information

Network operators can provide information such as network policies and regulations, BS locations, UE mobility and usage statistics, and so on. External sensors can be sensors on the device, such as gyroscopes and temperature sensors, or sensors within the network, such as cameras. Online databases can contain information such as terrain topology, road maps, census data, and so on.

6.3 Context-aware PLS framework

Context or situational awareness aided PLS will allow operators to utilize the network and shared resources more efficiently. This is undoubtedly vital, as the number or wireless-capable devices, wireless-dependent applications, and required data rates is exponentially increasing per coverage area. Thus, all aspects of a transmission, including the security, should be optimized. Additionally, it is a preemptive form of security and enhances security measures if an attack is predicted. This allows the legitimate user to avoid the attacks altogether. Figure 6.6 illustrates the fundamental building blocks of a context-aware PLS framework, with possible input parameters and output information.

6.3.1 Situational awareness

Situational awareness in wireless communication is the ability to perceive and comprehend the devices, objects, and phenomenon in the environment, understand the relation of these elements to each other, and predict their future states. The environment, devices, infrastructure, and physical phenomenon around the wireless device contains an innumerable amount of data. From this data, it is possible to infer critical details about the environment, intent/purpose of the devices, and the criticality of the data, and thus gain some level of situational awareness. The REM information can be used for risk identification and the optimal PLS method selection by providing situational awareness and information on resource availability. In Figure 6.6,

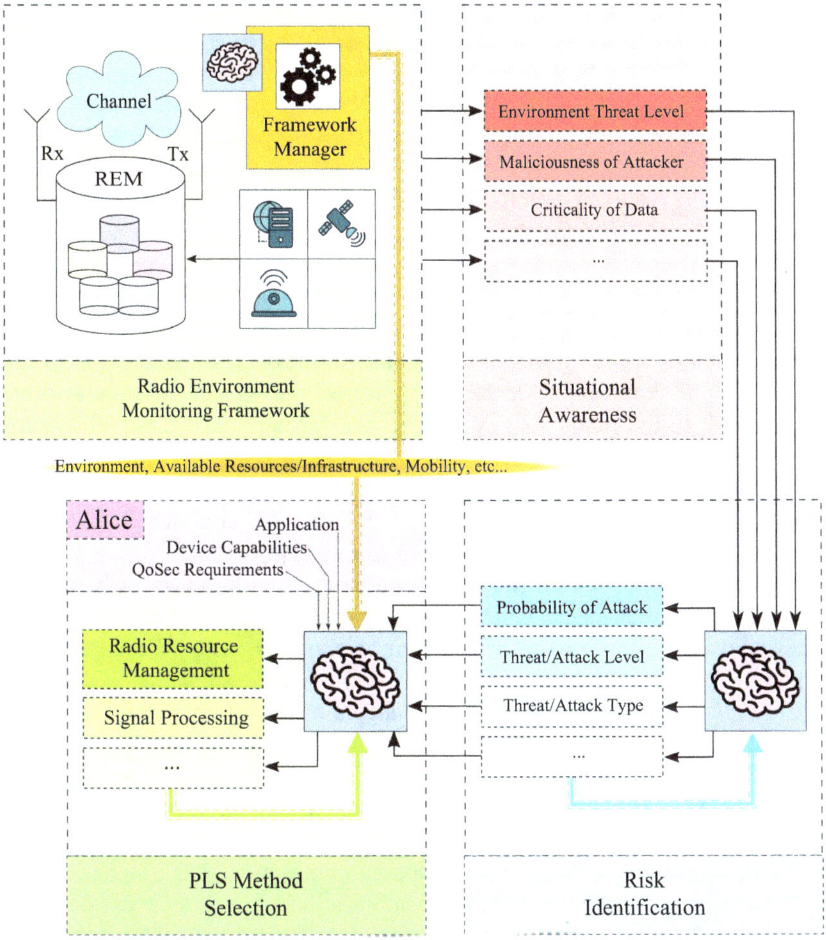

Figure 6.6 Context-aware PLS method selection framework with some exemplary parameters

three examples of situational awareness are given. These examples will be explained through realistic scenarios below.

1. *Environment threat level*: Wireless devices in an area are generally randomly scattered. For example, the devices within a cell are assumed random, in location, identity, and capability. However, cells or areas with a cell with high-criticality communications/UEs are likely to have more attackers/illegitimate nodes. These areas require more secure PLS techniques. Additionally, areas where attacks have been detected in the past or where a detected and identified illegitimate/compromised node is re-detected can also be assumed to have

a higher threat level. These areas, UEs/illegitimate node activities, and high criticality communications are either stored in the REM or can be inferred by processing the REM information.

2. *Maliciousness of attacker*: Attackers or illegitimate nodes may be pre-programmed devices performing one, constant function, such as indiscriminate jamming, or they may be more intelligent and have a discrete purpose, such as jamming the communication of a specific UE. The latter would entail detecting, identifying, and tracking the signals of the legitimate node and jamming only a specific frequency band. Such intelligent attackers/illegitimate nodes may be found in military scenarios. Alternatively, the illegitimate node may be after a critical data or want to gain an operation control and may execute man-in-the-middle or spoofing attacks. The maliciousness of the attacker/illegitimate node can be defined with respect to their goal and their dedication to achieve said goal. In the examples given above, the second attacker can be said to be more malicious because it not only disrupts the communication of the legitimate node, but also tries to overcome security measures, such as frequency hopping. Attackers/illegitimate nodes attacking high criticality communications are also more malicious, such as those aiming to intercept communication in a water treatment plant or power grid. REM can be used to detect, identify, and track the activities of attackers. Once they have been labeled with their maliciousness level, it will be easier to determine their potential targets before an attack takes place.
3. *Criticality of data*: This refers to the degree of damage or fall-out a potential attack on the data can cause. It is intuitive that the day-to-day conversation of a random person is of low criticality. Here, the privacy of the person is no doubt important. However, it is not as critical as data involving remote surgery or autonomous vehicles. Similarly, the day-to-day conversations of a person are less critical than their online shopping information. The usage statistics of the person (or UE), along with other information, such as location, duration of usage, mobility, etc., can be used to infer whether the data is of low criticality or high criticality.

6.3.2 Risk identification

Risk identification is the mapping of situational-awareness information, like those above, to indicators or measures of attacks and their possible repercussions. Because the relationships between the known information and the risks are non-linear and may be due to unobservable factors, AI and ML algorithms are better suited to perform this mapping. In Figure 6.6, three possible risk identifiers are given and are detailed below:

1. *Probability of attack*: Situational awareness can be used to estimate the likelihood of an attack along with the additional information of UE identity, communication service type, location, statistics of past attacks, etc. Higher probability of attack would mean that a more secure PLS method should be used and vice versa.

2. *Threat/attack level*: This refers to the severity of the attack. It can be mapped from the maliciousness of the attacker/illegitimate node and the success rate of the attacks. Higher threat/attack level would mean that a more secure PLS method should be used and vice versa.
3. *Threat/attack type*: The type or method of the attack expected to be executed is of utmost importance. Knowing this would allow the legitimate node to choose a tailored PLS method. This can provide a higher level of security as well as optimal allocation of resources. The threat/attack type can be mapped using the known illegitimate node information, type of data or communication service and the application, among other information.

6.3.3 PLS method selection

Once the risks to the communication are identified, the PLS method selection block can process this information to decide on the optimal PLS method. Additional information that may be required, such as environment, available resources, available infrastructure and UE mobility can be acquired from the REM. Application and QoSec requirements can be acquired from the UE. Once again, to avoid modeling complex and unobservable relations, AI and ML techniques can be used to map the risk identifiers, REM information, and UE information to the optimal PLS technique. Depending on the techniques, the output of this block could be resource allocation/management schemes and signal processing techniques.

Here, although the risks are identified through information from the REM, the additional information is used to select the appropriate PLS methods under the environment and mobility conditions of the UE. For example, knowledge of the propagation channel and channel diversity, such as independent multipath information, are required to estimate security metrics such as secrecy capacity, and can be used to determine the efficiency whether channel-based key generation. Similarly, node communication patterns in a network can help identify trustworthy nodes which can be used as relays.

6.4 Context-aware security in the literature

Context-aware security is newly gaining attention in academia and the sector, mainly due to the lack or immaturity of enablers, such as AI/ML techniques, REM, required protocols and interfaces, and suitable chips. However, the years leading up to the sixth generation (6G) standardization will no doubt produce interesting research in this area. Nevertheless, select works on producing or relying on context-awareness are presented below.

6.4.1 Social reputation and trustworthiness

Networks are increasingly becoming more local, decentralized, and ad hoc. Decision-making is also moving to the edge, rather than at a centralized controller [16]. Device-to-device (D2D) communication is also regaining interest for relaying or communicating in a controlled environment, such as IoT devices in a factory. The advantage of these trends is the possibility for autonomy, which brings less delays, back-haul throughput and simpler infrastructure [17]. However, the vulnerability of these networks to illegitimate users remains a drawback. Social reputation or trustworthiness factors are a measure of the reliability of the UE and form a trust-relationship between UE-UE or UE-BS. Generally, each UE starts out as trustworthy and has an assigned score. When a UE behaves in unexpected ways, their score begins to decrease, until a threshold after which they are deemed untrustworthy. Unexpected behavior could be a drastic change to the usage pattern, deviations from assigned locations or trajectories, repeated failure to deliver a packet, delivery of an altered packet, and so on. Such shifts in behavior could indicate a man-in-the-middle attack.

6.4.1.1 Securing VANET communications

An example could be the realization of vehicular ad hoc networks (VANETs). Autonomous vehicles are expected to become a part of the near future, but the consequences of communication hijacking or failure is one factor impeding their implementation. While in urban areas, security can be guaranteed to a degree, as they may be communicating through cellular technologies, this service cannot be provided in rural areas with limited cellular penetration. In this case, adjacent vehicles have to form their own network and communicate accordingly. Here, block-chain-based social reputation assignment is one proposed approach [18,19]. Any alteration to the transmitted packet changes the hash value of that packet, which is detectable. UEs which transmit packets with altered root hash values consequently lose their social reputation scores and are eventually phased out of the network, i.e.: they are not chosen for routing or their transmissions are ignored. A similar work is proposed in [20] where the communication statistics, i.e.: packet type and frequency, of a UE is monitored by its peers and reported as proof to the road side unit. If the road side unit finds the communication statistics of the UE to be different from those recorded by its peers, it is deemed as untrustworthy. These schemes can be seen in Figure 6.7. Here, V_3 and V_4 have lower social reputation because their traffic patterns have been determined as odd and malicious. Additionally, V_3 has forwarded a message with an invalid hash.

6.4.1.2 Securing relay/routing

Another example could be message routing in D2D communication. Here, the packet is transmitted from one device to another until it reaches its destination. The choice of devices to forward the packet to depends on the willingness of the device to participate. Rather than a fixed route of devices, a random route would be more suitable to prevent unwanted interception of the packet. However, the candidate UE may also be malicious. To this end, Ref. [21] proposes a trustworthiness model with direct and indirect components. The direct component is a score contribution from the UEs which have interacted with the candidate UE. The indirect component is a score based on the structural and social properties of the candidate UE. This is acquired

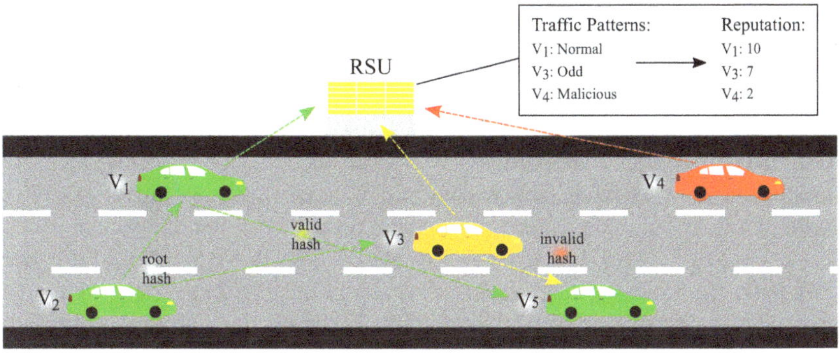

Figure 6.7 Social reputation-based authentication and routing for VANETs. Here, both the block chain hash and communication statistics schemes are illustrated

from the observations of the relay node (RN). In this manner, they can evaluate new candidate UEs.

6.4.2 Location, behavior, and mobility-aided authentication

REM contains information on network infrastructures, such as the locations of BSs and APs. In addition, UE localization methods have improved significantly in terms of accuracy and resolution. ML aided radio frequency (RF) fingerprinting based localization [22] and the 5G New Radio (5G-NR) resource block and symbol design for time-difference-of-arrival (TDOA) based positioning [23] are two examples of this. Furthermore, wireless sensing and integrated sensing and communication methods can be used to find the location of a UE or object. In addition, the REM also can contain mobility patterns of UEs.

6.4.2.1 Mobility Aided (Re)Authentication Overhead Reduction

Mobile devices may require frequent handovers, and in some cases, ping-pong handovers may occur if the UE is travelling along the intersection of two cells. The authentication process for UE-BSAP association may cause significant overhead, resulting in delays and dropped calls. The mobility patterns of the UE stored in the REM can be used as a form of authentication. For cellular networks, [24] proposes to classify the mobility of the UEs and perform authentication either solely at the cloud (centralized architecture) or the edge node (distributed architecture). In WLAN networks, authentication can take up to one second, which is problematic for UE requiring frequent handovers. Pre-authentication methods are present to mitigate this, but at the cost of increased overhead. Joshi *et al.* propose a mobility pattern based key sharing method which reduces the time for authentication by 20–50 ms and the overhead by 55%–70%. Specifically, they utilize the frequency of re-entrance to an AP coverage area and the time of last entrance to determine which

AP(s) are more likely to be visited next and distributes the pre-authentication key accordingly [25].

6.4.2.2 Location and behavior-aided BS authentication

False or rogue BS attacks are particularly critical since they occur before the device is associated with a legitimate entity. They aim to violate the users' privacy by eavesdropping on their communications. To prevent this, the 5G standards have adopted a device-assisted network-based false or rogue BS detection framework [26]. Here, received signaling messages are reported by devices to legitimate network entities. These entities analyze these reports and compare the results with their information on the network topology to detect anomalies. For example, false or rogue BS are known to transmit high power signals to encourage UE to associate with them. Murat [27,28] is one such framework that analyses report contents such as signalling message power, location, radio access type and fingerprints. Murat adds situational awareness to the framework by adding a mechanism that turns detection on or off based on a priority value. This value is determined by the susceptibility to attacks at a location and the time, or whether there is a high-profile event taking place.

6.5 Research directions

Context-aware PLS will surely become a critical part of future networks, as evidenced by the recent works in the literature and the 3GPP standards. However, there are many open research areas and shortcomings in current standards which need to be addressed.

6.5.1 Data/information collection and management

Wireless devices and networks have the potential to or are currently collecting extensive data from the users and their environment. The majority of this data is collected without the explicit permission of the user and may be sensitive. Thus, the data collected and stored must be well validated and the privacy of the user should be procurable. Another factor is that the indiscriminate collecting and storing of this data is unsustainable. Wireless networks collect much of the data for optimizing communication parameters, like associated cell, transmit power, and so on. Therefore, some data may not be necessary for context-aware PLS and may not need to be stored. Determining which data/information is useful is not straightforward. The complexity of the environment and communication statistics generally renders them difficult to model, as it is near impossible to determine the underlying factors for usage statistics and environment activities. However, efforts in this area are needed to at least ascertain a set of data that can be directly related to risk identification and PLS methods selection.

Managing the collection of this data is another issue. Age of information refers to the amount of time the information is valid or considered current. This varies for different types of data/information, scenarios/applications, and UEs. Acquiring most data incurs additional overhead, which should be avoided at all costs due to spectrum congestion. Therefore, when and which data should be collected/updated and how should be well studied.

6.5.2 *Validating the context or situational awareness*

While humans can understand context and make deductions based on observed, seemingly unrelated actions and events, this is on-going research in the AI/ML field. Currently, for most of the works explored in Section 6.4, a low level of context-awareness is acquired through rule-based reasoning, i.e.: if x location has signalling transmissions of y dB power or more, then there is a false BS. To attain a higher and more autonomous level of awareness, and thus fully realize the context-aware paradigm, more insightful algorithms need to be developed. Combining model-based reasoning and AI/ML methods are expected to provide better context awareness. To do this, the aid of data analysts and field experts is needed to better quantize the relationships between non-communication related topics, such as day-to-day activities, mobility patterns national and international events, natural disasters, and so on, with communication-related data. This is crucial, as mistakes or the inability to derive an accurate context can lead to under-protected communications and breaches of security.

6.5.3 *Unified understanding of risk identifiers and QoSec levels*

Till now, a consensus has not been reached on the levels of QoSec and mapping these levels to applications and solutions. While the term QoSec has been around since the 2000s, interest in it has recently reignited with the advent of context-aware PLS. As such, academia and industry are still trying to define these levels, let alone design frameworks. While there are generalized conceptual frameworks like the one depicted in Section 6.6, concrete methods are yet to be developed.

6.6 Conclusion

The development and adoption of communication technologies has taken an interesting turn, with not only pervading all aspects of human life, but taking on significant role in nearly all industries as well. With such a function, ensuring security is both imperative and extremely challenging. The plethora of different devices and capabilities, together with the resource limitations, compel different security levels with specific complexities and resource usages. Because the communication applications, device and resource limitations, attacks and threats are innumerable, intelligent mechanisms for security method selection should be developed. On this point, context or situation aware PLS techniques are a viable solution which can potentially support all applications, devices and attack scenarios, while ensuring more efficient resource allocation.

References

[1] Imran MA, Sambo YA, Abbasi QH. *Enabling 5G Communication Systems to Support Vertical Industries*. New York, NY: Wiley-IEEE Press; 2019.

[2] 3GPP. 3GPP TR 33.809: Study on 5G security enhancements against False Base Stations (FBS) (Release 16). 3rd Generation Partnership Project;2022.

[3] Yilmaz MH, Guvenkaya E, Furqan HM, *et al.* Cognitive security of wireless communication systems in the physical layer. *Wireless Communications and Mobile Computing*. 2017;2017:9.

[4] Furqan HM, Solaija MSJ, Türkmen H, *et al.* Wireless communication, sensing, and REM: a security perspective. *IEEE Open Journal of the Communications Society*. 2021;2:287–321.

[5] Chorti A, Barreto AN, Köpsell S, *et al.* Context-aware security for 6G wireless: the role of physical layer security. *IEEE Communications Standards Magazine*. 2022;6(1):102–108.

[6] ETSI. *Reconfigurable Radio Systems (RRS); Use Cases for Building and Exploitation of Radio Environment Maps (REMs) for Intra-Operator Scenarios*. European Telecommunications Standards Institute; 2013.

[7] Zhao Y, Le B, Reed JH. Network Support: The Radio Environment Map. In: Fette BA, editor. *Cognitive Radio Technology*. Burlington: Newnes; 2006. p. 337–363 (Chapter 11). Available from: https://www.sciencedirect.com/science/article/pii/B978075067952750012X.

[8] Gavrilovska LM, Atanasovski VM. Dynamic REM towards flexible spectrum management. In: *2013 11th International Conference on Telecommunications in Modern Satellite, Cable and Broadcasting Services* (TELSIKS). vol. 01; 2013. p. 287–296.

[9] Suarez-Rodriguez C, Jayawickrama BA, Bader F, *et al.* REM-based handover algorithm for next-generation multi-tier cellular networks. In: *2018 IEEE Wireless Communications and Networking Conference* (WCNC); 2018. p. 1–6.

[10] Du R, Xie H, Hu M, *et al.* An Overview on IEEE 802.11 bf: WLAN Sensing. arXiv preprint arXiv:220704859. 2022;.

[11] 3GPP. 3GPP, editor. 3GPP List of Work Items;. Available from: https://www.3gpp.org/DynaReport/WI-List.htm [cited 01.08.2022].

[12] Mahafza BR. *Radar Signal Analysis and Processing using MATLAB*. London: CRC Press; 2016.

[13] Ma Y, Zhou G, Wang S. WiFi sensing with channel state information: a survey. *ACM Computing Surveys*. 2019;52(3). Available from: https://doi.org/10.1145/3310194.

[14] He S, Chan SHG. Wi-Fi fingerprint-based indoor positioning: recent advances and comparisons. *IEEE Communications Surveys & Tutorials*. 2016;18(1):466–490.

[15] Zhang J, Han G, Sun N, *et al.* Path-loss-based fingerprint localization approach for location-based services in indoor environments. *IEEE Access*. 2017;5:13756–13769.

[16] He Y, Ren J, Yu G, *et al.* D2D communications meet mobile edge computing for enhanced computation capacity in cellular networks. *IEEE Transactions on Wireless Communications*. 2019;18(3):1750–1763.

[17] Islam T, Kwon C. Survey on the state-of-the-art in device-to-device communication: a resource allocation perspective. *Ad Hoc Networks*. 2022; 136: 102978. Available from: https://www.sciencedirect.com/science/article/pii/S1570870522001524.

[18] Lu Z, Liu W, Wang Q, *et al.* A privacy-preserving trust model based on blockchain for VANETs. *IEEE Access*. 2018;6:45655–45664.

[19] Lu Z, Wang Q, Qu G, *et al.* BARS: a blockchain-based anonymous reputation system for trust management in VANETs. In: *2018 17th IEEE International Conference On Trust, Security And Privacy In Computing And Communications/ 12th IEEE International Conference On Big Data Science And Engineering* (TrustCom/BigDataSE); 2018. p. 98–103.

[20] Oh H, Zou CC, Park S. An enhanced community-based reputation system for vehicular ad hoc networks. *International Journal of Engineering and Technology*. 2016;8:2272–2287.

[21] Suraci C, Pizzi S, Garompolo D, *et al.* Trusted and secured D2D-aided communications in 5G networks. *Ad Hoc Networks*. 2021;114: 102403. Available from: https://www.sciencedirect.com/science/article/pii/S1570870520307320.

[22] Zhu X, Qu W, Qiu T, *et al.* Indoor intelligent fingerprint-based localization: principles, approaches and challenges. *IEEE Communications Surveys & Tutorials*. 2020;22(4):2634–2657.

[23] Dwivedi S, Shreevastav R, Munier F, *et al.* Positioning in 5G networks. *IEEE Communications Magazine*. 2021;59(11):38–44.

[24] Abdullah F, Kimovski D, Prodan R, *et al.* Handover authentication latency reduction using mobile edge computing and mobility patterns. *Computing*. 2021;103(11):2667–2686. Available from: https://doi.org/10.1007/s00607-021-00969-z.

[25] Joshi T, Mukherjee A, Agrawal DP. Exploiting mobility patterns to reduce re-authentication overheads in infrastructure WLAN networks. In: *2006 Canadian Conference on Electrical and Computer Engineering*; 2006. p. 1423–1426.

[26] 3GPP. 3GPP TS 33.501: Security architecture and procedures for 5G system (Release 17). 3rd Generation Partnership Project; 2022.

[27] Nakarmi PK, Ersoy MA, Soykan EU, *et al.* Murat: Multi-RAT False Base Station Detector. CoRR. 2021;abs/2102.08780. Available from: https://arxiv.org/abs/2102.08780.

[28] Nakarmi PK, Norrman K. *Detecting false base stations in mobile networks* [online]; 2022. Available from www.ericsson.com/en/blog/2018/6/detecting-false-base-stations-in-mobile-networks.

Chapter 7
Signal domain physical modification for PLS

Salah Eddine Zegrar[1], *Ahmed Naeem*[1], *Haji M. Furqan*[1], *and Hüseyin Arslan*[1]

In wireless communication, the data is conveyed through transmission over a physical signal, occupying a bounded specific space–time–frequency block in hyperspace. The building block of this signal is a distinct waveform that conveys the data symbols. A successful reception or detection of a signal requires the receiver to know the exact location of the transmitted signal in space–time–frequency hyperspace. Moreover, a successful identification requires that the receiver knows the features of the transmitted signal in order to decode the data. The goal of physical layer security (PLS) is to ensure the successful detection and identification of the information at the legitimate receiver while preventing illegitimate receivers from doing that. Initially, this chapter presents an overview of the inherent security characteristics provided by different waveforms against detection and identification. Then, it investigates how securing the control channel signals can affect the secrecy capacity of the communication system. However, inherent security and securing the control signal cannot fulfill the security requirements of all wireless applications. Therefore, in order to provide the required level of security, PLS comes into play, where it includes designing different processes that affect the signal based on the parameters extracted from the observation plane to modify the transmissions to ensure secure communication as highlighted in Chapter 3 (Modification plan).

7.1 Waveform & security

In both civil and military applications, wireless networking is extremely crucial. However, securing information transfer through wireless networks from threats such as eavesdropping, jamming, and spoofing is still a difficult problem. Despite their widespread adoption, conventional cryptographic security mechanisms are unable to scale with increasingly decentralized and heterogeneous networks. So, PLS, which is the utilization of the inherent characteristics of the observation plane between transmission and reception nodes, is proposed as a complementary solution along

[1]Department of Electrical and Electronics Engineering, Istanbul Medipol University, Turkey

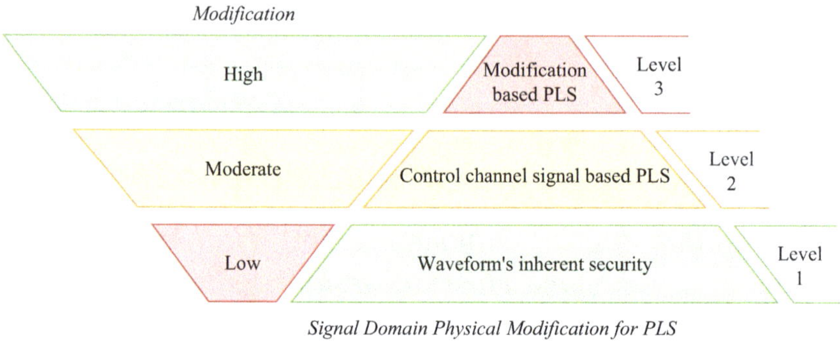

Figure 7.1 The three levels of signal domain physical modification for PLS with respect to the modification and level of security

with cryptographic security algorithms. PLS exploits the features of the observation plane, specifically randomness and uniqueness, etc., to establish a more secure link employed for effective communication. The vast majority of the work that is related to PLS can be classified under the domain of wireless signals. It is important to note that in the context of the modification plane, the term "wireless signal" refers to the entirety of the blocks that exist between the coded bit-stream of data and the antenna on the transceivers. In this chapter, we categorize signal domain physical modification for PLS into three levels with respect to the modification, as illustrated in Figure 7.1. The details of the categorization are as follows:

- *Low modification (waveform's inherent security)*: In this category, security is provided only by selecting the type of waveform without making any modifications to the signal. Specifically, the waveform, which interacts directly with the wireless channel, can also act as a crucial medium in providing security since the observation plane is solely dependent on the correct reception and detection of the waveform. It is preferable to have characteristics that render the waveform undetectable rather than securing it using well-established security techniques. Additionally, the waveform's location and shape in the time–frequency–space dimension directly affect how vulnerable it is to security risks like eavesdropping, jamming, and spoofing.
- *Moderate modification (control channel signal-based PLS)*: The level of complexity of signals that are physically transmitted continues to rise with next-generation wireless technologies beyond fifth generation (5G) and sixth generation (6G). For instance, in order to generate a physical signal, there is an increasing number of variables that can be altered in order to support different future applications, such as modulation order and type, waveform, numerology, frame duration, coding scheme, pilot ratio, subcarrier spacing, and multiple-input multiple-output (MIMO) configuration. Consequently, there is potentially an infinite number of distinct types of signals that can be produced depending on

factors such as the channel, application, and user context. This large number of permutations renders the control signal/channel just like an encryption key. This makes the data signal a totally random meaningless signal without the correct key. Thus, securing the control signal in wireless communication is equivalent to securing the key in asymmetric cryptography. PLS methods can be applied to secure the control channel signal. In this category, the control channel signal containing the data frame's specifications is secured using PLS methods, while the signal conveying the information is intact.

- *High modification (modification-based PLS)*: In cases where the waveform is known and detectable and/or the control signal is shared, the communicating nodes can make use of PLS algorithms to secure the wireless link in this category. To do so, the randomness of the channel is exploited, which serves as a main driver of the observation plane. The primary goal of this class of security measures is the provision of enhanced data decoding capabilities for the legitimate receiver in comparison to those available to the illegitimate attacker. The data decoding competency can be improved by either deliberately lowering the eavesdropper's performance or by increasing the quality of service (QoS) for the authorized receiver. This can be done by adapting, modifying, and optimizing the transmission parameters of the transmitted signal, by adding an interfering signal with the information signal, or by changing the physical features of the signal based on pre-shared or PLS-based generated keys.

7.2 Low modification: waveform's inherent security

The principal concept of PLS is the utilization of the inherent characteristics of the environment existing between transmission and reception nodes. Even though the propagation environment plays an important role and acts as a rich source of entropy, it is not the only factor impacting the establishment of a secure link. For instance, the randomness of the environment can be leveraged to the maximum extent with the help of the proper selection of the transmitted waveform. The waveform, which interacts directly with the wireless channel, can also act as a crucial medium in providing security. This information-carrying waveform acts as the first level of signal domain physical modification for PLS, since channel estimation is solely dependent on the reception and detection of the waveform, as shown in Figure 7.1. The traits that make the waveform undetectable by the attacker are more desirable than applying any security method. For instance, an undetectable plain waveform transmission is considered to be more secure in wireless communications than a detectable one despite being encrypted, since the data can be recovered from the finitely known encryption techniques. Moreover, the shape of a waveform in the time–frequency–space dimension has a direct impact on its susceptibility to security threats like eavesdropping, jamming, and spoofing.

In order to employ PLS algorithms, the observation plane (such as channel state information (CSI)) must be known at all legitimate nodes. To do so, PLS algorithms make use of the transmitted waveform by first detecting the domain of the data transmission (time, frequency, and space) and then by extracting the features of the received signal/waveform (modulation order, coding scheme, etc.). Finally, the features are extracted for the observation plane. Therefore, the inherent secrecy performance of waveforms for robust communication is centered on two criteria:

- *Signal detection*: It is defined as the procedure that determines the occupied region in time–frequency–space dimension and also identifies the underlying information parameters.
- *Signal identification*: It is defined as the extraction of features and hyper-parameters of the received signal, such as waveform type, subcarrier spacing, modulation type, and coding scheme, that change the statistical properties of the transmitted signal for a given waveform.

7.2.1 Signal detection

Once the signal is passed through the wireless channel, it is composed of noise and signals with varying characteristics and access techniques defined by their respective technologies, e.g., standards, air interfaces, etc. The unique traits exhibited by these signals have projections in time, frequency, and space dimensions and can be utilized to detect the signals. As explained earlier, signal detection is defined as the procedure that determines the occupied dimension of time–frequency–space dimension and identifies the underlying information such as bandwidth, carrier frequency, frame duration, transmission/reception direction, and other parameters related to air interfaces. In signal detection, the performance of single carrier/multi-carrier waveforms in terms of security is of prime importance. The merits and demerits of these waveforms are discussed in the following sections.

7.2.1.1 Signal detection for single carrier modulation

Spread spectrum (SS) is among the renowned digital communication schemes due to its distinct properties, which lead to a better and more secure communication network. One reason why it was first employed by the military is that it is difficult to intercept or detect. A definition of SS can be defined as: "a method of transmission, where the signal uses more bandwidth than it is required to transmit the information. In particular, at the transmitter, the signal is spread using special codes that are independent of the data, while at the receiver, the de-spreading and subsequent data recovery processes use synchronized reception with those codes." [1].

For instance, considering frequency hopping spread spectrum (FHSS) transmission, random or pseudo-random (PN) sequences are used to spread the data signal over a wide range of frequencies, as depicted in Figure 7.2. This will alter the carrier frequency, but no change will occur to the original bandwidth of the data [2]. The communication system's diversity is also increased as a result of the modulation and transmission of various data portions over different carrier frequencies. The employed PN sequence determines the order and sequence of the carrier frequencies. The simplest frequency hopping form is given by:

$$S_m = b_{\mathrm{m}} \cos\left(2\pi f_{\mathrm{m}} t\right) \mathrm{P}_{T_{\mathrm{b}}}\left(t - mT_{\mathrm{b}}\right), \tag{7.1}$$

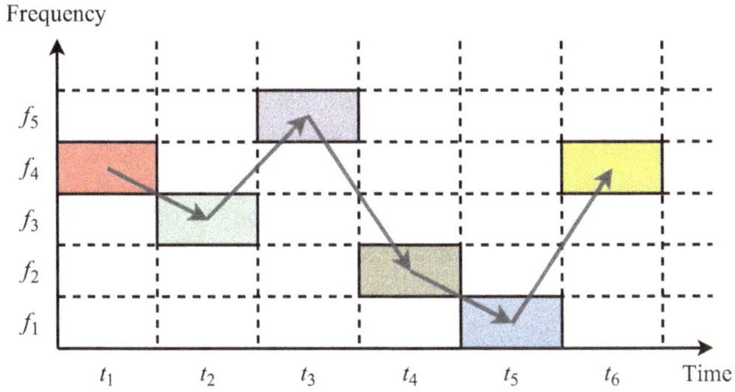

Figure 7.2 Frequency hopping spread spectrum

where b_m is the information sequence, P_{T_b} is the pulse shape, and f_m is one of the N frequencies selected to be the random frequency sequence, and the dwell time T_b (which is the amount of time that the data signal spends in each frequency hop). As a result, the data signals hop to a different frequency.

In SS systems, higher frequency code sequences are multiplied with the signals. This increases the occupied bandwidth of data to a wider bandwidth and extends power to a wider range as well, which makes SS signals resistant to narrow-band interference because of their large bandwidth and low power spectrum density, and also makes them able to be readily masked within the noise floor, preventing them from being detected by an unauthorized party. Additionally, due to the use of random codes known only between the transmitter and the receiver, SS has better security performance against eavesdropping.

7.2.1.2 Signal detection for multi-carrier modulation

In contrast to single-carrier transmission, where whole data is transmitted over a single carrier, multi-carrier transmission is a technique that transmits data by dividing it into several components and transmitting each of these components over multiple individual carriers. The individual carriers have narrow bandwidth, but the composite signal can have broad bandwidth [3]. Multi-carrier modulation has several advantages, including relative immunity to fading caused by transmission over more than one path at a time (multipath fading), less susceptibility than single-carrier systems to interference caused by impulse noise, and enhanced immunity to intersymbol interference (ISI). There are many forms of multi-carrier modulation techniques that are in use. Among these, some widely known schemes are orthogonal frequency division multiplexing (OFDM), filter bank multi carrier (FBMC), and generalized frequency division multiplexing (GFDM).

In commercial systems, OFDM is extensively used because of the numerous benefits it offers. It can achieve high data rates with manageable complexity and provides high bandwidth efficiency, which is essential for modern communication systems. Also, the parameters in OFDM are adaptive in nature depending on the environmental characteristics [4]. One of the main advantages of OFDM is that it is immune to multipath distortion. In OFDM, linear convolution of the transmitted signal with the channel impulse response (CIR) is converted to circular convolution by cyclically extending the symbols and adding a cyclic prefix (CP). This addition of CP makes the recovery of data symbols simple, using a single-tap frequency domain equalizer. With the usage of CP, the cyclo-stationary property of OFDM signals is exploited [5]. This redundancy introduced by CP is also useful to obtain time and frequency synchronization once the symbol duration and CP size are estimated. Moreover, the cyclo-stationary property of OFDM can be used for estimating CSI [6]. The CP can be used for blind detection by the attackers to extract transmission parameters and synchronization to the transmitted signal, which makes multi-carrier poorly secured transmission. Moreover, these systems are also affected by jammers because of their narrow band nature. It makes them vulnerable even to single-tone or partial-band jammers [7].

7.2.2 *Signal identification/feature extraction*

Following up with Section 7.2.1 and assuming that the signal conveying the data is well received and detected, a crucial component of wireless communications systems is the extraction of features from the received signal. For instance, blind receivers might benefit from techniques that can determine the modulation of received signals without the need for prior knowledge. In addition, making a distinction between single-carrier systems and multi-carrier systems is crucial for an attacker or eavesdropper. However, even if the modulation type is correctly identified, the information conveyed in the signal cannot be detected straightforwardly. This is due to the waveform hyper-parameters, such as modulation order and coding scheme, that change the statistical properties of the transmitted signal for a given waveform. The degree of difficulty of blindly extracting these features by the attacker denotes the inherent secrecy level provided by these waveforms which is investigated in the following sections for single-carrier and multi-carrier transmission systems.

7.2.2.1 Signal identification for single carrier modulation

One of the main categories of spread spectrum includes direct sequence spread spectrum (DSSS) that uses a PN sequence of positive and negative pulses at a very high repetition rate (chip rate) to spread the data bandwidth signal as in code division multiple access (CDMA) [8]. As shown in Figure 7.3, the spreading code is multiplied by the data signal before the signal is up-converted to the desired carrier frequency. The output shape of the spread signal is expressed as:

$$s(t) = a(t)d(t)\cos(2\pi f_m t + \theta), \tag{7.2}$$

where $a(t)$ is a sequence of pulses used to spread the data and $d(t)$ is the digital data. By using a "de-spreading" code that is identical to the spreading signal used at the transmitter, the spread signal is recovered at the receiver. Note that the PN sequence, which is a high-bit-rate binary sequence, has distinct characteristics like randomness, which makes them appear as noise to an attacker. However, legitimate devices

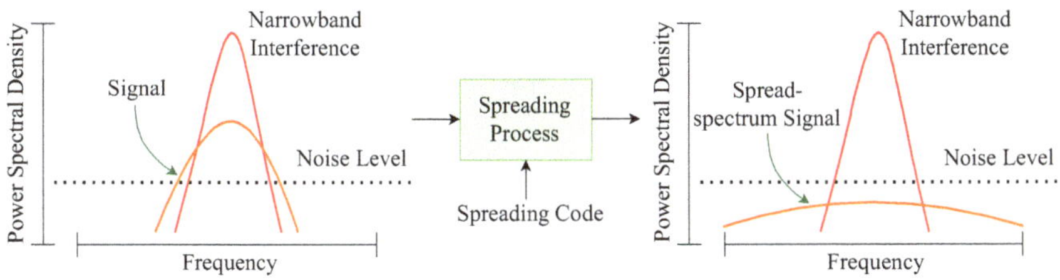

Figure 7.3 Direct sequence spread spectrum

can decode the transmission correctly using the secure pre-shared PN codes. Moreover, the inherent physical security of such systems depends on the use of relatively long PN scrambling sequences [9]. For instance, in [10], a method was proposed to enhance the PLS of CDMA systems by using Advanced Encryption Standard (AES) operation to generate the scrambling sequences. The AES specifies three key sizes (128, 192, and 256) so that the AES-CDMA method can raise the security level to guard against exhaustive-key-search attacks.

7.2.2.2 Signal identification for multi-carrier modulation

Multi-carrier signals, for instance OFDM signals, can be viewed as the combination of several different independent random signals, where sampling these large numbers of random variables (sampled signals) yields a Gaussian distribution based as per central limit theorem. This information can be used to categorize the incoming signals as single-carrier or multi-carrier by applying a normality test. Additionally, diverse techniques have been developed to estimate different parameters for OFDM systems. The majority of these techniques rely on the cyclo-stationary property of OFDM signals based on CP extension. The redundancy introduced by the CP can be used for detecting the transmission parameters [11], symbol duration [12], channel estimation [13], and synchronization [14].

In addition to using CP, numerous algorithms are proposed to estimate the parameters of an OFDM modulated signal and to blindly distinguish between different OFDM-based systems (such as WiFi, Wimax, 3GPP/LTE, and DVB-T). These algorithms rely on the normalized kurtosis, the maximum-likelihood principle, the matched filter, and the second-order cyclo-stationary property [15]. Additionally, with the advent of artificial intelligence (AI), many deep learning (DL)-based OFDM parameters identification schemes have been proposed to improve the classification accuracy in the dynamic channel. These DL-based models achieve high classification accuracy by determining the optimal OFDM parameters. Furthermore, they perform relatively good for various signal-to-noise ratio (SNR) values, under synchronization impairments, and in fading environments [16].

The ability to estimate all of the multi-carrier parameters makes OFDM like waveforms less secure and vulnerable to different security threats. However, even though the multi-carrier waveforms are less secure compared to spread spectrum, they are flexible enough to serve a wide variety of applications. This flexibility can also be advantageous in terms of incorporating them with the diverse PLS techniques, achieving both performance and secrecy gains. These techniques will be covered in Section 7.4 under modification-based PLS.

7.3 Moderate modification: control-signal/channel-based PLS

The complexity of physically transmitted signals keeps increasing with every new wireless generation technology. There are several variables (such as modulation and waveform) that can be changed to create a physical signal, as illustrated in Figure 7.4. Ultimately, an infinite number of different types of signals can be produced based on the channel, application, user context, etc. For instance, 5G New Radio (5G-NR) can operate in frequencies ranging from very low to very high (0.6–30 GHz). It provides an ultra-wide carrier bandwidth which can be up to 100 MHz at frequencies below 6 GHz, and up to 400 MHz at frequencies above 6 GHz. There are numerous physical channels in 5G-NR, for instance, there are downlink shared channel (PDSCH), broadcast channel (PBCH), and downlink control channel (PDCCH) for the downlink and three channels for the uplink: uplink shared channel (PUSCH), uplink control channel (PUCCH), and random access channel (PRACH). Also, many different reference signals and synchronization pilots are exchanged on both the downlink and uplink in the physical layer of 5G-NR. For downlink frame synchronization and transmission, the base station (BS) uses the primary synchronization signal (PSS) and secondary synchronization signal (SSS), which are used for frame/slot/symbol

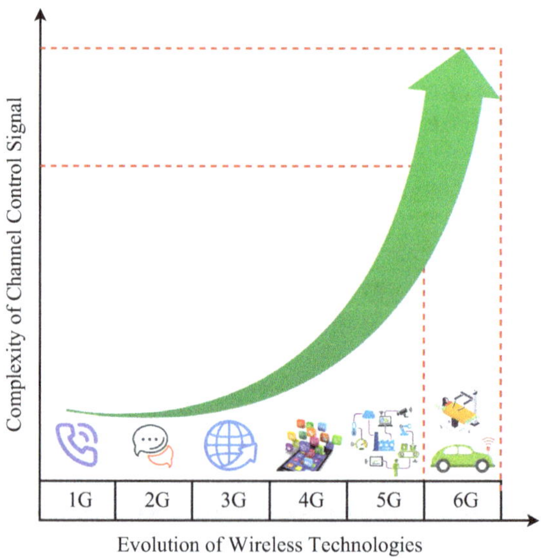

Figure 7.4 Relationship between the evolution of the wireless communication technologies and the complexity of the control signals

timing as well as conveying the physical cell identity. There are three forms of PSS and 336 combinations of SSS, where each PSS and SSS is composed of an m-sequence of length 127 that is mapped to a group of 127 subcarriers within the same OFDM symbol or in two different OFDM symbols. The Gold sequence, which is produced by fusing two orthogonal m-sequences, allows the user equipment (UE) to distinguish between various BSs on the same carrier at a low SNR.

Considering the frame structures of 5G-NR and Long Term Evolution (LTE), some significant changes have been made in 5G-NR. Similar to LTE, there are 14 OFDM symbols in each slot, each frame is 10 ms long, and there are 10 subframes in each frame. It is a significant change that the slots per subframe, which in LTE were always two, are now variable. More specifically, the number of slots in each frame depends on the subcarrier spacing, which varies from 15 to 240 KHz. Table 7.1 lists the various options along with the carrier frequencies for which they are intended. 5G-NR can address various radio spectrum, bandwidths, applications, and services thanks to its adaptable numerology. For example, macro coverage, small cell, indoor, and mmWave, respectively, call for subcarrier spacing of 15, 30, 60, and 120 kHz. Larger subcarrier spacing is also appropriate for ultra-reliable low-latency communication (uRLLC), whereas small subcarrier spacing is used to deliver high data rates. Mini-slot are made up of 2, 4, and 7 symbols, which can be set aside for very short transmissions. Slots may be combined for prolonged communication periods. Similar to LTE, a resource element (RE) consists of one subcarrier and one OFDM symbol, and the OFDM symbol contains PSS, PBCH, and SSS. In contrast to LTE, the resource block (RB) uses 12 subcarriers by 1 OFDM symbol. Table 7.2 displays

Table 7.1 Options in 5G NR for subcarrier spacing

Subcarrier spacing	Slots	Range	Min BW (MHz)	Max BW MHz)
15 kHz	1		4.32	49.5
30 kHz	2	< 6 GHz	8.64	99
60 kHz	4		17.28	198
120 kHz	8	> 24 GHz	34.56	396
240 kHz	16		69.12	397.44

Table 7.2 Number of RBs for different BWs and subcarrier spacings

Spacing	5 MHz	10 MHz	15 MHz	20 MHz	25 MHz	30 MHz	40 MHz	50 MHz	60 MHz	70 MHz	80 MHz	90 MHz	100 MHz
15 kHz	25	52	79	106	133	160	216	270					
30 kHz	11	24	38	51	65	78	106	133	162	189	217	245	273
60 kHz		11	18	24	31	38	51	65	79	93	107	121	135

the number of RBs as a function of system bandwidth and subcarrier spacing when operating below 6 GHz.

From the perspective of the coding schemes, 5G NR uses low-density parity-check (LDPC) and polar coding for the data channel and control channel, respectively. It has been proven that when used for error correction, LDPC codes are efficient for small data chunks. On the other hand, polar coding needs a lot of data and has performance that is very close to the Shannon limit. The modulation types that 5G NR supports include quadrature Phase Shift Keying (QPSK), 16 quadrature amplitude modulation (QAM), 64 QAM, and 256 QAM modulations which determine how many bits, whether they are useful bits or parity bits, can be carried by a single RE. With QPSK, each RE can be transmitted using just 2 bits, while 16 QAM, 64 QAM, and 256 QAM each require 4 bits, 6 bits, and 8 bits, respectively. The QAM modulation order is defined as 16, 64, and 256. The modulation and coding scheme (MCS) determines how many useful bits are present in each symbol. MCS is selected based on the block error rate (BLER) and radio condition. MCS is altered by gNodeB (gNB) based on the link adaptation algorithm. The modulations QPSK, 16 QAM, 64 QAM, and 256 QAM are supported by 5G NR for PDSCH. MCS indexes 29, 30, and 31 are set aside for re-transmission, and the MCS indexes (0–31) start at roughly 32. The 64 QAM Table, the 256 QAM Table, and the low spectral efficiency 64 QAM Table are the three tables for PDSCH MCS that are provided in 3GPP Specification 38.214. The use of massive MIMO to increase the capacity and coverage of wireless cells is another feature of 5G NR. The quantity of antennas and the number of layers are the main components of MIMO configuration. These elements are expressly mentioned in the radio resource controller (RRC) message used in LTE. However, the number of antennas and the number of layers in NR are not explicit RRC parameters. Instead, a few predefined tables defining the number of antenna ports and layers, which have 129 different options in downlink, are carried by both UE and gNB.

Considering various parameters in the aforementioned paragraphs, the physical signal will be very rich and cannot be interpretable without knowing those parameters. We can think of this as if a cryptographic key is applied to the physical signal and can only be extracted by the legitimate receiver who knows the information related to each individual parameter. For example, 5G signal is encrypted with a key of length:

$$L_{\text{key}} = \log_2(\#\ \text{parameter combinations}). \tag{7.3}$$

We can say 5G signal is more secure than LTE, and LTE is more secure than previous generations, and consequently, 6G will be more secure than all previous generations. So, for new wireless technologies (5G and beyond), securing the control channel/signal is sufficient enough to secure the transmitted data frame. Also, this emphasizes the importance and the need of securing the control signal against all security threats. Therefore, it is sufficient to secure only the control channel signal between the legitimate node to ensure secure transmission. This can be done by implementing various PLS algorithms as given in Section 7.4.

7.4 High modification: modification-based PLS

When the low and moderate modification-based PLS techniques fail to provide security, or when high secrecy levels are needed, then high modification-based PLS comes into play. In modification-based PLS, the whole transmitted signal, including the data-containing part, is modified. Specifically, to secure the wireless link using modification-based PLS, the randomness of the channel provided by the observation plane is exploited in bits, signals, or network domains. Comprehensively, the modification plane is the medium between the data bit stream and the transceiver's antenna. The main objective of this class of security measures is to provide more data decoding competence to a legitimate receiver as compared to a malicious attacker. The data decoding competency can be improved by either deliberately lowering the eavesdropper's performance or by increasing the QoS for the authorized receiver. This can be done by adapting, modifying, and optimizing the transmission parameters of the transmitted signal, by adding an interfering signal with the information signal, or by changing the physical features of the signal based on pre-shared or PLS-based generated keys.

7.4.1 Adaptation-based PLS

The fundamental principle of adaptation-based PLS is the modification and optimization of transmission parameters, such as location, channel conditions, etc., related to the legitimate receiver. Since adaptation is done for the legitimate link of Bob rather than Eve, Bob's signal-to-interference-plus-noise ratio (SINR) is higher as both experience different channel conditions. This method eliminates the need for additional processing for decoding data at a legitimate receiver. However, complete or partial CSI is essential by the transmitter for adaptation-based PLS. To improve the security performance, in addition to CSI, other feedback messages such as acknowledgment (ACK), negative acknowledgment (NACK) messages in the automatic repeat request (ARQ) process, received signal strength (RSS) indicator (RSSI), precoding matrix indicator (PMI), and rank indicator (RI) in multi-antenna systems can be useful.

This knowledge of CSI helps the transmitter to design different adaptive techniques, specifically, adaptive waveforms and pulse shaping, adaptive resource allocation, adaptive scheduling, adaptive interleaving, adaptive modulation and coding, adaptive power allocation, adaptive pre-coding, etc. based on the legitimate receiver's requirements. These adaptation techniques not only enhance the reliability and power efficiency of legitimate nodes but also strengthen the security at the physical level [17]. Since no additional processing is required at the decoding side, these techniques can be utilized in Internet of things (IoT) systems. Also, these techniques are supported by different duplexing mechanisms such as frequency division duplex (FDD), time division duplex (TDD), or hybrid duplexing. Despite numerous benefits, these techniques might not be very effective in the case of several cooperative attackers, each recording a distinct copy of the signal. However, these techniques can be improved by combining them with other security methods to maintain robustness against cooperative eavesdroppers.

The hyperspace domain can be utilized in designing adaptation-based algorithms since these can be implemented in time, frequency, and space domains. For instance, in [18], a time domain PLS scheme is proposed, where security performance is improved in addition to power and spectral efficiency. The main idea is to construct a channel shortening equalizer which modifies the signal such that the length of the effective CIR is made greater than CP for the illegitimate receiver while ensuring that it is less than or equal to CP for the legitimate party. This leads to loss of orthogonality at Eve due to ISI, which significantly degrades its ability to detect the signal. The frequency domain can also be utilized for adaptation mechanisms. For instance, Ref. [19] proposes a method known as OFDM with sub-carrier index selection (OFDM-SIS) and adaptive interleaving. The method utilizes the knowledge of good and bad sub-channels by dividing the OFDM block into small number of sub-blocks. At the authorized receiver, an ideal channel-based selection of the sub-carrier indices is used in each sub-block to improve the SNR at legitimate node. Finally, the space domain can also be utilized for adaptation mechanisms, which include MIMO, multiple-input single-output (MISO), single-input multiple-output (SIMO), distributed antenna system (DAS), coordinated multipoint (CoMP) systems, relays, reconfigurable intelligent surface (RIS)s [20], etc. Examples include adaptive antenna power allocation, transmit antenna selection, beamforming, pre-coding (zero forcing, geometric mean decomposition, minimum mean squared error, and generalized singular value decomposition), full/partial pre-equalization, relay selection, RIS phase optimization.

7.4.1.1 Secure beamforming

Beamforming is the process of transmitting data in a specific direction (angular domain) by utilizing different signal processing techniques, as shown in Figure 7.5.

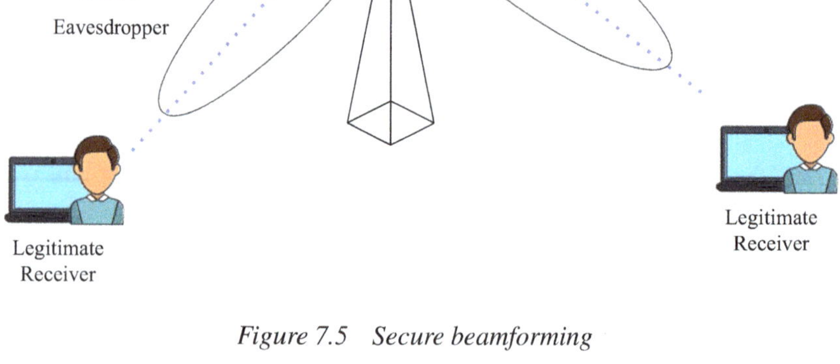

Figure 7.5 Secure beamforming

The main concept of using beamforming in PLS is to focus the beam towards the legitimate receiver while suppressing the beam in the other directions by varying the phase and amplitude of the signal at each antenna element. This will enhance the overall secrecy rate of the system

7.4.1.2 Signal design

One approach to secure data is to design the signal between a legitimate transmitter and receiver in such a way that its characteristics are unaware to the eavesdropper. Even though the eavesdropper is receiving the information, due to the peculiar design of the signal between legitimate parties, it is unable to decipher useful information from the received signal. This method is known as signal design [21] based PLS. In particular, by designing the signal structure in terms of the constellation, modulation scheme, etc., the link between Alice and Bob is secured by some sort of constellation or modulation code-book. If the constellation is considered as a design parameter, as in the case of [22], the mapping of points on the constellation and the order of the constellation are varied based on the channel phase. Since the eavesdropper is unaware of the modulation order or mapping of the points, the received signal appears to be cloudy or with severe distortions.

7.4.1.3 Effective channel modification

Instead of utilizing the channel state information directly to PLS algorithms, one approach is to take advantage of the effective channel observed at the receiver end. For example, in Ref. [18], an effective channel-based technique is presented, where a shortening filter is designed based on the channel of legitimate node such that the length of the effective CIR observed at illegitimate receiver is made greater than CP while ensuring that it is less than or equal to CP for the legitimate party. This leads to loss of orthogonality at Eve due to ISI, which significantly degrades its ability to detect the signal. This technique is computationally effective since there is no need for signal processing at the receiver node.

In addition to secure data, pilots can also be protected as information about the propagation environment can be extracted from them. For example, the authors in Ref. [23] decomposed the channel into all pass and minimum phase channels and used this to ensure security for both pilots and data.

7.4.2 Interference-based PLS

The main objective of interference-based PLS is to add an interfering signal with the information signal by a trusted party to degrade the performance of illegitimate receivers. This method utilizes the null space of the legitimate receiver, where an artificial noise is added such that it does not affect the performance of the legitimate receiver but acts as interference for eavesdroppers. The additional advantage of this strategy is that it is backward compatible because it does not add complexity and additional processing at the receiving end. This strategy is also applicable to FDD and TDD systems. In addition to maintaining confidentiality, the additional interfering signals can naturally offer additional advantages, such as a decrease in peak-to-average power ratio (PAPR), the mitigation of adjacent channel interference, and out of band emission (OOBE) [24].

Like every other scheme, this method also has trade-offs. For example, the transmitter uses some power resources for noise and needs to have CSI knowledge in order to inject an interfering signal, where channel estimation errors may also cause a slight performance degradation at the legitimate receiver. Furthermore, an increase in PAPR may result from improperly designed interference signals. The addition of artificially interfering-based techniques can be further divided into time, frequency, and space domains as follows:

- **Time domain interference addition:** The artificial noise-based techniques in the time domain involve adding the interference signal on top of the information signal in such a way that it causes significant degradation at the attacker while being completely canceled out at the legitimate node. For example, the work in Ref. [25] uses OFDM as the transmission scheme and adds a properly constructed interference to the transmitted signal such that when it traverses a frequency selective channel, the interference accumulates over the CP portion of the signal at the legitimate receiver. Therefore, legitimate receiver may automatically remove the noise when the signal is processed to remove the CP component of the OFDM signal, as illustrated in Figure 7.6. However, signal degradation is observed at Eve's end due to the spread of interference over whole signal at it.
- **Frequency domain interference addition:** The addition of an interfering signal is also possible in the frequency domain without degrading the performance at the desired receiver. For example, the author of Ref. [26] used faded sub-carriers in OFDM systems to add interference instead of data in order to deceive the eavesdropper, as depicted in Figure 7.7. Here, the eavesdropper is unable to distinguish between sub-carriers carrying data symbols and interference because it experiences a different channel than Bob.
- **Space domain interference addition:** As illustrated in Figure 7.8, there are numerous intriguing works in the literature in the space domain that are based on

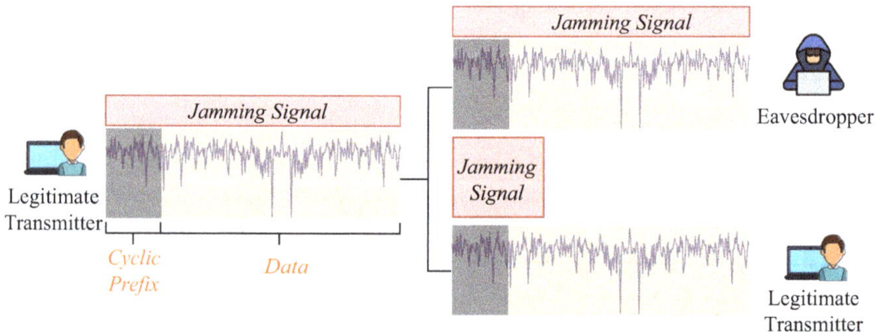

Figure 7.6 Time domain interference addition

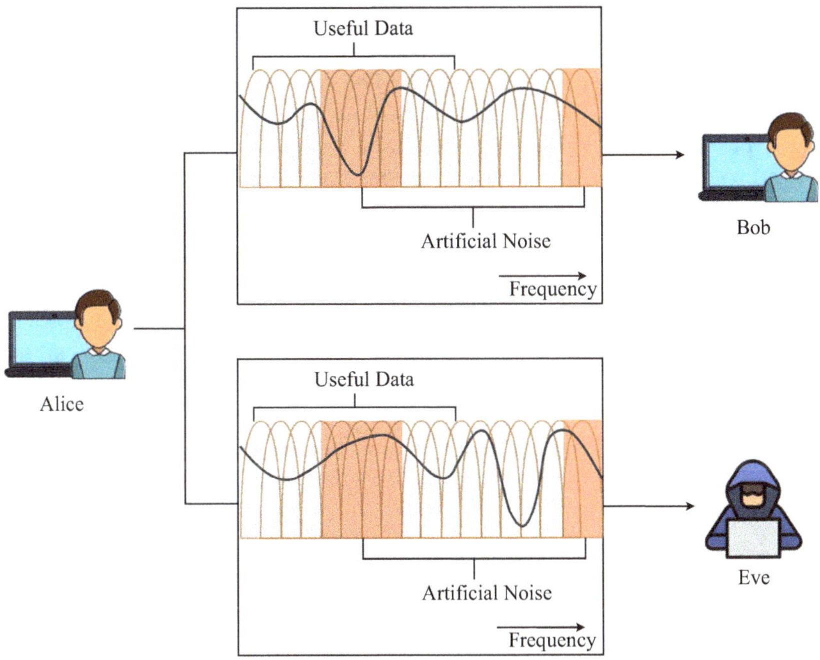

Figure 7.7 Frequency domain interference addition

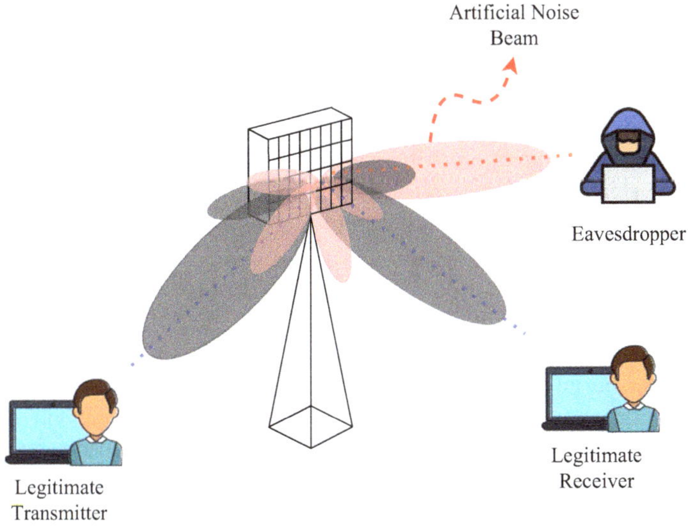

Figure 7.8 Space domain interference addition

the idea of interference [27], including MIMO, MISO, SIMO, CoMP, relays, etc. Goel and Negi provide an example in [27], where they develop an interference-based method using multiple antennas. Two requirements must be met for this technique to be effective: (1) the channel must have a null space, and to achieve this, the number of transmit antennas used must be greater than the number of legitimate receiver antennas; and (2) the number of antennas used at Eve must be lower than that at Alice in order to ensure that Eve cannot align the added noise.

Similar to artificial noise, Interference alignment (IA) is an effective method that can manage interference and provide PLS. In order to separate the intended signals from the interference and enable simultaneous transmission among multiple users with limited signal resources, the basic concept behind IA is to compress all interference signals into a constricted space at each receiver while enlarging the desired signal dimensions [28]. In the case of multiuser networks, IA can be used for PLS in such a way that the legitimate destination will receive the useful signals free from interference while the eavesdropper will receive a mixture of useful signals and interference.

7.4.3 Key-based modification at signal level

In wireless communication systems, the channel itself serves as the source of randomness as it varies randomly in time, space, and frequency. Using the principle of reciprocity and spatial decorrelated, key-based PLS algorithms can extract secret keys from the random characteristics of the radio channel [29].

Key-based PLS can be implemented in various ways. One such strategy is primarily based on quantizing the RSSI, where RSSI is a physical layer metric that measures the average received power of a communication signal over a predetermined period of time. RSSI has been widely used for key-based algorithms because they can be implemented in the majority of off-the-shelf devices and do not require significant hardware alteration. Also, due to the fact that the terminals are moving (or there are significant mobile objects or people in the vicinity), RSSI-based PLS benefits most from a time-variant scenario (i.e., fading). However, these methods are ineffective in scenarios where the subject has limited mobility [30]. On the other hand, some strategies make use of all the random features offered by the multipath channel, such as amplitude and phase of CIR/CSI.

The extracted key (from time, frequency, or spatial domains of the channel) can be applied at a bit level, similar to encryption-based techniques, or at a signal level (symbol level basis) to change the physical features of the signal. The application of a key (channel-based key or pre-shared key) at the physical layer refers to "physical layer encryption." In this context, the use of keys at the application layer for encryption is vulnerable to hacking and cracking using brute-force attacks. However, physical layer encryption is immune to such brute-force attacks. For example, one way of using physical layer encryption is to add secret key-based intentional ISI to achieve secrecy. The key determines the filter coefficients of the time-varying ISI filter such that the legitimate receiver can remove the ISI while the eavesdropper cannot be able to remove ISI [31]. Similarly, the work in Ref. [32] involves adding a random jamming signal to each symbol, which can be subtracted by a legitimate receiver only [32]. Another example is related to constellation rotation, where the signal is modified at the constellation level using pre-shared keys at the transmitter and receiver. For instance, the RSSI values are used to extract random sequences that are then used to rotate the symbols at the physical layer level, rather than transmitting a fixed constellation, in order to secure communication in the presence of an eavesdropper [33].

7.5 Conclusion

The waveform is the physical signal transmitted in the time–frequency–space domain. Modulating these waveforms allows them to carry unknown information from one point to another. Therefore, securing the communication can be equivalent to securing the waveform itself, either by hiding its physical features (making it robust against detection), by hiding the features of the modulation used (making it robust against identification), or by securing the control channel signals. PLS comes into play when security provided by inherent waveform and control signals is not enough or when higher security levels are needed. Signal domain physical modification for PLS includes the processes that alter the transmitted signal based on the parameters extracted from the observation plane to ensure secure communication.

References

[1] Viterbi AJ. *CDMA: Principles of Spread Spectrum Communication*. Boston, MA: Addison Wesley Longman Publishing Co., Inc.; 1995.

[2] Strasser M, Pöpper C, Čapkun S. Efficient uncoordinated FHSS anti-jamming communication. In: *Proceedings of the Tenth ACM International Symposium on Mobile Ad Hoc Networking and Computing*; 2009. p. 207–218.

[3] Bingham JA. Multicarrier modulation for data transmission: an idea whose time has come. *IEEE Communications Magazine*. 1990;28(5):5–14.

[4] Prasad R. *OFDM for Wireless Communications Systems*. Norwood, MA: Artech House; 2004.

[5] Al-Naffouri TY, Quadeer AA. Cyclic prefix based enhanced data recovery in OFDM. *IEEE Transactions on Signal Processing*. 2010;58(6):3406–3410.

[6] Ankaral ZE, Karabacak M, Arslan H. Cyclic feature concealing CP selection for physical layer security. In: *2014 IEEE Military Communications Conference*. New York, NY: IEEE; 2014. p. 485–489.

[7] Park J, Kang C, Hong D, *et al.* Effect of partial band jamming on OFDM-based WLAN in 802.11 g. In: *2003 IEEE International Conference on Acoustics, Speech, and Signal Processing, 2003. Proceedings* (ICASSP'03), vol. 4. New York, NY: IEEE; 2003. p. IV–560.

[8] Hara S, Prasad R. Overview of multicarrier CDMA. *IEEE Communications Magazine*. 1997;35(12):126–133.

[9] Shiu YS, Chang SY, Wu HC, *et al.* Physical layer security in wireless networks: a tutorial. *IEEE Wireless Communications*. 2011;18(2):66–74.

[10] Li T, Ren J, Ling Q, *et al.* Physical layer built-in security analysis and enhancement of CDMA systems. In: *MILCOM 2005—2005 IEEE Military Communications Conference*. New York, NY: IEEE; 2005. p. 956–962.

[11] Yucek T, Arslan H. OFDM signal identification and transmission parameter estimation for cognitive radio applications. In: *IEEE GLOBECOM 2007—IEEE Global Telecommunications Conference*. New York, NY: IEEE; 2007. p. 4056–4060.

[12] Li H, Bar-Ness Y, Abdi A, *et al.* OFDM modulation classification and parameters extraction. In: *2006 1st International Conference on Cognitive Radio Oriented Wireless Networks and Communications*. New York, NY: IEEE; 2006. p. 1–6.

[13] Heath RW, Giannakis GB. Exploiting input cyclostationarity for blind channel identification in OFDM systems. *IEEE Transactions on Signal Processing*. 1999;47(3):848–856.

[14] Speth M, Classen F, Meyr H. Frame synchronization of OFDM systems in frequency selective fading channels. In: *1997 IEEE 47th Vehicular Technology Conference. Technology in Motion*, vol. 3. New York, NY: IEEE; 1997. p. 1807–1811.

[15] Bouzegzi A, Ciblat P, Jallon P. New algorithms for blind recognition of OFDM based systems. *Signal Processing*. 2010;90(3):900–913.

[16] Park MC, Han DS. Deep learning-based automatic modulation classification with blind OFDM parameter estimation. *IEEE Access*. 2021;9:108305–108317.

[17] Solaija MSJ, Salman H, Arslan H. Towards a unified framework for physical layer security in 5G and beyond networks. *IEEE Open Journal of Vehicular Technology*. 2022;3:321–343.

[18] Furqan HM, Hamamreh JM, Arslan H. Enhancing physical layer security of OFDM systems using channel shortening. In: *Proc. IEEE 28th Annu. Int. Symp. Personal Indoor Mobile Radio Commun.* (PIMRC). Montreal, Canada; 2017. p. 1–5.

[19] Hamamreh JM, Basar E, Arslan H. OFDM-subcarrier index selection for enhancing security and reliability of 5G URLLC services. *IEEE Access*. 2017;5:25863–25875.

[20] Eddine Zegrar S, Afeef L, Arslan H. Reconfigurable intelligent surface (RIS): eigenvalue decomposition-based separate channel estimation. In: *2021 IEEE 32nd Annual International Symposium on Personal, Indoor and Mobile Radio Communications* (PIMRC); 2021. p. 1–6.

[21] Xiong T, Lou W, Zhang J, *et al.* MIO: Enhancing wireless communications security through physical layer multiple inter-symbol obfuscation. *IEEE Transactions on Information Forensics and Security*. 2015;10(8):1678–1691.

[22] Althunibat S, Sucasas V, Rodriguez J. A physical-layer security scheme by phase-based adaptive modulation. *IEEE Transactions on Vehicular Technology*. 2017;66(11):9931–9942.

[23] Zegrar SE, Furqan HM, Arslan H. Flexible physical layer security for joint data and pilots in future wireless networks. *IEEE Transactions on Communications*. 2022;70(4):2635–2647.

[24] Hamamreh JM, Furqan HM, Arslan H. Classifications and applications of physical layer security techniques for confidentiality: a comprehensive survey. *IEEE Communications Surveys & Tutorials*. 2018;21(2):1773–1828.

[25] Qin H, Sun Y, Chang TH, *et al.* Power allocation and time-domain artificial noise design for wiretap OFDM with discrete inputs. *IEEE Transactions on Wireless Communications*. 2013;12(6):2717–2729.

[26] Güvenkaya E, Arslan H. Secure communication in frequency selective channels with fade-avoiding subchannel usage. In: *2014 IEEE International Conference on Communications Workshops* (ICC). New York, NY: IEEE; 2014. p. 813–818.

[27] Goel S, Negi R. Guaranteeing secrecy using artificial noise. *IEEE Transactions on Wireless Communications*. 2008;7(6):2180–2189.

[28] Guo J, Zhao N, Yu FR, *et al.* A novel anti-jamming scheme for interference alignment (IA)-based wireless networks. In: *2015 IEEE/CIC International Conference on Communications in China* (ICCC). New York, NY: IEEE; 2015. p. 1–5.

[29] Hassan AA, Stark WE, Hershey JE, *et al.* Cryptographic key agreement for mobile radio. *Digital Signal Processing*. 1996;6(4):207–212.

[30] Zoli M, Barreto AN, Köpsell S, *et al.* Physical-Layer-Security Box: a concept for time-frequency channel-reciprocity key generation. *EURASIP Journal on Wireless Communications and Networking*. 2020;2020(1):1–24.

[31] Sheikholeslami A, Goeckel D, Pishro-Nik H. Artificial intersymbol interference (ISI) to exploit receiver imperfections for secrecy. In: *2013 IEEE International Symposium on Information Theory*. New York, NY: IEEE; 2013. p. 2950–2954.

[32] Sheikholeslami A, Goeckel D, Pishro-Nik H. Jamming based on an ephemeral key to obtain everlasting security in wireless environments. *IEEE Transactions on Wireless Communications*. 2015;14(11):6072–6081.

[33] Allen T, Cheng J, Al-Dhahir N. Secure space–time block coding without transmitter CSI. *IEEE Wireless Communications Letters*. 2014;3(6):573–576.

Chapter 8
Physical modification plane: cross MAC/PHY scheduling and resource allocation

Ahmad Jaradat[1] *and Hüseyin Arslan*[1]

This chapter discusses the strategies for providing physical layer security (PLS) using current scheduling and resource allocation (RA) algorithms. PLS could be a useful notion for resource management that focuses on security. Scheduling and RA have been used to accomplish a number of objectives, including improving spectrum efficiency (SE) and energy efficiency (EE), as well as increasing fairness. Information security can be efficiently achieved by carefully allocating the available resources based on the PLS concept. This chapter will review the scheduling and RA and their popular methods. Performance indicators and fundamental PLS optimization issues are then discussed. The secure RA and the prevalent PLS scheduling strategies are then covered. The importance of PLS-based scheduling in downlink transmission in wireless networks is demonstrated. Finally, the potential difficulties are discussed, and it is anticipated that the reader will understand PLS-based scheduling strategies and be motivated to use them in the future.

8.1 Introduction

The performance of the communication systems' secrecy could be improved by utilizing the scheduling strategies. Scheduling modes, or transmission modes, could be created with PLS restrictions, and this link is referred to as a security-aware RA dilemma in several research investigations. The development of a reliable detection system appropriate for wireless networks depends on the proper interaction and coordination between the various protocol layers. The foundational components of cross-layer architectures are such interactions. In the cross medium access control (MAC)/physical (PHY) layer security concept, the MAC operations, such as hybrid automatic repeat request (HARQ), scheduling, and RA, as well as the PHY parameters, can be simultaneously tuned to maximize secrecy while supporting the essential quality of service (QoS) needs of the authorized users. The authors of [1] provide an illustration of a study that relates to this new issue and demonstrates how

[1]Department of Electrical and Electronics Engineering, Istanbul Medipol University, Turkey

automatic repeat request (ARQ) functionality and artificial noise (AN) are used to provide eavesdropping protection.

The inclusion of PLS in the RA process for multi-antenna systems leads to some difficult optimization challenges; nonetheless, many degrees of freedom can be utilized to offer confidentiality and traditional rate advantages. The field's RA issues are theoretically challenging to resolve in an efficient manner. Also, the secrecy rate (SR) formula is a more complex function than the traditional transmission rate function, this is mostly because of their integer, combinatorial structure. A typical RA problem in wireless systems that use subchannels considers not only the binding of the subchannel group of users but also the design of the beamforming matrices within the subchannels, the adjustment of the power allocated to each subchannel and user, QoS guarantees, etc. Allowing transmission to only one user per subchannel is a typical assumption that makes the RA problem easier to understand.

Several methods are offered in [2,3] for a range of optimization goals under the assumption of one user per sub-channel. However, it is advantageous to apply spatial multiplexing inside each sub-channel in multi-user, multi-antenna systems. The authors of [4] have concentrated on RA to maximize system throughput, whereas the studies [5–7] have concentrated on optimizing a number of fairness-aware constrained issues. In general, it is a non-linear, non-convex problem to jointly discover the best sub-channel user binding and beamforming. The issue of throughput maximization is examined in [8], and several different aims are investigated in [9] using the same presumption.

The correlation between nearby subchannels can be used in a system that transmits simultaneously over a number of subchannels to lessen the computational complexity and signal overhead of the RA process. One strategy in this direction is to group a number of adjacent subchannels into a single entity and perform RA on a chunk basis. Obviously, chunk-based RA may only perform as well as sub-channel-based RA on a percentage basis. The bulk of existing works implement chunk-based RA using a single, representative value for each chunk, such as the median sub-quality channel's or the mean value of subchannels' quality within the chunk. However, in doing so, the system's frequency selectivity is disregarded, and the advantages of the built-in frequency diversity gain are lost.

The coherence bandwidth frequently exceeds the sub-channel bandwidth and is comparable to the chunk size in broadband wireless systems. As a result, there is a strong likelihood that a user will experience high (low) channel quality in part of the chunk and low (high) channel quality in part of the chunk. The performance difference between chunk-based RA and sub-channel based RA can be further closed by considering feature, which also reduces computational complexity and signal overhead. A new dimension that determines which user (or group of users) is served within each of the parallel channels is added to the linear processing design and the user selection problem in a system that transmits over a series of parallel channels. The main issue, often known as the RA problem, is combinatorial in nature and belongs to the non-deterministic polynomial (NP)-class of problems. Therefore, in theory, it cannot be solved optimally due to the fact that its size scales with regard to the quantity of users, the quantity of parallel channels, and the quantity of antennas at BS. The presentation of heuristic solutions that are very effective in terms of both performance and computing complexity is therefore of great interest [10].

8.2 Scheduling and resource allocation

The traditional definition of radio resource before fifth generation (5G) includes some factors such as time, frequency, and transmit power. In 5G networks, the resource concept incorporates several other factors. The management mechanism for the RA is very significant in wireless communication. It may include various network features such as scheduling, transmitter assignment, and broadcast rate management. The radio resource scheduling architecture is shown in Figure 8.1 [11]. The scheduling procedure can be divided into three parts depending on the goals of the tasks: prioritization, resource number determination, and RA. As a result, by combining these processes, many scheduling algorithms can be produced.

Prioritization is a scheduling phase that ranks the members who are eligible for radio resource service at the present transmission time interval (TTI) with an emphasis on user fairness and QoS fulfillment. The amount of resources that should be used for each user is determined by the resource number. The radio resource's time and frequency domain position for transmitting the control and shared channels with an emphasis on SE is determined by RA.

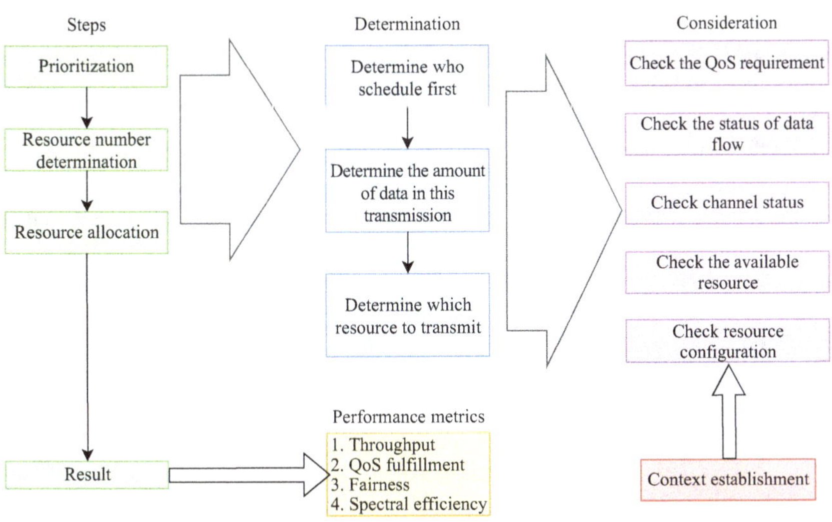

Figure 8.1 The architecture of radio resource scheduling [11]

Additionally, a number of user attributes are considered while scheduling tasks. The scheduler must first verify each user's status and do its best to meet the QoS requirement. Second, the data flow when data enters the radio link control layer buffer can be observed based on the type of application. Third, each user's channel quality and the associated resource block (RB) are different. Fourth, the resources that are accessible in each scheduling TTI vary based on the resources set aside for broadcasting, synchronization, and other capabilities. Finally, the scheduler needs to consider the context.

Scheduling and RA are the two key functional categories for radio resource management (RRM). The decision for the served user and the quantity of packets needed at the moment are typically included in the scheduling routines. The RA procedure includes assigning RB to the selected user and figuring out how many RBs are required to satisfy the user's QoS needs. Several performance measures, such as SE, scalability, computational complexity, QoS, fairness, goal latency, queue length, priority, and guaranteed bit-rate, have been researched for Scheduling and RA. Channel and QoS aware schedulers as well as channel unaware schedulers are typically used in scheduling approaches. RRM will achieve previously unheard-of degrees of complexity in 5G networks. Latency, control channel, HARQ, and radio channel are some of the variables that could have an impact on how the transmission resources are allocated to different users. The major design enablers for scheduling in future wireless systems should be complexity and scalability. The conditions for QoS must be met for users and applications, including SE, fairness, and energy consumption.

8.3 Popular scheduling and resource allocation algorithms

In [12], a thorough evaluation of traditional physical RB scheduling strategies for the RA of Long Term Evolution (LTE) systems is provided. The topic of user scheduling that is fair and reliable has been well researched in the literature. These scheduling methods include proportional fair (PF), round robin (RR), and best CQI (BCQI), among others. One of the most popular techniques for fair scheduling is PF. By taking advantage of favorable channel conditions and dynamically allocating resources to users, PF scheduling seeks to promote fairness. It has been established that the capacity limits make the PF algorithm not optimal.

Users are given resources by the RR scheduler without considering channel conditions, which ensures fairness. The processes in the RR cannot have priorities established, i.e., more significant jobs are not given a higher priority when using RR scheduling. The most popular channel-dependent scheduler is the BCQI scheduler. The channel differences between users are exploited in BCQI to maximize cell throughput at the expense of fairness. The efficient and dynamic RA is the key issue caused by the diverse 5G service requirements. The PF, RR, and BCQI schedulers, among the other classical schedulers described above, are not adaptable enough to handle the wide range of use cases in next-generation wireless networks.

The Third Generation Partnership Project (3GPP) specifications do not include the scheduling technique because it is implementation specific. The majority of schedulers' primary goal is to take advantage of differences in channel conditions between terminals and, ideally, plan transmissions to those terminals at those times. Therefore, most scheduling strategies require the following:

- Conditions of the terminal's channel.
- Priorities of various data flows and buffer status.
- The interference from the neighboring cells.

8.4 Performance metrics and basic optimization problems in resource allocation and scheduling for physical layer security

The selection of performance measures for secure transmission schemes is remarkably important. The following are several issues with PLS designs that are typically brought on by various performance requirements such as:

- The achievable SR/capacity is used to assess the transmission efficiency of secure transmission systems.
- Secrecy outage probability (SOP)/capacity is used to assess the reliability of secure transmission techniques.
- The minimal amount of power required to guarantee secure QoS.
- The EE of secure transmission techniques focuses on the number of secret bits that may be sent with a given amount of energy or the amount of energy needed to deliver a single secret bit.

The appropriate metrics, SR/capacity, SOP/capacity, power consumption, and secure EE, which are commonly used to evaluate the effectiveness of proposed secure transmission methods, were studied in order to investigate the above-mentioned issues. Particularly for system designs in varied application situations, these performance metrics are frequently employed as optimization goals.

8.4.1 Secrecy rate/capacity

The SR is a crucial parameter for evaluating the transmission performance of physical-layer secure techniques. The SR in PLS is the quantity of secret bits sent over a specific channel each second; it is highly reliant on the channel inputs. The Gaussian inputs and the achievable SR are commonly used to evaluate secrecy more conveniently and economically. The difference between the data rates for a legitimate channel and a wiretap channel while employing the Gaussian codebook is the attainable SR.

The lower bound of the secrecy capacity (SC) is the achievable SR. In actual designs, some technologies, such as secure beamforming or RA, can achieve a non-zero SR by purposely degrading the wiretap channel while boosting the quality of the legitimate channel. SC, which is the limit of the SR, is another statistic that is closely related to the SR. The SC is the maximum SR at which the source node's confidential messages may be securely and reliably delivered to the destination node, while unauthenticated users are prevented from learning anything of value during this transmission

The fading of wireless channels has not been considered when investigating the aforementioned SR and SC. However, the fading is a common problem that cannot be avoided. When channel fading is taken into account, it is important to evaluate the average capability of secure communication over fading channels. In this case, the ergodic SR or SC is a very suitable metric. In practice, the secrecy performance in fading scenarios is evaluated using the achievable ergodic SR because, in many cases, creating ergodic SC (ESC) may be computationally difficult. The attainable ergodic SR is the ratio of the ergodic rates of the legitimate channels and the wiretap channels with Gaussian codebooks, the latter of which is frequently more computationally effective. In secure transmission systems, the feasible ergodic SR has traditionally been taken into account as the optimization goal when channel fading is considered as the ESC's lower bound.

How much secrecy may be obtained for securely and reliably conveying secret data is the main concern for secure communication system designs. This issue can be represented as maximization of the achievable SR, which entails applying some techniques, such as RA, beamforming, cooperative diversity, or other optimization techniques, to maximize the achievable SR as much as feasible. The power limitation and bandwidth are the most crucial consideration for maximizing the achievable SR. As a result, one popular formulation of SR maximization on the available channels often seeks to maximize SR within the limitations of the maximum allowable power.

8.4.2 Secrecy outage probability/capacity

Secure transmission may fail as a result of channel fading and imperfect channel state information (CSI). Investigating how a secure transmission scheme behaves during secrecy outages is therefore particularly interesting. As a result, the SOP is a suitable metric to describe the likelihood that secure communication will not be possible. The probability that a secrecy outage event occurs is precisely what is meant by the phrase "SOP." Secrecy outage occurrences have two distinct definitions. According to one of them, which is more frequently acknowledged, a secrecy outage happens when the instantaneous SC is less than a given SR. A failure to achieve perfect secrecy does not necessarily follow from an outage based on this criterion. To be more precise, blackout events mean that the secure communication will undoubtedly stop since the legitimate channels cannot handle the SR. The likelihood that a sent message will not attain absolute secrecy is directly measured by the alternative concept of a secrecy outage. The price of protecting message transmission against eavesdropping is reflected in the rate differential.

The secrecy outage capacity, which is the highest SR that may be provided under a tolerable SOP, is another critical idea connected to the SOP. In other words, the capacity for secrecy outages is the highest SR that may be achieved while still having an acceptable SOP. The reliability of secure transmission, which is typically assessed by SOP, has also drawn growing attention for the optimal design in PLS. Without an outage, secure communication should ideally be implemented. As a result of this observation, it is expected to lower the probability of a secrecy outage. This puts the optimization problem of minimizing the probability of a secrecy outage under resource limitations.

8.4.3 Power/energy consumption

In cases when resources are few, such as battery-powered networks, power/energy usage is a crucial factor. The most significant concern is generally the long-term viability of secure communications in such networks. Power/energy cost is thus one of the key factors considered in PLS designs to increase network lifetime. The elements affecting power consumption in wireless networks are investigated before creating a secure transmission strategy with constrained power and energy. The output power of power amplifiers has a significant impact on how much power they use. The basic transmitter and receiver circuits, excluding power amplifiers, are included in the other circuit blocks. These include active filters, frequency synthesizers, mixers, intermediate frequency amplifiers, and others.

Development of power-efficient transmission techniques with the goal of reducing power consumption is motivated by the regime of limited resources. To achieve this, transmitter power levels need be changed in order to reduce transmission power consumption and still meet the desired QoS standards. Thus, the power adaption must be completed in order to satisfy the demands of power-efficient secure transmission. If there is no requirement for confidentiality, a higher message transmission rate can be attained. When secrecy is considered, secure coding will cause confidential message transmission rates to drop. Therefore, more power must be used to maintain a higher level of PHY layer secrecy.

8.4.4 Secure energy efficiency

The utilized efficiency of the system energy, or EE, in conventional communications without confidentiality restrictions, is a crucial parameter for developing green transmission plan designs. When security issues and energy constraints in wireless networks are considered jointly, energy-efficient secure transmission systems should operate in a private and environmentally responsible way. A suitable metric for evaluating the energy utilization efficiency of the system is therefore of utmost significance. It tries to measure both the energy used by all communication nodes and the network level performance toward the EE of the system/network level. In the design of transmission strategy, the EE of the system/network level is frequently used.

For evaluating the EE of emerging strategies aimed at PLS, two key metrics have been proposed. A measurement is the secure EE, which is the quantity of secret bits communicated per unit of energy usage. It is anticipated that using this metric to design energy-efficient secure transmission schemes will maximize the secure EE. The end result is that with a certain quantity of energy, as much private information as possible is conveyed. The popular metric for secure EE is the ratio of SR to total power consumption, and its unit is secret bits per Joule.

Generally, the optimization issue that results from using one metric is the dual problem of the problem that results from using the other metric. To ease the challenges of secure transmission systems, the best metric should consider all realistic cases. Since it is easier to measure the degree of proportionality between energy consumption and various degrees of load, the secret bits per Joule metric is more widely used. This statistic can depict dynamic network conditions that consider power usage and confidentiality restrictions under various load scenarios. The network EE can only be evaluated using the energy per secret bit metric at non-zero SR. Furthermore, it is clear that the model of power consumption and the secure EE metrics are intertwined. Transmission power is the component of power consumption in networks and traditional energy-efficient technologies. The secure EE needs to consider all network-wide power usage, including fundamental circuit power, transmission power, and signaling overhead.

PLS may use more power and energy to secure sensitive information from eavesdropping than traditional communication without a need for concealment. The SR function, which calculates the rate at which information leaks to eavesdroppers, can be used to confirm this discovery. This feature might make electricity and energy supply more demanding, especially in situations when those resources are few. The initial concern, driven by the demands of "green communication," is to send secret information with high secure EE as much as feasible when the limited power and energy become the major elements for securing communications. This drive increases PLS's secure EE maximization.

As the above-mentioned metrics and optimization issues are linked to the SR/capacity, which is based on information theory. The level of secrecy is measured through metrics that can be seen in real-world communication contexts, such as the secrecy gap, which is typically represented by bit error rate or packet error rate versus signal-to-noise ratio (SNR). This is another sort of performance metric for secrecy. The minimal necessary difference between the SNR of the legitimate receiver and eavesdropper for which secure communication is possible is represented by the secrecy gap, to be more precise. This metric has also been applied to the creation of a numerical evaluation of system designs.

8.5 Resource allocation for physical layer security

The orthogonal frequency-division multiple access (OFDMA) networks include multidimensional wireless resources such as frequency, time slot, and power. Thus, it is possible to purposely increase the distance between the legitimate and wiretap channel by utilizing secure RA. The resources in multi-antenna and multi-node wireless networks often refer to the degrees of freedom offered by numerous antennae and nodes. The fundamental problem of secure RA, however, is to make optimum use of the restricted resources to meet the performance metric requirements. The wireless resources in the OFDMA-based wireless network are depicted in Figure 8.2 [13].

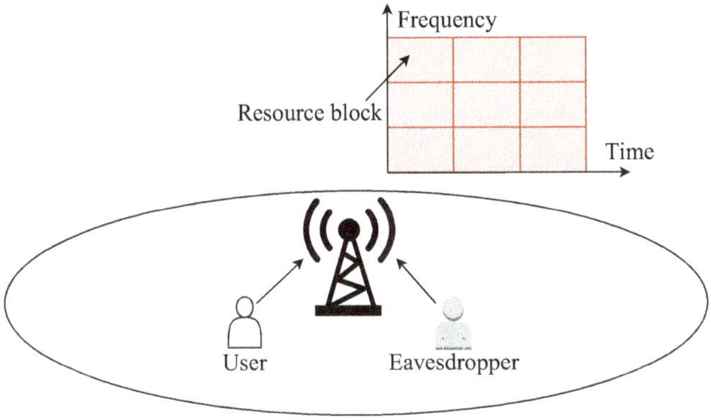

Figure 8.2 The multidimensional wireless resources in a OFDMA-based wireless network [13]

8.5.1 Literature on secure resource allocation

There has been a lot of discussion in the literature about two key problems with secure RA: subcarrier allocation and power allocation (PA) in multicarrier networks. The RA problem for ultra-reliable low-latency communication (uRLLC) in real-time wireless control systems is investigated in [14] based on uplink transmission. Convergence rate is controlled while optimizing bandwidth and transmission PA in uRLLC under the restrictions of communication and control. To design and resolve the optimization problem, the authors first convert the control convergence rate requirement into a communication reliability constraint. The wireless RA problem is then used in place of the co-design problem. An iterative approach is proposed to determine the best distribution of communication resources after it is established that the transformed problem is concave. On the basis of that, the ideal control convergence rate is determined to improve system performance as a whole.

In [15], two useful random subcarrier selection techniques are proposed for transmitting private messages by creating a random subcarrier set and performing a randomization process to ensure secure accurate wireless transmission per orthogonal frequency division multiplexing (OFDM) signal. The RA for a secure amplify-and-forward (AF) relay communication is examined in [16], where the optimization variable is defined for the source and the relay, respectively, to specify the status of communication on a particular carrier. The work given in [17] intends to assign sub-channels and allot power to optimize the max–min fairness across the users' in secure multi-user OFDMA downlink operations.

For maximizing the capacity for secrecy outages and reducing the likelihood of interception, respectively, the solutions of the ideal relay PA for a large-scale multiple-input multiple-output (MIMO) decode-and-forward (DF) relay network are derived in [18]. The findings in [18] are advanced in [19], where the alternating optimization approach is used to handle the joint node power and transmission time allocation problem's non-convexity by first maximizing over part of the variables and then the remaining variables. To maximize the minimal average SR of the total number of information receivers, authors in [20] have formulated an optimization problem to simultaneously design the trajectories and transmit power of base station (BS) of unmanned aerial vehicle (UAV) and UAV-Jammer.

8.5.2 Optimization problems in secure resource allocation

Secure RA has been thoroughly studied for many uses as a promising means of enhancing the performance requirements of PLS. As was previously described, four optimization problems that are connected to the associated performance indicators are typically solved to create designs for secure RA.

8.5.2.1 Maximization of achievable secrecy rate

To increase the achievable SR, several research concentrate on building secure RA algorithms. The traditional method for increasing the SR in multicarrier systems is to distribute the finite amount of power and subcarriers equally among all transmission nodes. In many situations, this goal typically results in mixed integral programming, which has been studied in numerous works [16,17,21–24]. In this field, several particular formulations geared at certain scenarios have also been investigated. The study [21] discussed a secure RA method for a downlink OFDMA-based network with the coexistence of normal and secure users who do not have any important messages and are unconcerned about security. In [23,24], jamming and AN-aided RA for sum SR maximization are examined, with the former focusing on OFDMA-based two-way relay wireless sensor networks and the latter on OFDMA systems with joint secrecy information and power transmission. The work described in [17] intends to assign subchannels and allocate power to optimize the max-min fairness criterion over the users' SR for evaluating the fairness of RA in secure multi-user OFDMA downlink works. Additionally, Ref. [25] investigates robust secure RA in relay-assisted cognitive radio (CR) networks considering the unpredictability of CSI.

8.5.2.2 Minimization of secrecy outage probability

A secure RA is a good strategy for reducing the likelihood of a secrecy outage. The outage-optimal subcarrier allocation is addressed for a typical secure OFDMA downlink system to reduce the probability of each user's secrecy outage while ensuring that each user has an equal chance of accessing each subcarrier. Additionally, the outage-optimal power distribution is thoroughly investigated. In the circumstances of secure wireless information and power transfer in [26–28], the reduction of the SOP is also increased. In [26], the dual decomposition and alternating optimization approaches are coupled to improve the transmission PA and power splitting ratio for the AN signal in order to reduce the outage risk for the transmission of delay-limited secret information. In [27], the best location of the energy harvesting node is optimized using PLS concerns to achieve the lowest SOP. The problems of SOP minimization and average harvested energy maximization in wireless information and power transmission systems are addressed in [28] using an optimization framework of goal SR and PA ratio. Another optimization strategy linked to secrecy outage performance is the minimization of the SOP and the maximizing of the secrecy outage capacity. An OFDMA DF relay network's packet data rate, secrecy data rate, power, and subcarrier allocation policies are created using the dual decomposition and gradient approach to maximize the average secrecy outage capacity.

8.5.2.3 Minimization of power consumption

The creation of secure RA algorithms can help reduce the power consumption of secure communications. To be more precise, we can meet various requirements for safe QoS while using the least amount of energy possible by allocating resources optimally. The wiretap channel can be harmed by the use of jamming or AN signals, but the overall power consumption also rises. The best possible power distribution between the intended information and jamming/AN signals is therefore crucial for power conservation. The ideal PA between transmitted information and AN is created in a multiple-input single-output (MISO) system to reduce transmission power while maintaining a specific probability of secrecy. The PA approach is optimized, which considers a multi-user MISO network with friendly jamming, to reduce the total power provided to the information and jamming signals while ensuring secure QoS requirements. A closed-form solution to reduce the transmission power is derived while considering a non-orthogonal multiple access system.

8.5.2.4 Maximization of secure energy efficiency

The EE of secure communications is effectively enhanced with the usage of secure RA. The number of channels used, the transmission length, or the quantity of energy used can all affect how much the SC costs. The majority of secure EE maximizing formulations are non-convex, making them extremely challenging to implement. To address these issues, non-convex optimization techniques including the penalty function method, fractional programming, difference convex (DC) programming, alternating optimization, etc. are introduced. To be more precise, the Dinkelbach algorithm can solve the parameterized polynomial subtractive form so that the secure EE function can be transformed using fractional programming. The non-convex constraint of SR can be removed using the penalty function method by putting the constraint into the objective function. Sometimes an optimization problem is non-convex or impossible to solve when applied to all variables, but it is tractable when applied to some variables first, then the rest. These qualities make using alternating optimization advantageous. DC programming is an efficient method for addressing optimization problems where it iteratively solves a string of convex sub-problems.

8.6 Scheduling for physical layer security

Here, the popular PLS scheduling strategies are reviewed in this part. The RA issue in multi-user OFDMA broadband systems is one that has attracted a lot of attention.

The literature on physical RA in a multi-user OFDMA system is always predicated on the ideal scenario in which the legitimate user's channel quality is better than the eavesdropper's, and it pays little attention to physical RA techniques that would increase safety in unfavorable communication environments.

In the OFDMA network, there are two different kinds of users coexisting. The first category of users is the secondary user which requests non-zero SR and has confidential communications to communicate with BS. The second group is NU, which only conducts routine data transmission and is unconcerned with PLS concerns. All secure and normal users are true members of the network who are totally sincere with the BS.

When secure and normal users coexist in an OFDMA network, the RA problem is significantly different from those in networks without a secrecy restriction. First off, only the subcarrier set on which this secure user has the best channel conditions out of all the users' subcarriers are legal to be assigned to this secure user. This is so that any other user on the same network might potentially eavesdrop in on each secure user. As a result, a non-zero SR on a subcarrier can only be achieved if the secure user on that subcarrier has the highest channel gain out of all the users. Second, when the QoS provided by normal users is considered, assigning a subcarrier to an secure user even though it has the best channel conditions on that subcarrier may not be the best course of action for the system. This is because subtracting two logarithmic functions yields the achievable SR.

In the current and future wireless networks, secrecy or secret message exchanges between users and the BS are typically required. Therefore, it is crucial to incorporate PLS into the multi-user OFDMA systems RA problem. Independent parallel channels have been investigated, and it has been demonstrated that the system's SR equals the average of the secrecy capacities attained on each individual channel. An OFDM-based broadband system's power and subcarrier allocation has the goal of optimizing the sum SR.

The literature on RA for OFDM-based relay PLS algorithms is summarized in the list below. The authors of [29] considered secure RA and developed DF OFDM relaying systems, and the sum SR is increased in an OFDM relay system in [22]. Numerous significant research findings on the RA of OFDMA networks have been published in the last ten years, including [2,30–34], where the main differences in the problem formulation are the optimization objectives and constraints.

8.6.1 Physical layer security-based scheduling in downlink networks

In the current communication systems, a downlink network is crucial. Figure 8.3 depicts the basic downlink scheduling scenario in which the user sends a CQI report based on downlink reference symbols. The scheduler assigns resources per RB based on QoS, CQI, and so on. RA is sent in conjunction with data. Due to the significance of this network in communication systems, numerous efforts have been made to improve SE in various downlink network types. In a typical downlink network made up of a single BS and a number of receivers, the BS takes advantage of numerous antennas and opportunistic receiver scheduling as its main methods for

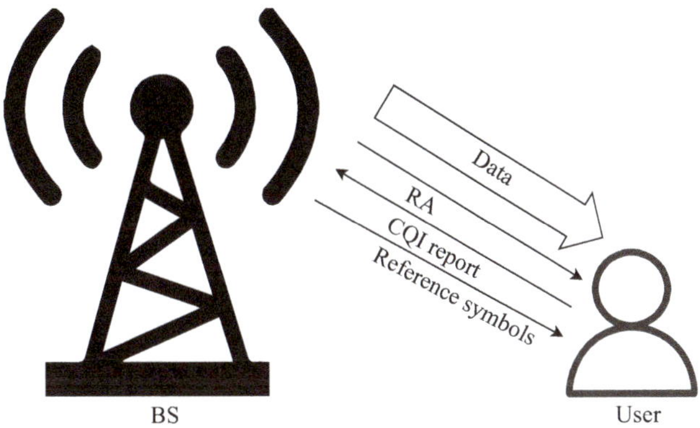

Figure 8.3 Basic downlink scheduling concept

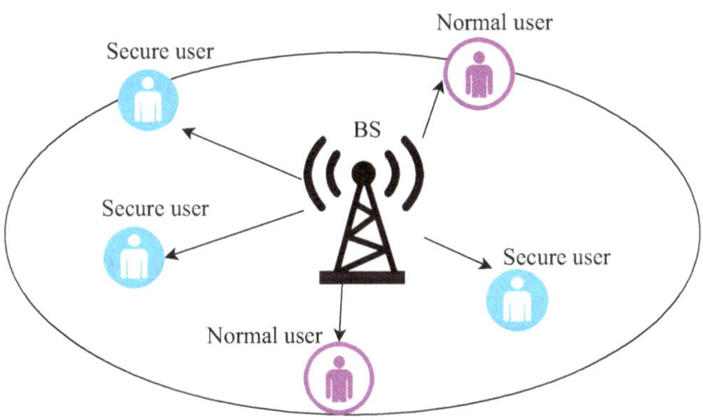

Figure 8.4 The OFDMA downlink network with two types of users: secure and normal users [35]

enhancing SE. Furthermore, similar technologies are used to increase the rate that may be reached by the legitimate user and to stop eavesdroppers from obtaining any information.

Figure 8.4 shows the OFDMA downlink network with two types of users; secure and normal users [35]. A BS with M antenna, K legitimate users with N_t antennas, and N eavesdroppers with N_e antennas constitute the network. It is assumed that the BS is aware of the perfect or imperfect CSI of the K legitimate users, but is unaware of the CSI of the eavesdroppers.

Positive secrecy can only be obtained, according to PLS ideas, when the main channel between the BS and the legitimate user is in better condition than the eavesdropping channel between the BS and the eavesdroppers. The main channel condition cannot always be superior to the eavesdropping channel condition due to

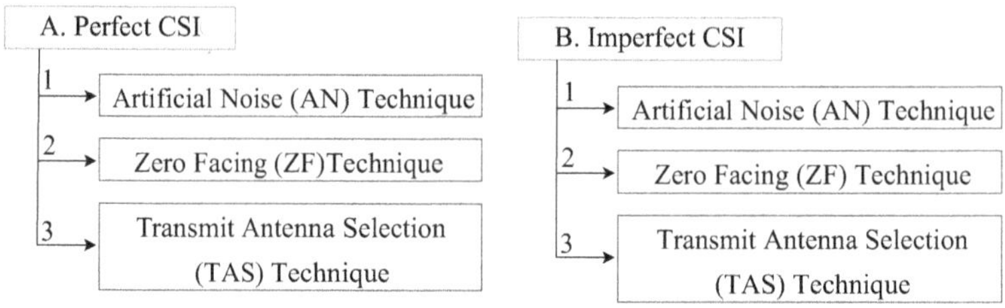

Figure 8.5 PLS-based downlink scheduling techniques based on CSI availability

the randomization of the channel condition. The BS's difficulty in obtaining perfect CSI also contributes to some of the challenges. We will go over the scheduling strategies in the following to maximize the secrecy performance to the CSI availability at the transmitter. Figure 8.5 shows the existing CSI-based PLS scheduling techniques.

(A) *Perfect CSI*: Here, we primarily focus on transmission techniques that exploit perfect CSI of legitimate receivers for various sorts of eavesdroppers. Because of the estimation error and feedback latency, the perfect CSI is not the best assumption, but it provides the performance upper bound and allows us to concentrate on the problems brought on by eavesdroppers. Thus, a lot of PLS research relies on the ideal CSI of legitimate receivers. By adapting the transmission beam to the channel realizations and scheduling the user who has a good effective channel gain, the BS can increase the SR under the perfect CSI assumption. It can also be used to determine the codeword length and the best time to transmit data.

If the BS is outfitted with a single antenna, the best strategy to optimize the SR in a downlink network when everyone else is regarded as a potential eavesdropper is to schedule the user with the highest SNR. The SR can then be calculated as the difference between the achievable rate of the primary connection to the scheduled user and that of the secondary link to the users with the second-highest SNR. Since the achievable rate of the link to the user with the second-largest SNR converges to that to the scheduled user due to the opportunistic user scheduling at the BS, the SR of the network drops as the number of users increases. Eventually, given a network with many of users, the SR approaches zero.

(1) *Artificial Noise (AN) Technique*: The multi-antenna at the BS offers more options for increasing the SR. It enables the BS to broadcast AN in addition to performing beamforming to advance the information signal, preventing information from being extracted by eavesdroppers. The transmit signal of the user k is $\mathbf{x_k}$, in other words, a linear combination of an information signal and AN as follows:

$$\mathbf{x}_k = \sqrt{(1-\alpha)P}w_k s + \sqrt{\alpha P}\eta_k \tag{8.1}$$

where $P = Tr[x_k^H x_k]$ represents the transmit power and α denotes the power portion between the information signal and the noise. $w_k \in C^{M\times 1}$ represents the beamforming vector, which is given as $w_k = h_k^H/||h_k^H||$; and it is known as maximum ratio transmission (MRT). η_k denotes the AN vector, which is intended to be orthogonal to the scheduled user's channel but not to the eavesdroppers channels. Therefore, the noise vector guarantees that no channels, aside from those used by eavesdroppers, will degrade.

The SR of the network is greatly influenced by the types of eavesdroppers in terms of whether they can share the received signal between themselves when the BS only has statistical knowledge of the link to the eavesdroppers and just the CSI of the ties to legitimate users. Sharing information has a similar impact on performance as having numerous antennas at BS. Since the mutual information between the BS and the eavesdroppers grows as the number of cooperating eavesdroppers increases, the BS is required to allocate more power on the AN transmission in the case of information-sharing eavesdroppers than in the case of non-cooperating eavesdroppers. Additionally, as the number of information sharing eavesdroppers rises, the SR decreasing speed with respect to user count monotonically grows. The non-adaptive equal PA between the information signal and the AN, on the other hand, produces a SR performance close to the optimal PA in the case of information non-sharing eavesdroppers [36]. Opportunistic user scheduling might be viewed as a useful strategy if the downlink network's goal is to maximize the network security rate.

The user with the highest channel gain is scheduled by BS in conventional networks in the absence of eavesdroppers to increase network throughput. However, if eavesdroppers are also considered in the network, the opportunistic user scheduling rule and the accompanying SR characteristics alter dramatically. Since there is always a chance that some of the eavesdroppers have a better channel state than the scheduled user, it is difficult to ensure a positive SR. The codeword length in the slow fading channel is less than the channel coherence time. Instead of using the SR, we may use the outage SC as a performance parameter to account for secure communication failures. The highest SR that may be maintained while keeping the chance of a secrecy outage below a specific amount is known as the outage SC ε [36]. ESC is a better performance metric in the fast fading channel, when the codeword length is significantly more than the channel coherence time.

(2) *Zero Forcing (ZF) Technique*: An efficient method to prevent signal reception at the eavesdroppers is the zero forcing (ZF) scheme, which broadcasts the information signal in the orthogonal direction of the subspace spanned by the eavesdroppers' channels. But in essence, it calls for a precise CSI at the BS. The ZF reliably transmits the message to the users with a relatively low interference level when the CSI of both users and eavesdroppers is completely known to the BS with a suitable number of transmit antennas.

Furthermore, BS uses both techniques when it has incomplete or imperfect knowledge of the eavesdroppers' CSI. The ZF approach is employed to maintain secrecy among legitimate users and convey messages with the least amount of interference and AN to promote secrecy. Even though there is no CSI of the eavesdropper accessible other than its statistical data, BS can relay the information to any legitimate user with a SR proportionate to the transmit power if the number of antennas at eavesdropper is $N_e = M_K$. However, the SR becomes insensitive to the power if eavesdropper is aware of the global CSI and has enough antennas to cancel out the AN $N_e >= M_K$ [37].

In many cases, the BS is unable to collect the CSI of the eavesdroppers because they may not be willing to provide the CSI feedback. If an eavesdropper's CSI is unavailable, the BS creates beamforming vectors and schedules users based only on their CSI, much like in a typical network without any eavesdroppers.

(3) *Transmit antenna selection (TAS) technique*: The TAS is a straightforward transmission method that makes use of a single radio frequency network. As a result, we may decrease the hardware size, computational complexity, and implementation cost. To increase the SR with a single antenna, the BS chooses an antenna with a high main channel gain and a low eavesdropping channel gain.

(B) *Imperfect CSI*: Imperfect CSI at the BS is assumed here. At the CSI of the legitimate users, there are a few estimation inaccuracies. It is challenging to use a perfect channel estimation in real-world situations since a better estimation needs more time, bandwidth, and energy. The BS can follow the CSI in a number of ways. The BS in time division duplex (TDD) systems can calculate the CSI using the channel reciprocity. Since there is no channel reciprocity in frequency division duplex (FDD) systems, CSI is derived from the feedback. Three types of imperfect CSI exist: statistical CSI, in-deterministic imperfect instantaneous CSI, and deterministic imperfect instantaneous CSI. The user's estimation and estimation error are added to represent their actual channel gain. The estimation error is unknown to the transmitter and is unrelated to the error estimated by the transmitter. It is impossible to fully align the beam vectors for the information signal and the AN to the subspace and null space spanned by the main channel, if there is an error in the CSI at the BS. Then, AN obstructs the legitimate user's received signal.

(1) *AN technique*: In a network with one transmitter, one legitimate receiver $K = 1$, and N non-cooperating eavesdroppers, imperfect CSI affects the power allocation between the information signal and the AN. The power share allotted to the AN rises as N_e rises and converges to a constant amount at high SNR. Additionally, more power needs to be given to the AN as the estimation error rises. This is because the channel estimate error affects the legitimate user but not the eavesdroppers while receiving information signals, and it has no negative effects on the eavesdroppers when using AN.

(2) *ZF technique*: ZF is a successful method to prevent signal reception at prospective eavesdroppers in the case where $M > K$ and everyone else besides the scheduled user are thought to be potential eavesdroppers. However, it essentially calls for precise CSI at the BS. The rate of network secrecy reduces quickly as the CSI error's magnitude rises. We might think about the regularized channel inversion (RCI) as a solution to reduce the performance decrease caused by the insufficient CSI in the ZF. RCI behaves as ZF with $\varphi = 0$ if the BS is aware of the users' perfect CSI. In contrast, RCI consistently beats ZF if the BS has an imperfect CSI [37]. The greatest option for maximizing network SR for $M > K$ is not always nulling out the received signal at the eavesdroppers with RCI. Even if it results in AN reception at the scheduled users due to the imperfect CSI, AN transmission beats RCI in some networks with non-cooperative eavesdroppers.

(3) *TAS technique*: If the CSI is obsolete, the BS cannot achieve the diversity order proportional to the number of broadcast antennas via transmit antenna selection (TAS). The BS cannot accomplish the diversity order proportional to the number of broadcast antennas through TAS if the CSI is obsolete.

8.7 Challenges, recommendation and future directions for physical layer security in scheduling

Here, we highlight some challenges, recommendation, and future directions for secure RA and scheduling as follows:

- Among many secure scheduling challenges is satisfying the diverse users requirements in the next-generation wireless networks where secure scheduling methods are needed.
- So far, only a few works on cross secure MAC/PHY scheduling algorithms have been revealed. Therefore, there is a need for improved PLS algorithms that consider the implications of various scheduling problems.
- The range of services necessary to serve various user requirements is a distinguishing aspect of 5G and beyond networks, but they are sometimes in conflict with one another.
- The number of transmission parameters and related optimization problems would multiply as a result of the increased heterogeneity of sixth generation (6G) in terms of deployment scenarios, mobility management, frequency spectrum, application needs, and device capabilities. If we ensure that the eavesdropper is operating below the QoS needs that each service in actuality has that are unique from the others, practical service-based secrecy can be guaranteed. In light of the real-world QoS needs of genuine services like uRLLC,massive machine-type communication (mMTC), video, etc., a potential research field would be to redesign the current secure RA and scheduling algorithms from a much more realistic perspective.
- Furthermore, there should be more efficient adaptive RA methods that can change the transmission parameters to suit user needs and channel conditions, which make correct detection hard for the attacker.

8.8 Conclusion

Despite the fact that PLS is a well-studied issue, there has been a gap in defining a generic framework addressing secure RA and scheduling. This chapter studied the modification plane by highlighting ways for developing PLS using existing scheduling and RA algorithms. A number of goals have been achieved through the use of scheduling and RA, including enhancing SE, EE, and fairness. The main criteria employed in security research to evaluate the efficacy of a scheduling solution are SC and SOP. By carefully arranging the available resources based on the PLS idea, information security can be efficiently achieved. This chapter has revisited the scheduling, RA, and their widely used approaches. Then, the fundamental PLS optimization problems and performance measures have been covered. The popular secure RA and PLS scheduling techniques have then been discussed.

Acknowledgment

This work was supported by the Scientific and Technological Research Council of Turkey (TÜBİTAK) under Grant 5200107, with the cooperation of Turkcell Technology and Istanbul Medipol University.

References

[1] Hamamreh JM, Arslan H. Joint PHY/MAC layer security design using ARQ with MRC and null-space independent PAPR-aware artificial noise in SISO systems. *IEEE Transactions on Wireless Communications*. 2018;17(9):6190–6204.

[2] Wong CY, Cheng RS, Lataief KB, *et al.* Multiuser OFDM with adaptive subcarrier, bit, and power allocation. *IEEE Journal on Selected Areas in Communications*. 1999;17(10):1747–1758.

[3] Lo ES, Chan PW, Lau VK, *et al.* Adaptive resource allocation and capacity comparison of downlink multiuser MIMO-MC-CDMA and MIMO-OFDMA. *IEEE Transactions on Wireless Communications*. 2007;6(3):1083–1093.

[4] Tang J, Cumanan K, Lambotharan S. Sum-rate maximization technique for spectrum-sharing MIMO OFDM broadcast channels. *IEEE Transactions on Vehicular Technology*. 2011;60(4):1960–1964.

[5] Maciel TF, Klein A. On the performance, complexity, and fairness of suboptimal resource allocation for multiuser MIMO–OFDMA systems. *IEEE Transactions on Vehicular Technology*. 2009;59(1):406–419.

[6] Henarejos P, Perez-Neira AI, Tralli V, *et al.* Low-complexity resource allocation with rate balancing for the MISO-OFDMA broadcast channel. *Signal Processing*. 2012;92(12):2975–2989.

[7] Chen J, Swindlehurst AL. Applying bargaining solutions to resource allocation in multiuser MIMO-OFDMA broadcast systems. *IEEE Journal of Selected Topics in Signal Processing*. 2012;6(2):127–139.

[8] Chan PW, Cheng RS. Capacity maximization for zero-forcing MIMO-OFDMA downlink systems with multiuser diversity. *IEEE Transactions on Wireless Communications*. 2007;6(5):1880–1889.

[9] Perea-Vega D, Girard A, Frigon JF. Dual-based bounds for resource allocation in zero-forcing beamforming OFDMA-SDMA systems. *EURASIP Journal on Wireless Communications and Networking*. 2013;2013(1):1–16.

[10] Huang H, Papadias CB, Venkatesan S. MIMO Communication for Cellular Networks. New York, NY: Springer Science I& Business Media; 2011.

[11] Tseng SC, Liu ZW, Chou YC, et al. Radio resource scheduling for 5G NR via deep deterministic policy gradient. In: *2019 IEEE International Conference on Communications Workshops* (ICC Workshops). New York, NY: IEEE; 2019. p. 1–6.

[12] Richart M, Baliosian J, Serrat J, *et al.* Resource slicing in virtual wireless networks: a survey. *IEEE Transactions on Network and Service Management*. 2016;13(3):462–476.

[13] Wang D, Bai B, Zhao W, *et al.* A survey of optimization approaches for wireless physical layer security. *IEEE Communications Surveys & Tutorials*. 2018;21(2):1878–1911.

[14] Chang B, Zhang L, Li L, *et al.* Optimizing resource allocation in URLLC for real-time wireless control systems. *IEEE Transactions on Vehicular Technology*. 2019;68(9):8916–8927.

[15] Shen T, Zhang S, Chen R, *et al.* Two practical random-subcarrier-selection methods for secure precise wireless transmissions. *IEEE Transactions on Vehicular Technology*. 2019;68(9):9018–9028.

[16] Jindal A, Bose R. Resource allocation for secure multicarrier AF relay system under total power constraint. *IEEE Communications Letters*. 2014;19(2): 231–234.

[17] Karachontzitis S, Timotheou S, Krikidis I, *et al.* Security-aware max–min resource allocation in multiuser OFDMA downlink. *IEEE Transactions on Information Forensics and Security*. 2014;10(3):529–542.

[18] Chen J, Chen X, Gerstacker W. Optimal power allocation for a massive MIMO relay aided secure communication. In: *2015 IEEE Global Communications Conference* (GLOBECOM). New York, NY: IEEE; 2015. p. 1–6.

[19] Chen J, Chen X, Gerstacker WH, *et al.* Resource allocation for a massive MIMO relay aided secure communication. *IEEE Transactions on Information Forensics and Security*. 2016;11(8):1700–1711.

[20] Zhou X, Wu Q, Yan S, *et al.* UAV-enabled secure communications: joint trajectory and transmit power optimization. *IEEE Transactions on Vehicular Technology*. 2019;68(4):4069–4073.

[21] Wang X, Tao M, Mo J, *et al.* Power and subcarrier allocation for physical-layer security in OFDMA-based broadband wireless networks. *IEEE Transactions on Information Forensics and Security*. 2011;6(3):693–702.

[22] Jeong C, Kim IM. Optimal power allocation for secure multicarrier relay systems. *IEEE Transactions on Signal Processing*. 2011;59(11):5428–5442.

[23] Zhang Y, Shen Y, Wang H, *et al.* On secure wireless communications for IoT under eavesdropper collusion. *IEEE Transactions on Automation Science and Engineering*. 2015;13(3):1281–1293.

[24] Zhang M, Liu Y, Zhang R. Artificial noise aided secrecy information and power transfer in OFDMA systems. *IEEE Transactions on Wireless Communications*. 2016;15(4):3085–3096.

[25] Mokari N, Parsaeefard S, Saeedi H, *et al.* Secure robust ergodic uplink resource allocation in relay-assisted cognitive radio networks. *IEEE Transactions on Signal Processing*. 2014;63(2):291–304.
[26] Xing H, Liu L, Zhang R. Secrecy wireless information and power transfer in fading wiretap channel. *IEEE Transactions on Vehicular Technology*. 2015;65(1):180–190.
[27] He B, Zhou X. On the placement of RF energy harvesting node in wireless networks with secrecy considerations. In: *2014 IEEE Globecom Workshops* (GC Wkshps). New York, NY: IEEE; 2014. p. 1355–1360.
[28] Yu H, Guo S, Yang Y. An optimization framework of target secrecy rate and power allocation for SWIPT system. In: *2016 IEEE Global Communications Conference* (GLOBECOM). New York, NY: IEEE; 2016. p. 1–6.
[29] Ng DWK, Lo ES, Schober R. Secure resource allocation and scheduling for OFDMA decode-and-forward relay networks. *IEEE Transactions on Wireless Communications*. 2011;10(10):3528–3540.
[30] Jang J, Lee KB. Transmit power adaptation for multiuser OFDM systems. *IEEE Journal on Selected Areas in Communications*. 2003;21(2):171–178.
[31] Tao M, Liang YC, Zhang F. Resource allocation for delay differentiated traffic in multiuser OFDM systems. *IEEE Transactions on Wireless Communications*. 2008;7(6):2190–2201.
[32] Hui DSW, Lau VKN, Lam WH. Cross-layer design for OFDMA wireless systems with heterogeneous delay requirements. *IEEE Transactions on Wireless Communications*. 2007;6(8):2872–2880.
[33] Mokari N, Javan MR, Navaie K. Cross-layer resource allocation in OFDMA systems for heterogeneous traffic with imperfect CSI. *IEEE Transactions on Vehicular Technology*. 2009;59(2):1011–1017.
[34] Song G, Li Y. Cross-layer optimization for OFDM wireless networks—Part II: algorithm development. *IEEE Transactions on Wireless Communications*. 2005;4(2):625–634.
[35] Zhou X, Song L, Zhang Y. *Physical Layer Security in Wireless Communications*. London: CRC Press; 2013.
[36] Zhou X, McKay MR. Secure transmission with artificial noise over fading channels: achievable rate and optimal power allocation. *IEEE Transactions on Vehicular Technology*. 2010;59(8):3831–3842.
[37] Li N, Tao X, Wu H, *et al.* Large-system analysis of artificial-noise-assisted communication in the multiuser downlink: ergodic secrecy sum rate and optimal power allocation. *IEEE Transactions on Vehicular Technology*. 2015;65(9):7036–7050.

Chapter 9

Physical layer security in distributed wireless networks

Muhammad Sohaib J. Solaija[1], Hanadi Salman[1], Haji M. Furqan[1], and Hüseyin Arslan[1]

According to the framework presented in Chapter 3, the *network* domain represents the various physical nodes present in the propagation environment and how their transmissions can be modified or adapted to enable physical layer security (PLS). Specifically, these nodes may include relays, transmission points (TPs), and reconfigurable intelligent surfaces (RISs) in cooperative communication, coordinated multipoint (CoMP) and smart radio environment paradigms, respectively. All three of these are discussed as potential solutions against eavesdropping, jamming, and spoofing attacks.

9.1 Cooperative communication for physical layer security

As a wireless signal traverses the environment, it experiences various impairments including pathloss, shadowing, blockage, and fading. One way of combating the channel losses is to transmit multiple copies of the signal to provide *diversity*. The diversity benefits of multi-antenna technologies such as single-input multiple-output (SIMO), multiple-input single-output (MISO), and multiple-input multiple-output (MIMO) are well established to the extent of being standardized for various wireless technologies. However, in certain scenarios, it is not feasible for devices to have multiple antennas. This may be due to size, cost, or energy constraints as in the case of wireless sensors. The cooperative communication paradigm enables such resource-constrained single-antenna devices to enjoy the diversity benefits otherwise limited to MIMO systems [1]. This is achieved by the cooperation of multiple (single-antenna) nodes for transmission/reception of wireless signals. Essentially, multiple nodes located between the source and the destination offer their antennas to form a virtual MIMO system [2]. Before jumping into the PLS methods utilizing cooperative communication, we would provide an overview of the general system model considered in such systems in the following subsection.

[1]Department of Electrical and Electronics Engineering, Istanbul Medipol University, Turkey

9.1.1 General system model in cooperative communications

The cooperative communication model comprises of three (types of) terminals, i.e., source, relay, and destination. The information/data originates at the source and travels to the destination with the help of (one or more) relay(s). In general, cooperative transmission is split into two phases. The first phase, also known as the broadcast phase, involves the transmission of information from the source. The second phase, referred to as multiple-access*, includes the relayed transmissions [3]. Figure 9.1 illustrates the various combinations in terms of source and relay transmissions, where the solid line represents the broadcast phase while the dashed line depicts the multiple access phase. The first model (top left) shows the most common scenario where the source transmits to both receiver and destination in the first phase. In the second phase, both relay and source transmit to the destination. The second scenario (top right) involves the source's transmission to the relay in the first phase, while both the relay and the source transmit to the destination in the second phase. In the third case (bottom left), source broadcasts to both the relay and destination in the first phase and only the relay transmits to the destination in the second phase. The last (bottom right) scenario represents the case commonly known as *multi-hop* communication, where the source transmits to relay in the first phase and relay transmits to the destination in the second phase. Since there is only a single path for the signal, this case does not provide any diversity but can be used to improve the communication range. Here it should be noted that there might be multiple relays in the vicinity of a

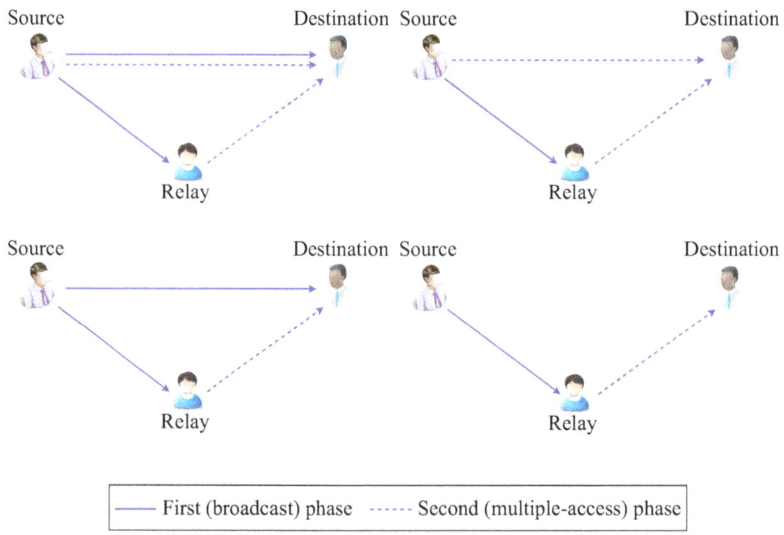

Figure 9.1 Illustration of the different system models considered in cooperative communication scenarios

*This terminology is reused due to its presence in the literature [3]. In essence, this refers to the phase that includes retransmission by the relay node(s).

source-destination link, and it is also possible for relays to have full duplexing (simultaneous transmission and reception) capability.

As mentioned earlier, cooperative communication is more popular in scenarios where devices are resource-constrained such as *ad hoc* and wireless sensor networks (WSNs). Apart from limited antennas, these systems also possess less sophisticated authentication procedures compared to cellular networks, rendering them more vulnerable to security breaches such as eavesdropping or spoofing due to the presence of malicious nodes. In the following subsections, we will look at the cooperative solutions for the different attacks.

9.1.2 Cooperative solutions against eavesdropping

In general, solutions against eavesdropping are divided into two categories, i.e., *cooperative relaying* and *cooperative jamming*. As mentioned earlier, cooperative communication comprises two stages, i.e., broadcast and multiple access. In the latter, the relay transmits the message using one of amplify-and-forward (AF), compute-and-forward (CTF), compress-and-forward (CF), or decode-and-forward (DF) strategies [4]. In AF, the relay transmits an amplified version of the signal received from the source. While AF is simple to implement, it results in the amplification of noise in addition to the intended signal. In CF, the relay compresses the signal before retransmitting it. CTF leverages structured (linear) codes in a multiuser scenario where the relay decodes the messages using linear combinations of the noisy signals before retransmitting them to the destination. The destination is able to decode the transmitted messages if a sufficiently large number of linear combinations are received. A DF relay first decodes the signal before re-encoding and transmitting it to the destination.

Since cooperative communication does not boast an authentication mechanism similar to or as sophisticated as cellular systems, there is the possibility of having untrusted relays as part of the system which further compounds the eavesdropping problem. Consider the case where one of each, i.e., source, destination, (untrusted) relay, and eavesdropper is present. In such a case, DF relaying is usually avoided so that the relay is not required to decode the transmission. A theoretical CTF approach leveraging nested lattice codes for multi-hop scenario is studied in [5], where the nodes can only receive transmissions of their neighbors. A MIMO system for AF relaying protocol is considered in [6], where the source beamforming vector and relay's beamforming matrix are jointly optimized. The results indicate that this scheme is particularly suitable when the relay has high transmit power and a large number of antennas are located closer to the source compared to the relative distance between source and destination. An interesting approach to limit eavesdropping capability for both AF and DF relays is given in [7], where instead of achieving perfect secrecy the goal is to ensure that only parts of the data are secure at the physical (PHY) layer. Essentially, this approach involves transmit beamforming at the source such that the rate at the relay is lower than a certain ratio of the rate achieved at the destination. As such, this approach is devised to be used in conjunction with upper layer techniques rather than being used in isolation.

In the case of trusted relays, an approach is to optimize the relay's power allocation such that any leakage is minimized while ensuring a minimum desired signal-to-interference-plus-noise ratio (SINR) at the destination. In theory, the power can be limited on a system or per-relay basis. However, in practice, the latter is more relevant since the power for any node is limited itself [8]. A subcarrier allocation approach for AF multi-hop system is presented in [9], however, it requires the channel state information (CSI) of all links at the source which is not a suitable assumption in real networks. The selection of appropriate relays for security can be considered an extension of the resource allocation in the spatial domain. This is illustrated in Figure 9.2, where the relay close to the eavesdropper (Relay 2) is not used in the multiple-access phase. The biggest challenge in this regard is to ensure reliable transmission at the destination while ensuring security. This tradeoff is studied in [10]. Optimal relay selection for AF/DF protocols is discussed in [11] in a multi-hop scenario, where both destination and eavesdropper are outside the range of the source. In the broadcast phase, all relays receive the source's transmission. Later, only the optimally selected relay re-transmits the signal for the destination. The selection of optimal relay, however, requires not only the CSI of the legitimate link but also the link between relay and eavesdropper. While this system cannot be considered practical, it does give a benchmark for other studies.

The other popular approach for mitigating eavesdropping in these systems is the use of cooperative jamming. This requires additional power and carefully designed interference/noise signals to ensure that the performance of the destination node is not degraded. It is possible to employ a dedicated helper/jamming node that only serves to degrade the eavesdropper's reception. If a dedicated helper node is not available, either the source or the destination are required to send the jamming signal but this requires the presence of multiple antennas (for the former) or full-duplexing capability (for the latter) [12]. Here it should be pointed out that full-duplexing is

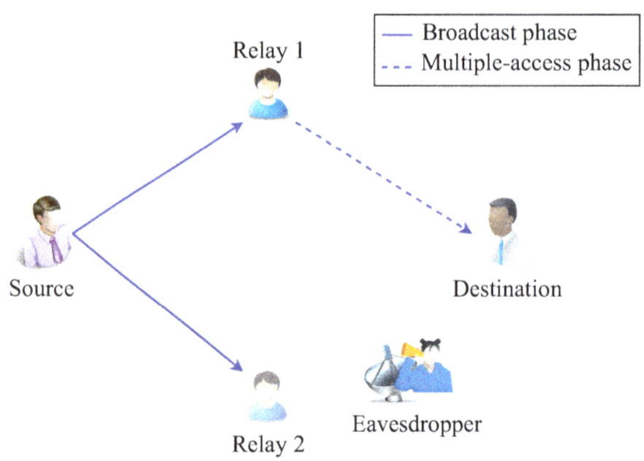

Figure 9.2 Illustration of relay selection against eavesdropping. Relay 2 is not used in the multiple-access phase due to the eavesdropper's presence in its vicinity

incredibly difficult to achieve in practice, particularly for low-end devices like the ones normally expected in cooperative communication networks. A possible solution for this exploiting the fast Fourier transform (FFT) operation in orthogonal frequency division multiplexing (OFDM)-based systems is presented in [13], where the destination transmits a known jamming signal during the cyclic prefix (CP) portion of the received signal in the broadcast phase. Even though it only exists for a small portion of time, due to fast Fourier transform (FFT) it spreads to all subcarriers in the frequency domain where the data is present in the case of OFDM communication.

9.1.3 Cooperative solutions against jamming

The case of jamming in cooperative communication is particularly challenging since a successful jamming attack on either link can drastically decrease the overall system performance. The impact of jamming on a two-hop AF system is discussed in [14]. The authors conclude that legitimate communication is optimal when the performance in both broadcast and multiple-access phases is similar. For the jammer, on the other hand, the goal is to unbalance the two phases' performance. In cooperative communications, the jamming attacks can be combated using the other (trustable) nodes in the vicinity. Figure 9.3 illustrates how the relay (Relay 2) close to the jammer can be deactivated in the multiple-access phase and a different relay (Relay 1) can be used to successfully transmit the signal. This is studied for a line-of-sight (LoS) scenario in satellite to unmanned aerial vehicle (UAV) communication in the presence of an aerial jammer in [15], where a two-stage mitigation method is proposed. In the first stage, the satellite transmits to the different UAV groups using beamforming. During this stage, the jammed and relaying UAVs are selected based on the successful reception and decoding of the satellite signals.

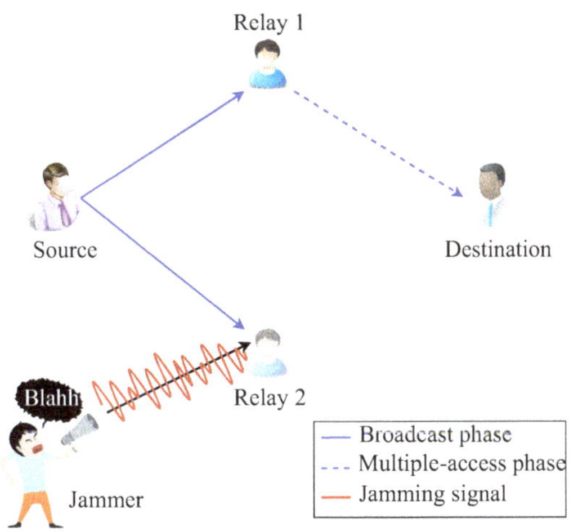

Figure 9.3 Illustration of relay selection against jamming. Relay 2 is not used in the multiple-access phase due to the jammer's presence in its vicinity

Later, the relaying UAVs uses DF protocol to retransmit the signal to the jammed UAVs. Similarly, the jamming of the links between UAVs in a swarm configuration is studied in [16] where the relay UAV's frequency, location/motion, and antenna direction is optimized to combat jamming using reinforcement learning. In general, the ability of UAVs to be moved instantaneously and deployed dynamically is highly advantageous in mitigating jamming attacks.

9.1.4 Cooperative solutions against spoofing

In general, spoofing may be mitigated using authentication mechanisms that leverage the properties of the channel or the device itself. The former includes measurement such as channel impulse response (CIR), channel frequency response (CFR), and received signal strength indicator (RSSI) while the latter comprises of impairments like in-phase/quadrature imbalance (IQI) and carrier frequency offset (CFO) [17]. In cooperative communications, it is possible to get multiple observations of these properties since we have more nodes involved in the communication. This potentially allows for more robust authentication, as exemplified in [18]. As in any PHY authentication scheme, a training stage is required where the destination is able to learn the properties of the legitimate source. The spoofer can also utilize these transmissions to learn about the latter. In the authentication phase, the destination compares the observed channel against a threshold obtained from the training phase and classifies the transmission as legitimate or otherwise. In order to leverage the multiple observations, however, it is necessary to have trusted relay nodes present in the system. In other cases when there is information available about the channel of the attacker such as its statistics, it is possible to mitigate spoofing using proper relay selection, as shown in [19]. In that work, the relay is selected such that the ratio of instantaneous signal-to-noise ratio (SNR) between legitimate and illegitimate links is maximized. The information about the spoofer's channel, if available, can also be used to further reinforce the authentication mechanism.

9.1.5 Challenges for physical layer security in cooperative communication

Having gone through some of the techniques by which cooperative communication can be secured against various threats, we will now take a look at some of the general challenges faced in realizing these techniques:

- *Trustworthiness of relays:* As mentioned earlier, the PLS methods vary significantly depending on the trustworthiness of the relays present. However, a more objective method of determining this trustworthiness is required rather than just relying on previous interactions with the said node.
- *Mobility of relays:* A significant amount of work in cooperative PLS relies on the given network topology and (relative) locations of the different nodes. However, in scenarios such as vehicular communication, this assumption cannot be satisfied. In these cases, it is important to devise strategies that either do not rely

on specific assumptions about the node locations or can autonomously adapt to such changes.

- *Incentive for cooperation:* Most works in this domain assume that relays cooperate without any incentive. However, in real networks, this is hardly the case, particularly for scenarios such as cooperative jamming where a significant amount of extra power is consumed by helper nodes. As such, modeling the interaction between jammer and other nodes can serve as the first step toward optimizing the incentive(s) for a friendly jammer.
- *Channel correlation:* A significant number of PLS approaches assume that the relay-destination link experiences independent fading compared to the relay-attacker link. However, considering the small separation between the nodes, this is not a realistic assumption. For instance, in the case of cooperative jamming where the jamming signal is designed considering the specific link, it is important to study the impact of possible channel correlation.

9.2 CoMP-aided physical layer security

The diversity of services, necessary to serve varied user requirements, is a distinguishing feature of fifth generation (5G) and beyond networks. However, the requirements of these services are frequently conflicting in nature, emphasizing the importance of a CoMP system. The initial goal of CoMP in fourth generation (4G) is to increase the performance at the cell edge. The collaboration between TPs can be used to assist the different requirements and enabling technologies of future generations. However, a few studies have turned at CoMP for wireless communication security. This section provides a summary of the evolution of coordination in cellular networks, with a particular emphasis on the development of CoMP-assisted PLS solutions against eavesdropping, jamming, and spoofing attacks.

In Third Generation Partnership Project (3GPP) Rel-11, CoMP was introduced as solution of inter-cell interference coordination (ICIC) and enhanced ICIC (eICIC) failure. The basic idea of CoMP is to allow the coordination among different TPs through ideal backhaul. As a result, CoMP brings spatial domain as another resource allocation dimension. This helps to improve spectral efficiency along with interference mitigation. Although different levels of coordination have been proposed in the standardization discussions, 3GPP recognizes three main types of downlink coordination based on required backhaul bandwidth and scheduling complexity. Figure 9.4 depicts downlink CoMP schemes.

- *Joint transmission (JT)*: In this scheme, the coordinated TPs share CSI/scheduling information as well as user data. This type of coordination works well but requires low latency and high backhaul bandwidth due to the exchange of user data between multiple TPs to convert interference signals into usable signals to provide coherent or non-coherent services to a single user. The concept of coherent transmission refers to the combination of precoding design with synchronous transmission to realize coherent coupling. However, it is not necessary

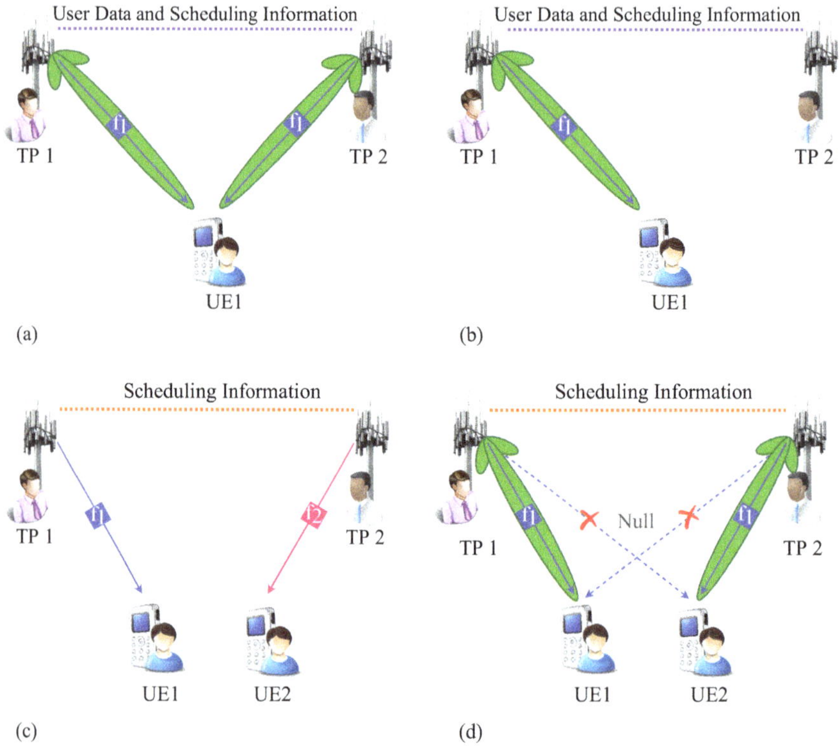

Figure 9.4 Illustration of CoMP schemes. As part of JT and DPS, user data, channel, and scheduling information are required to be shared. In contrast, CS/CB only requires channel and scheduling information to be exchanged. (a) JT scheme, (b) DPS scheme, (c) CS scheme and (d) CB scheme

to precode data from multiple TPs in non-coherent JT, since each TP precodes its own data.

- *Dynamic point selection (DPS)*: As a special type of JT, the user data is transmitted from only one TP at a time, and the serving TP is changed dynamically as resource availability and channel conditions change. In each transmission, the best serving cell is determined using fading conditions. This required that user data is accessible through multiple TPs, as with JT.
- *Coordinated scheduling (CS)/Coordinated beamforming (CB)*: User data is not shared between the coordinated TPs, so access to user data is restricted to one TP, but scheduling and beamforming design are exchanged among all coordinated TPs. A CS/CB reduces interference that may occur to a user served by a neighboring TP by dynamically preventing transmission of a particular time-frequency resource. This process is also known as *dynamic point blanking*. Additionally, coordinated power control or coordinated beamforming can also be involved

under CS/CB mechanisms, by dynamically changing the transmit power or transmission direction for a specific set of resources. Due to their reduced data exchange, CS/CB requires a lower backhaul bandwidth compared to JT.

9.2.1 CoMP-assisted solutions against eavesdropping

In addition to enhancing the received signal power for a legitimate receiver, CoMP also increases the risk of being eavesdropped on. For security enhancement, CoMP can complement other cooperation techniques such as coordinated jamming and relay [20]. As an additional benefit, PLS can also be performed using the multiple communication links in CoMP systems. Suppose an underwater communication scenario involves scheduling (and controlling the power of) multiple distributed TPs so that the received signal is clean and non-overlapping at the legitimate receiver, while packets from the different TPs overlap and interfere with the eavesdropper [21]. As a consequence of the spatial distribution of TPs, the directional modulation† limitation can also be addressed, since the eavesdropper lies in the same direction as the legitimate receiver [22]. The correct constellation is reinforced at the desired user location, but interference is observed at the eavesdropper location since multiple TPs use directional modulation. With this approach, location-based security can be maintained in sparse environments, which can pose challenges from a PLS perspective. It is then extended to the multi-user case in a multipath environment [23], in which only the intersection of coordinated transmission paths can recover the transmitted message.

9.2.2 CoMP-assisted solutions against jamming

Wireless communications are prone to jamming attacks. Typically, a jamming signal interrupts data reception over a legitimate transmission link, threatening the security of communications severely. Additionally, in CoMP scenarios with multiple transmission links, users will experience more significant performance degradation from external malicious jamming and internal co-channel interference. The spread spectrum/frequency hopping techniques conventionally used to mitigate jamming are more challenging to implement in CoMP networks due to the more rigid synchronization requirements [24]. Contrary to this, beamforming and directional modulation strategies are significantly more effective in preventing jamming.

9.2.3 CoMP-assisted solutions against spoofing

Authentication is essential in securing communication and validating the identity of the communicating node. Wireless channel or radio frequency (RF) impairments can be used to accomplish this. Consequently, these features can help detect malicious users and fake nodes. At present, there are no multipoint-based transmission strategies for authentication/spoofing detection. In order to address this gap, it is vital to

†Directional modulation is a multiple-antenna security technique where a distorted constellation is observed in all directions except the legitimate one.

develop techniques capable of extracting intended features from the signals obtained via spatially separated wireless links.

9.2.4 Technical limitations of CoMP deployment

From the experience of implementing and testing CoMP networks, the following key challenges have become apparent.

- *Synchronization:* Coordinated transmissions need to be synchronized in time and frequency in order to avoid intersymbol interference (ISI) as well as inter-cell interference (ICI) [25]. Due to the fact that CoMP requires the users to be synchronized with multiple TPs, the old timing advance technique will not work. This is due to synchronizing to one TP would actually affect the synchronization of the other TPs. It, therefore, restricts the scenarios that can use CoMP to those that guarantee that the guard interval between the arrived signal replicas at the TP would be less than the time difference between them. ISI and ICI are caused when the time difference of arrivals at the TP of the users' signals exceeds the CP's length. Thus, CoMP is not applicable to scenarios with large inter-site distances between TPs [26].
- *Backhaul:* As far as practical implementation is concerned, an ideal backhaul with zero delay path links is not available. In order to address the limited backhaul issue, the quantization of CSI and user data is used, as well as minimizing the number of connected users, which leads to a significantly reduced amount of throughput and an increase in end-to-end latency. Optical backhaul technology is extensively used today, but it may not have the capacity to support the increasing densification of future networks. Millimeter wave (mmWave) has been discussed as a means of integrated backhaul/fronthaul and access operation. In this case (especially self-backhauling), however, radio resource management becomes critical, necessitating methods that are flexible and adaptive when using radio resources.
- *Clustering:* With each CoMP strategy, TPs work together to choose the best channel for data sharing by minimizing interference levels to a minimum. clustering involves categorizing users/TPs according to SINR value or application requirements. Thus, the cluster size should not reach the threshold level, as this results in a performance delay. A systematic method of clustering, either statically or dynamically, can be used to achieve more advanced and efficient clustering results [27].
- *Complexity:* Whenever a scheduling algorithm is processed, computational complexity is always a concern. This not only delays the signal overhead but also decreases the network performance overall. There are several significant enhancements occurring, the single carrier frequency division multiple access (SC-FDMA) scheme suffers from increased complexity as MIMO equalization and additional subcarriers are enabled [28]. To reduce complexity, we need to design more improved computational algorithms that use fewer reference parameters.

9.3 RISs for secure and smart environments

The conventional wireless system design optimization is based on the consideration of the wireless communication environment as an uncontrollable factor that usually harms the communication performance and needs to be mitigated and compensated. However, RIS empowered smart radio environment envisions the wireless channel as a controllable entity.

In particular, an RIS consists of low-cost passive reflecting elements arranged in the form of a planar array, where each of these reflecting elements can be smartly controlled to modify the amplitude and/or phase of incoming electromagnetic waves [29]. This capability can be used to add different signals to enhance or weaken their overall strength by adding them constructively/destructively. Thus, RIS can be employed for improving the signal-to-noise ratio, capacity, coverage, PLS, and so on.

9.3.1 RIS-assisted PLS solutions against eavesdropping

From the PLS against eavesdropping point of view, the capability of RIS in terms of controlling the amplitude and/or phase of incoming electromagnetic waves can be exploited to degrade the listening capability of an eavesdropper while enhancing the signal quality at legitimate nodes, thus enhancing the overall secrecy rate, as shown in Figure 9.5. This needs information related to CSI of both the legitimate receiver and the attacker for optimal performance. However, when the CSI of the legitimate node is available only, secure communication can still be ensured based on the channel, location, and requirements of the legitimate node. The RIS-assisted protected zone concept is another interesting concept for providing secure communication against eavesdropping attacks in the presence of unknown attackers [30]. In particular, a secure region is created around the legitimate node by masking the

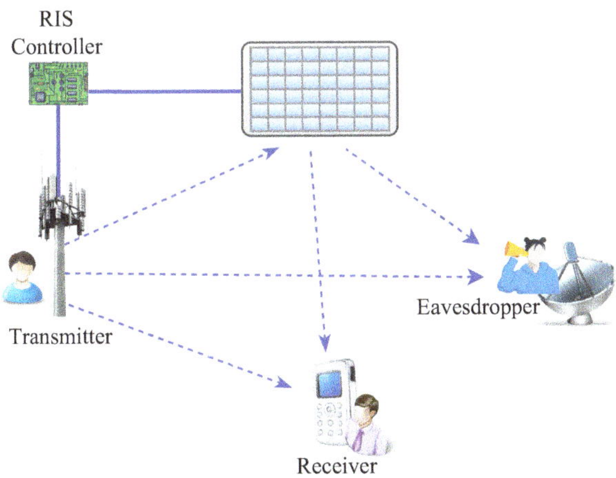

Figure 9.5 RIS-assisted security solutions for communication

signal corresponding to legitimate nodes using RIS based on location, geometry, angle of arrival (AoA), angle of departure (AoD), and the required level of security. Finally, one of the interesting advantages of RIS is its ability to provide secure communication in cases where conventional PLS techniques fail. For example, when the legitimate node and attacker are in the same direction and have a correlated channel, most of the PLS techniques such as directional modulation, beamforming, and other directional-based security will be ineffective. However, RIS can enable secure communication even in such challenging situations by providing an alternate path between the transmitter and legitimate receiver. Alternatively, RIS can be used in conjunction with artificial noise-based security techniques to redirect these signals towards the attacker to enhance the overall security.

In [29], a survey of RIS-assisted PLS approaches is presented along with various system models and scenarios such as single antenna and multi-antenna systems, single and multi-user systems, single or/and multiple attackers cases) along with different methodologies and optimization problems for enhancing overall security. Some of the important works are discussed in the aforementioned discussion. In [31], the authors employ joint active/passive beamforming at the transmitter and RIS to protect the communication. It is shown that such a system can provide security in an LoS even if the Eve channel is better/stronger compared to a legitimate node's channel. The authors in [31] discussed joint beamforming for RIS in terahertz (THz) and mmWave bands to enable secure communication. The effect of artificial noise on overall security in the presence of multiple eavesdroppers is presented in [32]. In [33], secure communication is provided using RIS in a device-to-device (D2D) scenario, where there is no direct path between the users. Finally, the impact/effect of RIS on average secrecy capacity and secrecy outage in a vehicular scenario is presented in [34] with a different configuration for RIS.

RIS can also enable covert communications by hiding the ongoing communication of legitimate nodes from the illegitimate user. For example, in [35], a case is considered where an eavesdropper aims to listen to the communication while a warder intends to detect the communication. In such system models, secrecy capacity needs to be maximized at the legitimate receiver while at the same time, received power at the warden needs to be minimized. In [36], the authors used adversarial perturbations to degrade the detection of deep neural network (DNN)-powered illegitimate receivers without affecting the performance of the legitimate receiver. Another interesting work is proposed in [37], where a RIS-assisted transmitter is proposed in which communication symbols are embedded in the radar waveform in a covert manner. In [38], a hybrid RIS/relay is presented, where a joint reflection coefficient/relay selection problem is formulated to ensure covertness from a warden while ensuring reception of the transmission at the legitimate receiver.

9.3.2 RIS-assisted solutions against jamming

The manipulative capability of RIS in terms of controlling the amplitude and/or phase of incoming waves can be used to weaken the effect of jamming signals at the legitimate nodes. For example, as shown in Figure 9.6, Bob will receive a signal from the legitimate node, a jammer, and a reflected version of the signal from RIS.

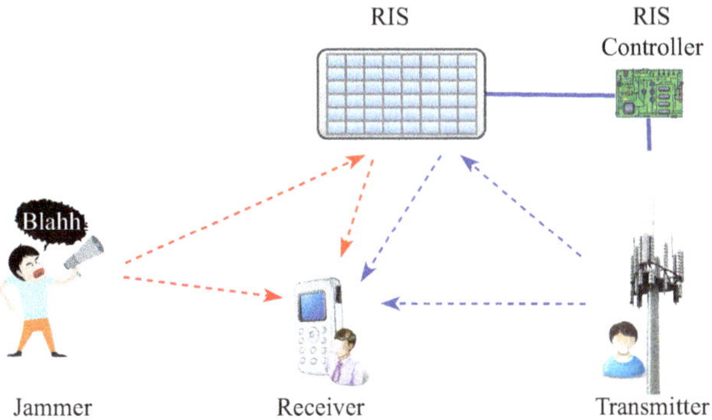

Figure 9.6 RIS-assisted anti-jamming solutions

Here, the RIS modifies the phases of its elements in such a way that the original jamming signal and reflected jamming from RIS are added destructively at Bob. Moreover, RIS can also provide an alternative path for legitimate transmission to tackle jamming attacks. For example, in [39], the authors employ joint optimization of power allocation and beamforming against a multi-antenna jammer to ensure quality of service (QoS) requirements of the different users. In particular, reinforcement learning-based joint optimization of reflecting beamforming and anti-jamming power allocation is employed.

9.3.3 RIS-assisted attacks against PLS

Besides introducing many novel properties with RIS to enhance overall system performance, including capacity, security, and interference management, RIS can also be exploited by the attacker to degrade the performance of PLS algorithms [40]. In particular, the RIS-involved channel is the superposition of the direct link and the RIS-induced link, where an attacker can destroy the reciprocity property between communicating nodes using RIS. This attack is challenging to detect because RIS uses the transmitted signal of the legitimate node for attacking, thus not leaving any energy fingerprint. Moreover, RIS can also be used to absorb signals, attenuate their strength, and generate fake multipaths, which leads to misrepresentation of the environment. Another possible attack is similar to a predictable channel attack, where different manipulations by the RIS result in predictable channel variations at legitimate nodes. This affects the performance of the channel-based PLS algorithms.

9.3.4 Challenges, recommendations, and future research directions

As explained earlier, RIS has great potential in terms of enhancing the security of wireless communication. However, there are a lot of challenges that need to be addressed before its full potential can be realized. The details are as follows:

- *Effect of CSI availability and imperfect CSI:* Most of the RIS-assisted PLS techniques are based on the assumption of perfect availability of CSI at the transmitter. However, practically, only imperfect CSI is available. Additionally, it is also assumed that the CSI of the attacker is available while designing different RIS-assisted PLS techniques. However, the attacker's CSI is available only if it is an active node or it is one of the licensed users. Consequently, both of the aforementioned issues related to CSI need to be considered while designing different RIS-assisted PLS techniques.

- *Flying RIS systems for PLS enhancement:* Recently, UAVs equipped with a RIS have received too much attention due to their ability for adaptive altitude, changeable location and direction, easy deployment, power-efficient beamforming, and so on. These capabilities can be exploited for enhancing PLS by adapting the transmission based on the location, channel conditions, and requirements of the legitimate node [41,42]. Additionally, RISs-equipped UAVs can also be used as a mobile-friendly jammer with ground TPs and active UAVs to improve overall security performance.
- *Effect of RIS physical design and deployment:* The performance of RIS-assisted PLS algorithms also depend on the physical design and deployment of RISs which need to be considered while designing different algorithms. The physical design includes (but is not limited to) the number of RISs, their distribution, size, geometric shape, and orientation. Moreover, it also includes the shape, size, and distribution of individual elements in each RIS. The optimal way to deploy, adjust, and associate different RIS for designing efficient PLS algorithms is still an open issue. Possible solutions can be stochastic geometry-based and machine learning (ML)-based solutions for such deployment.
- *Effect of non-continuous phase shifter:* It should be noted that although continuous phase shifts and unit modulus is assumed for RIS-reflecting elements. However, in practice, only RIS reflecting elements can offer discrete phase shifts and non-unit modulus, which need to be considered while designing RIS-assisted PLS techniques.
- *Optimization problems to enhance PLS:* As explained earlier, there are a lot of benefits of RIS in terms of enhancing PLS. However, RIS-based security systems are complex in terms of design and complexity as compared to conventional systems [43]. Data-driven tools such as deep learning and machine learning (ML) can provide solutions to reduce the complexity of such systems and support the self-optimizability and flexibility of such systems.
- *PLS in LoS environments:* Providing PLS in case of LoS transmission scenarios is quite challenging and some of PLS techniques, such as conventional beamforming, artificial noise (AN)-based MIMO techniques, etc., [44] fail to provide secure communication in such cases. In such a case, RIS can still enable secure communication by providing additional paths between legitimate nodes [45].

9.4 Conclusion

This chapter provides an overview of how the different nodes in the network can be leveraged to improve the security of wireless transmissions, primarily using macro-diversity. The diversity is used to improve the legitimate link's quality while degrading that of the illegitimate one. In general, spatially distributed techniques (particularly related to signal modification) are useful in mitigating eavesdropping and jamming attacks, with limited works addressing spoofing attacks. An interesting point to note here is that even though the integration of the different types of nodes (relays, TPs, and RISs) was not explicitly discussed, it could very well be required in the next-generation networks.

References

[1] Nosratinia A, Hunter TE, Hedayat A. Cooperative communication in wireless networks. *IEEE Communications Magazine*. 2004;42(10):74–80.

[2] Cui S, Goldsmith AJ, Bahai A. Energy-efficiency of MIMO and cooperative MIMO techniques in sensor networks. *IEEE Journal on Selected Areas in Communications*. 2004;22(6):1089–1098.

[3] Chakrabarti A, Sabharwal A, Aazhang B. Cooperative Communications. In: Fitzek FHP, Katz MD, editors. *Cooperation in Wireless Networks: Principles and Applications – Real Egoistic Behavior is to Cooperate!* New York, NY: Springer; 2006. p. 29–68.

[4] Jameel F, Wyne S, Kaddoum G, *et al.* A comprehensive survey on cooperative relaying and jamming strategies for physical layer security. *IEEE Communications Surveys & Tutorials*. 2018;21(3):2734–2771.

[5] He X, Yener A. End-to-end secure multi-hop communication with untrusted relays. *IEEE Transactions on Wireless Communications*. 2012;12(1):1–11.

[6] Jeong C, Kim IM, Kim DI. Joint secure beamforming design at the source and the relay for an amplify-and-forward MIMO untrusted relay system. *IEEE Transactions on Signal Processing*. 2011;60(1):310–325.

[7] Kaliszan M, Mohammadi J, Stańczak S. Cross-layer security in two-hop wireless Gaussian relay network with untrusted relays. In: *International Conference on Communications* (ICC). New York, NY: IEEE; 2013. p. 2199–2204.

[8] Hui M, Piming M. Convex analysis based beamforming of decode-and-forward cooperation for improving wireless physical layer security. In: *14th International Conference on Advanced Communication Technology* (ICACT). New York, NY: IEEE; 2012. p. 754–758.

[9] Aman W, Sidhu GAS, Jabeen T, *et al.* Enhancing physical layer security in dual-hop multiuser transmission. In: *Wireless Communications and Networking Conference* (WCNC). New York, NY: IEEE; 2016. p. 1–6.

[10] Zou Y, Wang X, Shen W, *et al.* Security versus reliability analysis of opportunistic relaying. *IEEE Transactions on Vehicular Technology*. 2013;63(6):2653–2661.

[11] Zou Y, Wang X, Shen W. Optimal relay selection for physical-layer security in cooperative wireless networks. *IEEE Journal on Selected Areas in Communications*. 2013;31(10):2099–2111.

[12] Zhao R, Tan X, Chen DH, *et al.* Secrecy performance of untrusted relay systems with a full-duplex jamming destination. *IEEE Transactions on Vehicular Technology*. 2018;67(12):11511–11524.
[13] Solaija MSJ, Furqan HM, Ankaralı ZE, *et al.* Cyclic prefix (CP) jamming against eavesdropping relays in OFDM systems. In: *Wireless Communications and Networking Conference* (WCNC). New York, NY: IEEE; 2022. p. 1976–1980.
[14] Zheng G, Jorswieck EA, Ottersten B. Cooperative communications against jamming with half-duplex and full-duplex relaying. In: *77th Vehicular Technology Conference* (VTC Spring). New York, NY: IEEE; 2013. p. 1–5.
[15] Yu J, Gong Y, Fang J, *et al.* Let's work together: cooperative beamforming for UAV anti-jamming in space-air-ground networks. *IEEE Internet of Things Journal*. 2022;9(17):15607–15617.
[16] Peng J, Zhang Z, Wu Q, *et al.* Anti-jamming communications in UAV swarms: a reinforcement learning approach. *IEEE Access*. 2019;7:180532–180543.
[17] Solaija MSJ, Salman H, Arslan H. Towards a unified framework for physical layer security in 5G and beyond networks. *IEEE Open Journal of Vehicular Technology*. 2022;3:321–343.
[18] Senigagliesi L, Baldi M, Gambi E. Authentication at the physical layer with cooperative communications and machine learning. In: *Joint European Conference on Networks and Communications & 6G Summit* (EuCNC/6G Summit). New York, NY: IEEE; 2022. p. 71–76.
[19] Liu J, Wang X, Tang H. Physical layer authentication enhancement using maximum SNR ratio based cooperative AF relaying. *Wireless Communications and Mobile Computing*. 2017;2017:1–16.
[20] Wang HM, Xia XG. Enhancing wireless secrecy via cooperation: signal design and optimization. *IEEE Communications Magazine*. 2015;53(12):47–53.
[21] Wang C, Wang Z. Signal alignment for secure underwater coordinated multipoint transmissions. *IEEE Transactions on Signal Processing*. 2016;64(23):6360–6374.
[22] Yusuf M, Arslan H. Secure multi-user transmission using CoMP directional modulation. In: *82nd Vehicular Technology Conference* (VTC2015-Fall). New York, NY: IEEE; 2015. p. 1–2.
[23] Hafez M, Yusuf M, Khattab T, *et al.* Secure spatial multiple access using directional modulation. *IEEE Transactions on Wireless Communications*. 2017;17(1):563–573.
[24] Xiao L, Dai H, Ning P. Jamming-resistant collaborative broadcast using uncoordinated frequency hopping. *IEEE Transactions on Information Forensics and Security*. 2011;7(1):297–309.
[25] Qamar F, Dimyati KB, Hindia MN, *et al.* A comprehensive review on coordinated multi-point operation for LTE-A. *Computer Networks*. 2017;123:19–37.
[26] Kotzsch V, Fettweis G. On synchronization requirements and performance limitations for CoMP systems in large cells. In: *8th International Workshop on Multi-Carrier Systems & Solutions*. New York, NY: IEEE; 2011. p. 1–5.

[27] Guidolin F, Badia L, Zorzi M. A distributed clustering algorithm for coordinated multipoint in LTE networks. *IEEE Wireless Communications Letters*. 2014;3(5):517–520.

[28] Cicalo S, Tralli V, Perez-Neira AI. Centralized vs distributed resource allocation in multi-cell OFDMA systems. In: *73rd Vehicular Technology Conference* (VTC Spring). New York, NY: IEEE; 2011. p. 1–6.

[29] Almohamad A, Tahir AM, Al-Kababji A, *et al.* Smart and secure wireless communications via reflecting intelligent surfaces: a short survey. *IEEE Open Journal of the Communications Society*. 2020;1:1442–1456.

[30] Wu Q, Mei W, Zhang R. Safeguarding wireless network with UAVs: a physical layer security perspective. *IEEE Wireless Communications*. 2019;26(5):12–18.

[31] Qiao J, Alouini MS. Secure transmission for intelligent reflecting surface-assisted mmWave and terahertz systems. *IEEE Wireless Communications Letters*. 2020;9(10):1743–1747.

[32] Guan X, Wu Q, Zhang R. Intelligent reflecting surface assisted secrecy communication: is artificial noise helpful or not? *IEEE Wireless Communications Letters*. 2020;9(6):778–782.

[33] Khoshafa MH, Ngatched TM, Ahmed MH. Reconfigurable intelligent surfaces-aided physical layer security enhancement in D2D underlay communications. *IEEE Communications Letters*. 2020;25(5):1443–1447.

[34] Makarfi AU, Rabie KM, Kaiwartya O, *et al.* Reconfigurable intelligent surfaces-enabled vehicular networks: a physical layer security perspective. arXiv preprint arXiv:200411288. 2020;p. 1–10.

[35] Altun U, Basar E. RIS enabled secure communication with covert constraint. In: *55th Asilomar Conference on Signals, Systems, and Computers*. New York, NY: IEEE; 2021. p. 685–689.

[36] Kim B, Erpek T, Sagduyu YE, *et al.* Covert communications via adversarial machine learning and reconfigurable intelligent surfaces. In: *Wireless Communications and Networking Conference* (WCNC). New York, NY: IEEE; 2022. p. 411–416.

[37] Du H, Kang J, Niyato D, *et al.* Reconfigurable Intelligent Surface-Aided Joint Radar and Covert Communications: Fundamentals, Optimization, and Challenges. arXiv preprint arXiv:220302704. 2022; p. 1–8.

[38] Hu J, Shi X, Yan S, *et al.* Hybrid Relay-Reflecting Intelligent Surface-Aided Covert Communications. arXiv preprint arXiv:220312223. 2022;p. 1–5.

[39] Yang H, Xiong Z, Zhao J, *et al.* Intelligent reflecting surface assisted anti-jamming communications: a fast reinforcement learning approach. *IEEE Transactions on Wireless Communications*. 2020;20(3):1963–1974.

[40] Li G, Hu L, Staat P, *et al.* Reconfigurable intelligent surface for physical layer key generation: constructive or destructive? *IEEE Wireless Communications*. 2022;29(4):1–12.

[41] Long H, Chen M, Yang Z, *et al.* Joint trajectory and passive beamforming design for secure UAV networks with RIS. In: *Globecom Workshops* (GC Workshops). New York, NY: IEEE; 2020. p. 1–6.

[42] Zhang Q, Saad W, Bennis M. Reflections in the sky: millimeter wave communication with UAV-carried intelligent reflectors. In: *Global Communications Conference* (GLOBECOM). New York, NY: IEEE; 2019. p. 1–6.
[43] Wang D, Bai B, Zhao W, *et al.* A survey of optimization approaches for wireless physical layer security. *IEEE Communications Surveys & Tutorials*. 2018;21(2):1878–1911.
[44] Hamamreh JM, Furqan HM, Arslan H. Classifications and applications of physical layer security techniques for confidentiality: a comprehensive survey. *IEEE Communications Surveys I& Tutorials*. 2019;21(2):1773–1828.
[45] Cui M, Zhang G, Zhang R. Secure wireless communication via intelligent reflecting surface. *IEEE Wireless Communications Letters*. 2019;8(5):1410–1414.

Chapter 10

Physical layer security for Internet of Things networks

Muhammet Kırık[1], Nusaibah A. Abusanad[1], and Hüseyin Arslan[1]

In this chapter, the physical layer security (PLS) is interrogated for Internet of things (IoT) networks. As a result of their nature, IoT networks consist of a massive number of devices. These devices are produced in compact sizes and powered with small batteries rather than a grid network. Thus, they are able to have high portability but have low computational capability. Because of these features of IoT devices, the conventional encryption/decryption-based cryptographic security methods used in the classical wireless devices fail to satisfy the security requirements of IoT networks. The main reason for this is that the requirements of these security methods are too demanding for the processing power of IoT devices and stringent to manage when these devices are deployed in a massive number. Therefore, in this chapter, the possible advantages of the PLS in IoT networks are explained. In this manner, the IoT architecture is explained and common attack types in IoT networks are reviewed. Then, the unique features of IoT networks from the PLS perspective are presented along with the possible challenges on the way. A literature review is conducted to propound the current PLS techniques used in IoT networks and the future directions are signified. As a result of this chapter, it is expected that PLS techniques in IoT networks are clarified to the reader and their future applications are encouraged.

10.1 Introduction

The increasing number of connected objects that have the availability to be located in every imaginable area from small apartments and offices to huge factories, agricultural fields, and even cities makes the realization of the Internet of Things (IoT) concept inevitable. Within the wide application range of the IoT, many of the daily tasks that require idle human efforts, such as heating, ventilation, and air conditioning control in smart homes, electronic health records from wearable devices, and etc., are aimed to be handled remotely with low-cost and instant reaction. Not only

[1]Department of Electrical and Electronics Engineering, Istanbul Medipol University, Turkey

in these simple tasks, IoT also has the capability to play a significant role in those disaster management cases where human lives are in danger or could be in danger if the required precautions are not taken beforehand. Such cases could be exemplified as the earthquake early warning systems that analyze the seismic waves under the earth surface to define the possible destructive events before they occur; the search and rescue operations that deploy small IoT sensors and devices to localize the victims and rescue them in the areas where human interaction is impossible or extremely dangerous such as building wrecks, high mountains, and caves; and the fire detection systems that immediately detect the fire with flame and gas detectors, warn the authorities and take immediate action with automated fire extinguishers before public servants reach to the scene of an incident.

In order to achieve these varieties of tasks, proliferation of IoT enabler sensors, devices and networks is the key factor. Considering the fact that the number of connected devices has increased from 10 billion to 50 billion between the years 2005 and 2020 and is expected to reach up to 80 billion by 2030 [1], it appears like the proliferation of the usage of IoT in every aspect of societies' lives will keep being a trend in the future as it is today. Furthermore, within the advent of the new radio (NR) fifth generation (5G) communication technologies, requirements of IoT networks are anticipated to be substantially satisfied thanks to the nature of 5G that ensures ultra-reliable low-latency communication (uRLLC), enhanced mobile broadband (eMBB), and massive machine-type communication (mMTC).

IoT aims to connect every imaginable object to each other by equipping them with transceivers, making them environment-aware by integrating sensors, and making them interactable from distance by using controllers not only with humans but also with other devices in the environment that are connected in the network. In this manner, mMTC is a key enabler to momentarily collect data from the environment and users by deploying billions of sensors, process and storing these data, and controlling the physical objects according to these obtained data in a harmony with the other surrounding IoT devices. With this enabler feature of 5G, it is anticipated that the growing IoT networks and applications may shift the scope of the currently known concept of internet to be machine to machine (M2M)-based rather than being purely centralized for humans.

The purpose of M2M communication phenomena under the IoT paradigm is to create an environment for human beings, where all the objects around them work in a harmony without requiring any additional information input provided by humans, but functioning only with the acquired information that is collected from the sensors and detectors mounted on these objects. By creating such a communication network among the physical objects, it is aimed to create a more convenient and comfortable world for humans, where their surroundings adjust their environment to the best conditions without they even realize.

However, getting used to the comfort of IoT technologies and purely depending on them by allowing the sensors to constantly collect private data and store them in the cloud, and allowing these devices to make physical changes in the environment based on these collected data may cause some serious issues in terms of security.

Vulnerabilities in the IoT network may lead the stored data to be stolen, changed, or used to make the connected devices to function in undesirable ways or not to function at all. In the individual case, these vulnerabilities might result in the user being threatened by using the stolen data or degrading the life quality by disfunctioning IoT devices around the individual. On a bigger scale, the infiltration of IoT networks by malicious attackers may cause many chaotic problems. For instance, in the smart cities, attackers might leak into the traffic control network to break the synchronization of traffic lights and lock the avenues of the city. Moreover, in e-health applications of the IoT, these malicious attacks may create some fatal consequences. For example, in those scenarios where the patient's data are reached by unauthorized third persons and altered without the knowledge of the health advisor, the applied treatment plan based on this corrupted data might cause some irreversible damage to the patient and even cause the patient's death. Thus, when all these scenarios are considered, the notion of security in IoT networks is a vital concern to deeply study.

Due to the broadcast nature of IoT networks, they can be threatened by malicious attackers with many different types of attacking techniques. In order to have full comprehension of the IoT security and be able to produce robust solutions against the security threats in IoT networks, these different types of attacks must be classified. In the following paragraphs, these security attacks are explained in detail to increase the comprehensibility of the following sections.

One of the common types of security attacks is eavesdropping. In eavesdropping, the aim of the malicious attacker is to infiltrate the communication system to reach the private data of the legitimate user. This infiltration by the eavesdropper could be operated passively or actively. In the case of passive eavesdropping, the malicious attacker captures the transmitted data of the legitimate user in silence for later analysis. However, in the case of active eavesdropping, the malicious attacker pretends like a legitimate node of the network and sends quests to the transceiver to retrieve data.

Another type of security attack in wireless networks is jamming. In jamming, the aim of the malicious attacker is to create an interference between the legitimate nodes and corrupt the integrity of the communication between the transmitter and receiver. Even though in jamming attacks the privacy of the legitimate user is not violated, they might still cause some serious issues such as denial of service (DoS), which obstructs the legitimate user to access the communication network.

Lastly, a commonly used type of network attack is spoofing. In spoofing attacks, the malicious attacker aims to take the control of the communication network between the legitimate nodes. In order to do this, the spoofer first waits for the legitimate transmitter to finish its data transmission, then starts to transmit the malicious data by pretending like it is the continuation of the legitimate signal. In these attacks, the spoofer can completely replace the transmitted data, or partially change a specific portion of it.

All these types of attacks may deeply affect the security of users both in the digital world and the physical world. Figure 10.1 illustrates some possible security attack examples against the IoT network of a smart home. As it can be seen from

Figure 10.1 Exemplification of the possible security attacks in a smart home

Figure 10.1, an attacker may reach the illumination information of the house by eavesdropping on the owner's data, and using this data to determine the time interval when the lights are off. By acquiring this data, the eavesdropper may estimate the time when the owner is asleep or not at home to break into the house for possible theft. Another example given in Figure 10.1 is a jamming attack performed against the smart lock system of the smart home. In this case, the jammer might interrupt the communication between the lock and the owner and obstruct the owner to get in or get out of the house for long hours. Lastly, Figure 10.1 shows a spoofing attack scenario. In this scenario, an alarm is set to turn off the power of a smart oven after a period of time that the owner defined. In such a situation, the spoofer may take over the control of the alarm, increase the specified time without the knowledge of the owner, and cause a possible fire in the house.

In order to avoid these security attacks, traditionally, Shannon's an encryption/decryption-based cryptographic security methods are used at the upper layers of the communication systems [2]. However, there are several reasons that the obsolete security methods of Shannon are impractical in today's conditions, especially for IoT networks. The first reason is that, within IoT technologies, the massive number of devices connected to the Internet creates a dynamic and heterogeneous network, which causes a challenge in terms of the management of the secret key generation for each device and distribution of them at different layers [3]. Second, due to the increasing computational capabilities of malicious attackers, there is a high risk that the generated secret keys are easily broken by using advanced techniques against IoT networks [4]. Lastly, since the IoT sensors and devices are produced in compact sizes to simplify their integration with the physical objects, their power, storage, and computational capabilities are not sufficient to satisfy the requirements of complex cryptography algorithms [5].

In this regard, PLS is an enviable solution to eliminate the inherent problems of cryptography-based security techniques. PLS techniques aim to provide perfect secrecy against the malicious attackers by exploiting the channel characteristics between the legitimate nodes and the random noise in the communication environment [6]. By doing so, PLS creates an availability to provide secrecy without deploying highly complex encryption/decryption-based techniques. Moreover, by exploiting the unique channel state information (CSI) between the transmitter and the legitimate receiver, PLS can offer more robust protection against the malicious attackers without using a secret key, which ensures the security between the legitimate nodes regardless of the malicious attackers' computational capabilities. Thus, the usage of PLS techniques in the current and future IoT networks is a strict requirement to create perfectly secured smart environments.

However, since the IoT concept is still in the early development phase and there are many unfilled gaps in terms of its security concerns, the usage of PLS in IoT networks is a challenging task to fully comprehend. Furthermore, the conventional PLS techniques that are usually deployed in the base station networks do not fully match with IoT networks' requirements because of their unique characteristics, and require an appropriate adaptation. Therefore, characteristics of the IoT in terms of security must be carefully classified for the active usage of PLS in these networks.

To make this classification, in this chapter, we comprehensively analyze the unique security requirements of IoT networks, present the deficiencies of the cryptography-based security techniques and the possible PLS alternatives to eliminate them, survey the commonly used PLS techniques in the literature and their possible usage scenarios in IoT networks, investigate the challenges of the PLS in IoT networks, and identify the future of the IoT security to pave the way for the future research.

The organization of this chapter is given as follows. In Section 10.2, the general architecture of IoT networks is explained. In Section 10.3, possible types of attacks that could be encountered in IoT networks are covered. In Section 10.4, the unique features of IoT from PLS perspective are investigated and challenges in front of the PLS for the IoT are identified. In Section 10.5, a literature review is conducted and the popular PLS techniques used for the IoT in the literature are reviewed. In Section 10.6, the future of PLS in IoT networks are envisioned and the possible research directions in this domain are identified to pave the way for the communication society. Lastly, in Section 10.7, the chapter is concluded.

10.2 IoT architecture

Conventionally, the structure of IoT is divided into three different layers. These layers can be listed as the perception layer, network layer, and application layer, which are shown in Figure 10.2. Each of these layers brings its own unique security requirements. Thus, the identification of the specific characteristics of these layers is crucial to apply effective PLS techniques in IoT networks.

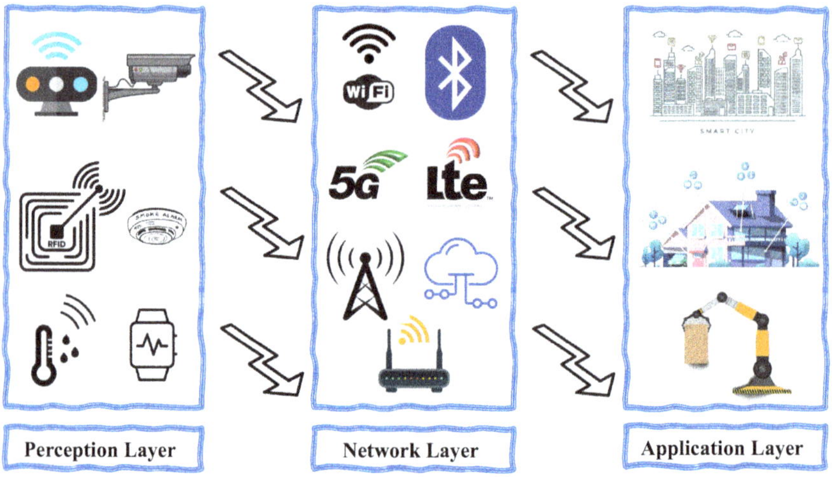

Figure 10.2 Illustration of the three different layers of IoT architecture

10.2.1 Perception layer

Also known as the physical layer, sensor layer, or acquisition layer since the environmental data is collected at this layer by deploying sensors, detectors, and actuators. In [7], the perception layer is correlated with the sensing receptors of the human body such as eyes, ears, nose, tongue, and skin. In the same way, as the human body detects its surroundings by using its sensing receptors, the perception layer of IoT uses sensors and detectors to collect the data from the environment. Then, the acquired data at this layer is transferred to the network layer.

In order to achieve the data collection at the perception layer, the most commonly used heterogeneous IoT network technologies are radio frequency identification (RFID) systems and wireless sensor network (WSN). RFID devices are aimed to be located on every object for identification purpose, which requires them to be produced in massive numbers as small and cheaply as possible. Thus, their interior hardware capabilities are not as advanced as the other wireless devices, which causes the RFID devices to be highly vulnerable to security attacks. Similar to the RFID devices, other types of IoT sensors such as cameras, voice capturing devices, and detectors serving for a variety of tasks suffer from the same issues. Because of their compact structure, they cannot handle complex cryptography-based security algorithms and even if they do, the advanced technologies of the malicious attackers put the perception layer of IoT networks in great danger. Therefore, PLS should be considered a vital part of not only the specific sensors or devices but also the whole perception layer of IoT networks.

10.2.2 Network layer

This layer is where the collected data from the sensors and detectors at the perception layer is processed and transmitted to IoT devices, and control operators through the Internet. Since the network layer is the enabling layer in the IoT architecture to provide connectivity, this layer includes the connection devices such as routers and gateways in addition to different types of connectivity technologies i.e., WiFi, Bluetooth, LTE, etc. Therefore, the security of these network devices and technologies at the network layer in terms of eavesdropping, jamming and spoofing is a crucial phenomenon. In order to avoid the security attacks from the malicious individuals and create a robust network layer in an IoT network, possible security threats must be characterized and possible solutions must be studied.

10.2.3 Application layer

At this layer, the received data from the network layer is exploited to provide the required IoT applications for the end user. The analysis and processing of the data that is acquired at the perception layer is operated at this layer to provide service. The main function of the application layer is to execute the IoT applications based on the collected data. These applications may vary from smart phones, smart watches, and smart glasses to smart homes, smart factories and smart cities. Since the application layer plays a critical role in the IoT applications and their executions in the physical world, the security concerns at this layer must be clarified and advanced solutions must be studied to eliminate these concerns.

10.3 Different attack types in IoT

10.3.1 Denial of service attacks

The main purpose of the DoS attacks is to create an inaccessibility to IoT devices by overloading the network and block the communication between them. In order to achieve this, the malicious attacker floods the network with undesired data, which creates a redundant traffic in the network. By doing so, the capabilities of the network is pushed to its limits. Thus, the legitimate device have a stringency to operate in the network and eventually, the service of the legitimate device is disrupted. DoS attacks can be performed from one source or multiple sources. The types of DoS attacks where the malicious attacker exploits different sources to flood the network are named as distributed denial of service (DDoS) attacks. Even though the DoS/DDoS attacks are not specifically performed against IoT networks, due to the inadequate capabilities of IoT devices the huge data traffic between these devices, their usage in IoT networks is more appealing for the attackers.

10.3.2 Denial of sleep attacks

IoT devices are compact devices that do not require too much power to operate. Therefore, their power supplies are generally provided by small batteries rather than being connected to the grid network. However, in order to provide the continuity of IoT devices' service, these batteries are required to be changed or recharged in certain periods. Any power shortage in these devices without completing the expected life-time period will lead IoT networks to be interrupted until the electric power is provided again. In order to prolong the life time of the battery-powered IoT devices, one method is to put these devices in sleep while their usage is unnecessary. However, the sleep cycles of IoT devices must be well protected against the malicious attackers since the subtle attacks in the network layer against the sleep time of IoT devices may cause them to lose their functionality.

In this regard, one of the popular attacks is the sleep deprivation attack. In this attack type, the malicious attacker keeps the victim unnecessarily busy by pretending like a legitimate node. The main purpose of this attack type is to avoid the IoT device to go to sleep mode until the victim is eventually shut down due to the power shortage.

Another type of denial of sleep attacks is the barrage attack. Similar to the sleep deprivation attack, the main purpose of this attack is to drain the energy source of the IoT device. However, in the barrage attacks the requests are sent to the legitimate node consecutively at much higher data rates. In comparison with the sleep deprivation attacks, barrage attacks are easier to detect since during the barrage attack the victim is active, whereas in the sleep deprivation attack, the victim is on standby.

10.3.3 Routing attacks

In routing attacks, the main purpose of the malicious attacker is to change the routing path of the legitimate data by deploying adversary nodes [8]. In order to achieve routing attacks, there are several different ways to redirect the routing path of the transmitted data from the adversary's perspective.

One type of routing attack is the sinkhole attack where the malicious attacker aims to attract the nearby nodes by emitting a false routing path information. After redirecting the routing path, the adversary obtains the ability to make changes on the data that passes through the illegitimate routing path or let only a specific portion of the data to be transmitted. As a result, the sinkhole attacks prevent the data to reach to the legitimate receiver, while the transmitter is believed that the data is reached to the legitimate receiver successfully.

Another type of routing attack is the wormhole attack, which can cause great damage in the network layer if it is combined with the other attacks such as the sinkhole attack. A wormhole in communication provides an out of band shortcut connection between the nodes, which leads the data packages to be transmitted faster in comparison with the standard nodes. In the wormhole attacks, the malicious attacker acquires the data by infiltrating the network from a specific node, channels the acquired data to another node that is situated in a distant location by using a wormhole, and retransmits the data back into the network [9]. As a result, the nodes in the network gain the perception that the wormhole node is in a nearby location albeit it is in multiple hops away. Consequently, the malicious attacker acquires supremacy against the other nodes by deploying a wormhole node, which can be exploited in variety of ways such as creating a DoS or altering the routing path of the data by performing a sinkhole attack.

10.3.4 Sybil attacks

In Sybil attacks, the adversary tries to infiltrate the communication by producing fake identities and manipulating the nodes in the IoT network [10]. Sybil attacks could be dangerous since they have the capability to spam the legitimate nodes redundantly, and violate the privacy.

10.3.5 Man in the middle attacks

In man in the middle (MITM) attacks, the malicious attacker aims to break the legitimate connection between two nodes and be located between these two as an illegitimate one. By operating this attack, the adversary guarantees that every transmitted data from one legitimate node to another passes through the malicious node. In this regard, an analogy based on a letter carrying postman from one person to another is given in [11]. If in such a scenario the postman is considered as an adversary, impacts of the MITM attacks can be expressed better. If the postman opens the letter in the middle before delivering it, he/she obtains the capability to change the context of the letter, delay the delivery time, or not even deliver it at all. Furthermore, the malicious postman can also get a new letter from the second person who received a letter from the first person based on corrupted data and deliver it back to the first user. Even though this scenario is an analogy the applicability of the MITM attacks in real-life scenarios is quite similar. If the attacker can break the link between the legitimate nodes and locate in the middle, this means that the whole communication between the two nodes is under the control of the adversary without the knowledge of the legitimate nodes.

10.4 Unique features and challenges of IoT from PLS perspective

10.4.1 Mobility

Since IoT devices are built in compact sizes and they are powered without connecting to a grid network, their movement and relocation are possible without disrupting the continuity of service. Such a capability that allows the sensors, detectors, and actuators to serve not only statically but also in movement emerges a unique subbranch of the IoT, termed as the Internet of Mobile Things (IoMT). Even though the concept of mobility in wireless communication is a familiar notion with many different applications such as cellular communication, vehicular communication, and non-terrestrial communication, the increased sensor devices, their interconnectivity, and reliance on each other within the emergence of IoT technologies has created a need for the mobility feature of IoT devices to be examined more vigorously to identify the characteristics of WSN in many different aspects [12]. Within the introduction of the mobility concept in WSN, the IoT applications have gained a whole new dimension in terms of location, connection, data acquisition, power consumption, etc. [13].

However, the mobility feature of IoT brings its own issues along with its many advantages. One of the main issues of IoT when the mobility is involved is the increased security vulnerabilities [14]. In order to boost the robustness of IoT networks against security threats in mobile applications, PLS could be a gratifying solution since the cryptography-based solutions fail to provide security in mobile IoT scenarios. Albeit the cryptography-based security techniques are not sufficient in IoT networks in general, the reason why they are not desirable especially in the mobility scenarios is due to the fact that the cryptographic key management in IoT networks is a challenging task to achieve. Because of the movement in IoT devices, the connection between the IoT nodes and devices is not easy to maintain for a long time. The connection is prone to be broken in accordance with the movement from one node, and reconstructed with another node, which causes a stringency for the IoT network to generate a secret key for each handover [15,16]. Furthermore, because of the spontaneous network disconnection and reconnection among the IoT access points in the mobility scenarios, the continuous switching between the nodes results in the authentication procedure to be repeated for each handover, which causes a latency in the authentication and puts the security of the system in a great danger [17]. However, this issue could be handled by deploying some advanced PLS techniques such as RF fingerprint, which exploits the unique characteristics of the device or the wireless channel to extract a fingerprint and will be explained in Section 10.5.3.

Another critical concern in the mobility applications of IoT networks is the battery power characteristics of the devices, which must be identified in many different aspects such as the battery duration, the next availability for recharge, and the energy requirement of the IoT application. All these concerns must be identified before letting the IoT device to move away from its charging station and after that, the communication between the device and IoT nodes must be protected against any possible attacks that might consume the power of the device unexpectedly fast such as the sleep deprivation attacks. Such a protection requires not merely the security of the data but also the security of the whole communication between the device and node. Thus, the usage of PLS is especially critical in the mobile IoT applications since it has the capability to protect the whole communication environment, whereas the cryptography-based security techniques can only provide a limited security for the transmitted data.

10.4.2 Low computational capability

Due to the targeted application scenarios of the IoT, the nature of the employed devices requires them to be in a compact size. This results in the components of the devices such as processor, storage unit, and battery to be built in smaller sizes than usual, which causes a limitation in the computational capability. In those cases where the cryptography-based techniques are considered as the main method to provide security, the limited computational capabilities of IoT devices can be a great challenge to get over because of the fact that the encryption/decryption procedures of the secret key in the cryptography-based security techniques are too overwhelming in terms of the processing power [18].

Another critical issue regarding the cryptography-based security solutions is the computational latency, which may cause the IoT network to perform abysmally in terms of the reaction time. Since the processing and the storage is limited in IoT devices, even if they can handle the encryption and decryption procedures in their hardware, the process will take too much time and the main tasks of the device will deprive the required processing power due to the fact that the processing power is directed to the secret key management. This will result in a distortion in the IoT device to react immediately based on the data that is sent from the network. When the nature of the IoT and its applications are considered it can be inferred that any delay in the IoT applications will degrade the quality of service and the end users' satisfaction.

Last but not least, in IoT networks where millions of devices serve different purposes in different environments from smart assistants, smart homes, and small healthcare services to smart factories and smart cities, the usage of the cryptography-based security techniques confronts a failure since the secret key distribution and management for massive number of components are not possible with these low-cost IoT devices [19].

In order to eliminate these problems without increasing the capabilities of IoT devices and dependently increasing their production costs, PLS could be a valid solution and it can provide security without compromising the quality of service. Since the PLS does not deploy secret keys in its security protocols, it saves IoT devices from the burden to handle complex and time consuming cryptographic procedures. This leads IoT devices to operate more resiliently without pushing their hardware limits and serve to their user without any additional latency that is caused by the encryption/decryption processes. Moreover, since the PLS algorithm exploits the channel and noise characteristics of the environment, it has the capability to identify each IoT device by exploiting their unique channel characteristics rather than assigning a secret key for each of them.

10.4.3 Uplink/downlink incompatibility in terms of hardware

Due to the application scenarios of IoT technologies, IoT devices are built as compact as possible. Therefore, IoT devices are not capable of being equipped with multiple antenna elements. On the contrary, the downlink equipment of IoT networks, namely the base stations and other types of access points, are usually composed of a legitimate fusion controller that can be equipped with multiple antennas. Hence, the degree of freedom (DoF) in the space domain, which normally comes with multiple antenna utilization cannot be available between the IoT nodes and devices. As a result of this incompatibility, the benefits of multiple-input multiple-output (MIMO) such as spatial diversity cannot be applicable on IoT networks anymore [20]. When the nature of the PLS, where the security is provided by utilizing the unique characteristics of the channel between the two nodes is considered, the incapability of exploiting the spatial diversity and obtaining a sparse channel during the communication affects the PLS to be properly applied in IoT networks.

10.4.4 Channel state information accuracy

CSI defines the properties of the channel in a wireless communication link. It is mainly categorized into two main classes perfect and imperfect CSI. While the perfect CSI provides complete information about the channel properties of the communication link, the imperfect CSI is concerned with the characterization of the statistical information such as the average channel gain and line-of-sight (LoS). Acquiring a perfect CSI is extremely hard to be achieved in some cases, especially in large-scale networks. Therefore, the channel quality of the legitimate devices cannot be perfectly predicted in some cases. Moreover, the CSI accuracy comes with signaling overhead, power consumption, and sacrifice of spectral resources, especially in the case of dense IoT deployment scenarios, which is contradictory to the inherent specifications of IoT. The reason for this is that IoT devices must transmit pilot symbols in short time intervals to acquire CSI, which causes IoT devices to consume an excessive amount of power and spectrum. Furthermore, in those cases where passive eavesdropping is conducted in the IoT network, the acquisition of the eavesdropper's CSI becomes difficult as well since the eavesdropper is external to the IoT system and remains completely silent. However, albeit all these challenges, once the accurate CSI of the IoT device is obtained, PLS is capable of achieving the perfect security in IoT networks [21,22].

10.4.5 Scalability

Scalability defines the ability of a device to adapt to the environment and meet the changing requirements at varying times and that is a primary concern in large-scale heterogeneous networks [23]. The ubiquitous nature of IoT networks allows millions of devices per square kilometer to exchange information, making scalability an important concern in the context of the PLS for IoT networks [24]. The vast majority of the current PLS schemes proposed in the literature are developed for small-scale networks. Therefore, the appropriate adaptations of PLS techniques must be developed for IoT networks to satisfy their unique requirements in terms of scalability.

10.5 Popular PLS techniques for IoT against eavesdropping, spoofing, and jamming

As it is discussed in the previous section, implementation of the PLS in IoT networks is a necessity. Therefore, there have been many studies in the literature throughout the years to satisfy the security requirements of IoT networks by applying advanced PLS techniques. In this section, a review of the literature is conducted and some of these techniques that have been implemented for IoT networks are discussed.

10.5.1 Beamforming

Beamforming is a signal processing technique used to maximize the received power of the desired user by creating a narrower beam while the interference power to other users is kept minimized. The implementation of beamforming requires the usage of directional antennas as well as the CSI at the transmitter, which increases the system complexity [25]. Since the data transmission in beamforming is conducted

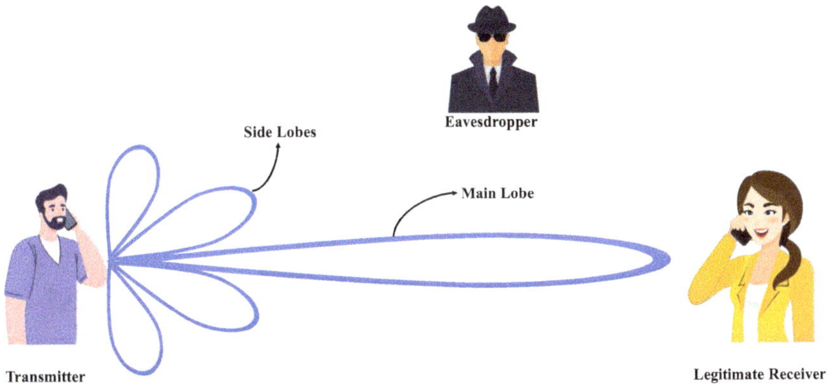

Figure 10.3 PLS scheme based on beamforming

by deploying a narrow beam, the amount of information leakage from the legitimate transmitter to illegitimate receivers is lower as shown in Figure 10.3. However, the usage of beamforming in PLS applications suffers from a lot of limitations such as the coexistence of the illegitimate receiver alongside the legitimate receiver in the same beamspace, information leakage in the side lobes, and uplink/downlink incompatibility in terms of hardware. PLS schemes that adopt beamforming showed that the secrecy rate of IoT networks can be significantly enlarged. Thus, the security performance can be improved by increasing the signal-to-noise ratio (SNR) at the fusion center. Therefore, the utilization of beamforming as a PLS technique in IoT networks must be well studied and deeply investigated.

10.5.2 Compressive sensing

Compressive sensing is a relatively new signal processing technique that aims to capture and recover a signal by solving an undetermined linear system. The signal can be captured and recovered by using fewer samples compared to the conventional sampling theorems. Compressive sensing requires two conditions to ensure successful implementation, which are sparsity and incoherence. In the former, the signal should have a sparse representation in a specific domain. In the latter, the measurement matrix should be incoherent in the signal domain [26]. In compressive sensing, a measurement matrix is used to encrypt and compress the message. Essentially, the measurement matrix generated for the compressive sampling process is used as a key for encrypting sparse messages. Transmission secrecy can be guaranteed if the measurement matrix is unknown to the eavesdropper. The compressive sensing process can be modeled as

$$C = \phi x = \phi \Psi s, \tag{10.1}$$

where ϕ represents the sensing matrix and Ψ is the sparsifying matrix. Consequently, any signal, x, can be represented as a linear combination of the element of Ψ with the vector s. where $s = [s_1, s_2, ..., s_N]^T$. So that x can be represented as

$$x = \sum_{i=1}^{N} \Psi_i s_i. \tag{10.2}$$

To implement a compressive sensing scheme effectively, measurement matrices with less coherence are strongly desired. Since the measurement matrix is used to sample only those components that perfectly represent the signal, optimizing the measurement matrix to achieve more incoherence is a necessity. It is worth mentioning that compressive sensing schemes do not require additional power. However, the measurement matrix must be shared. To provide secrecy, the measurement matrix should be randomly generated. Thus, the random measurement matrices such as Gaussian matrix, circulant matrix, and other special random matrices can be potentially applied for securing wireless communications. IoT networks can be easily integrated with compressive sensing thanks to the two unique features that IoT networks possess. First, by using the appropriate transform the sparsity can be easily explored for IoT networks since they have real data that can be represented by a sparse signal. Second, the vast majority of IoT networks exhibit sporadic transmission. In sporadic transmission, IoT devices transmit their data intermittently, meaning that not all devices transmit at the same time. Thus, at each transmission slot, only a small number of IoT devices are active. In this way, the activity of IoT devices can be observed and the sparsity can be explored [27].

IoT devices consume energy when performing those main tasks, which are data sampling, data processing, data transmission, and data reception. Using compressive sensing in IoT networks reduces energy consumption because of the lightweight encryption that compressive sensing provides during signal processing. In [28], the proposed scheme based on compressive sensing employs a learning phase where it compresses and decompresses the data at different rates and then computes the error at each rate. In other words, the sparsity level is tightly linked with reconstruction error. As a result, the proposed scheme increases the performance of IoT devices in terms of processing, memory, and energy compared to other schemes.

Optimizing the measurement matrix improves the effectiveness of the compressive sensing model, which reflects positively on the PLS schemes. Thus, a robust compressive sensing scheme can be guaranteed by optimizing the measurement matrix. Regarding that, a PLS scheme was proposed for IoT networks based on exploiting the circulant matrix to improve the efficiency of the PLS scheme by utilizing a binary resilient function and guarantee the security [29]. Another PLS scheme was proposed based on orthogonal frequency division multiplexing (OFDM), which is considered as a dominant waveform in 5G. Regarding the proposed scheme, the generated key is changed according to the changes on OFDM frames. In this scheme, the secret key management is simplified because it is generated from compressive sensing without a pre-sharing requirement [30].

10.5.3 RF fingerprinting

Radio frequency (RF) fingerprinting exploits the imperfection between the RF chains of legitimate and illegitimate devices. This imperfection occurs during the manufacturing process. Therefore, RF fingerprints are only dependent on the hardware characteristic of devices and independent of the digital modulation of the signal. Consequently, no overhead is loaded on IoT networks through RF fingerprinting-based identification process [31]. The real-time process of identifying a wireless device based on RF fingerprinting goes through three stages. The first stage is capturing the signal, the second stage is extracting the unique features of the captured signal, and the final stage is identifying and classifying RF fingerprints.

By enhancing the identification accuracy of RF fingerprint, the PLS based on RF fingerprinting can be built more robust. Thus, improving the identification accuracy of RF fingerprinting is important. To achieve this accuracy, RF fingerprints should be universal, time-invariant, environment-invariant, and collectible. Identification techniques are mainly categorized into two main categories, which are transient-based and steady-state-based identifications. In the transient-based, the identification is dependent on the switch of the transmitter that changes between turning on and off before the data transmission begins. A challenge of transient-based detection is to distinguish the channel noise and the exact position of the signal's starting point. In the steady-state-based identification, in order to identify a modulated signal, unique features of the signal need to be extracted such as frequency error, synchronized correlation, I/Q origin offset, magnitude error, and phase error [32].

Effective PLS techniques in IoT networks should be easily achieved and implemented since the low-cost IoT devices cannot handle complex methods. Thus, in [33], an inexpensive approach to implement RF fingerprinting identification was proposed using software defined radio (SDR) instead of expensive oscilloscopes. It is observed that such a scheme is more suitable for a commercial IoT deployment since the low-cost and low-complexity phenomena are crucial for IoT networks.

10.5.4 Cooperative jamming

The cooperative jamming (CJ) technique is based on broadcasting an artificial noise (AN) to block eavesdropping without affecting the legitimate receiver's performance, as shown in Figure 10.4. In IoT networks, better security can be achieved by increasing the AN level. Hence, to implement effective CJ-based security techniques, energy consumption should be taken into consideration. Depending on the AN source, the CJ schemes are classified into two categories. These categories are self-cooperative jamming and non-cooperative jamming. When self-cooperative jamming is used, AN is generated by a transceiver, while when non-cooperative jamming is used, AN is generated by a jammer [34].

According to the channel quality and its effect on power allocation for both noise signal and information signal, a scheme was proposed based on CJ to achieve a secure downlink transmission for IoT networks [35]. The proposed scheme is implemented based on the channel quality. When the channel quality is good, more power should be allocated to the noise signal and when the secrecy rate is high more power

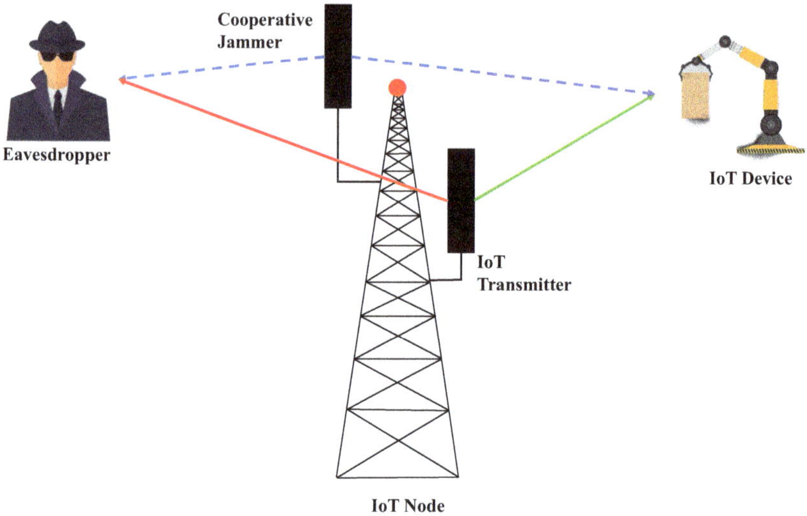

Figure 10.4 Cooperative jamming

should be allocated to the information signal. To improve the security performance, the vast majority of PLS based on CJ schemes usually employ continuous jamming signals to maximize the secrecy capacity. However, continuous jamming leads to a high energy cost for cooperative users. To overcome this problem, an intermittent jamming scheme was proposed in [36] where the jammer switches between transmission of jamming signals and being silent to guarantee the security requirement and energy conservation respectively. Accordingly, intermittent jamming is a feasible scheme to satisfy the requirements of secure performance and energy constraints. The intermittent jammer can strike a trade-off between the jamming effectiveness and energy savings by appropriately adjusting the duration of transmission and sleeping along with increasing the lifetime of devices in energy-hungry IoT networks.

Even though PLS schemes for IoT based on CJ have been successfully studied and implemented, these schemes should be extended in the future to overcome the drawbacks of existing schemes. For example, PLS schemes based on CJ cannot contribute to the enhancement of system security in the presence of a large number of eavesdroppers since PLS based on CJ assumes that the average legitimate channel is better than the wiretap channel, whereas multiple eavesdroppers may achieve a better wiretap channel than a legitimate one [37]. Moreover, the performance of CJ cancellation could be affected due to the Doppler shift, which causes the mismatch between a cooperative node and a receiver [38].

10.5.5 Spread spectrum

Spread spectrum fundamentally aims to spread narrow band signals over a wide bandwidth to reduce the effectiveness of jamming, eavesdropping, and spoofing attacks in addition to interference. Direct sequence spread spectrum (DSSS), frequency hopping spread spectrum (FHSS), and chirp spread spectrum (CSS) are some of the popular spread spectrum techniques used in the literature to provide security in IoT networks. In DSSS, the information signal is modulated by a periodic pseudo-random sequence before transmission, which creates a wideband signal with a low probability of interception. On the other hand, FHSS can be defined as a transmission technique where the carrier frequency jumps from one frequency to another, which is governed by pseudo-random sequences. In CSS, the chirp signal is used which has time-variant frequency. When the frequency increases, up-chirp is obtained and when it decreases, the down-chirp is obtained [39]. In general, CSS is used to provide data security in long-range communication systems, which makes it a suitable candidate to be used in long-range (LoRa) networks.

However, the conventional spread spectrum techniques have their restrictions. For instance, FHSS requires continuous synchronization between the transmitter and the receiver during the transmission. Additionally, it is not collision-free, which means that two nodes may use the same frequency at the same time. Hence, to avoid collision, a large bandwidth is required in FHSS. To solve this issue, a new frequency hopping scheme is proposed for IoT networks in [40]. The proposed scheme depends on sharing the encrypted hopping sequence information between legitimate devices before any transmission to secure the transmission against jamming attacks. The proposed scheme can be counted as a cost-effective scheme since it waives the need for bandpass filters. Another study proposed in [41] provides a jamming-resistant scheme for IoT networks under the presence of an untrusted relay and a jammer that has multiple antenna capability. Unlike the traditional FHSS technique, the proposed scheme exploits the local observations of the two-hop channels to generate a shared secret key.

Since the vast majority of existing pseudo-random hopping patterns tread channels with a similar condition, without considering the likelihood of a jamming attack based on the quality of the channel, in [42], a scheme called probabilistic hopping pattern where the usage probabilities of the channels are changed with the different channel qualities is proposed. In this study, channels with a lower probability of jamming attacks are utilized with a high usage probability and vice versa. Another security scheme has been proposed based on exploiting the instantaneous channel phase between the legitimate devices to impose a secret frequency shift on the transmitted frequency chirp [43]. According to the proposed scheme, the transmitted data is secured against eavesdropping as long as the channel phase is not revealed.

10.5.6 Bit flipping

Bit flipping technique is proposed in the literature to secure the communication between massive sensor nodes and legitimate fusion center (FC). In bit flipping technique, nodes are divided into two groups based on their channel quality. By doing so, the nodes with low channel quality and high channel quality are separated from each other. Regarding this classification, while the nodes with low channel quality are exploited to send the bit-flipped data, i.e., the faulty data, to interfere with the eavesdropper's FC, the nodes with high channel quality are utilized to send the information-carrying data. By applying such a technique, the SNR of the eavesdropper becomes much lower than the SNR of the legitimate user, which yields a significant performance degradation at the eavesdropper's side [44]. The main drawbacks of this technique are considered to be the high power consumption and waste of spectral resources, which is caused by the fact that the nodes with low channel quality need to transmit faulty data to confuse the eavesdropper. To overcome these drawbacks and prevent the eavesdropper attacks, a PLS technique was proposed in [45] based on clustering algorithms and bit flipping. In this study, it is proven that the proposed scheme has the capability to successfully reduce the energy consumption and enhance the spectral efficiency.

10.5.7 Noise aggregation

In order to apply advanced PLS techniques in wireless networks, the unique characteristics of the channel must be exploited. One way to exploit the channel characteristics of a network is to create an AN. By creating such noise that is specifically designed based on the legitimate nodes in a communication system, the infiltration of malicious attackers into the communication between two legitimate nodes can be avoided since the malicious attackers' channel is not matched with the intentionally produced AN. However, the usage of an AN at the physical layer causes the communication devices to consume extra power. When the nature of IoT devices is considered, it is not too difficult to infer that these devices do not have such a luxury to sacrifice an extra power to produce an AN to provide security. Therefore, the technique called noise aggregation is proposed in the literature to solve this issue.

The fundamental logic behind noise aggregation relies on the idea that forming part of the transmitted data in a way that it acts like an AN only to illegitimate receivers [46]. In order to achieve such a technique, first, the data is separated into packages and each package is assigned a number. After that, the odd-numbered packages are directly transmitted toward the receiver, whereas the even-numbered packages are XORed with the decoded versions of the odd-numbered packages before they are transmitted. This procedure is illustrated in Figure 10.5. Since the channels of the legitimate receiver and eavesdropper have their own unique characteristics, it is obvious that the decoded version of the odd-numbered packages is going to be different at the receiving ends of the legitimate receiver and the eavesdropper. This difference results in the even-numbered packages being difficult to be resolved at the eavesdropper's receiving end because these packages are affected by the decoded versions of the legitimate receiver's odd-numbered packages, which the eavesdropper cannot estimate. Consequently, the security between the transmitter and the legitimate receiver is provided by exploiting the natural channel characteristics of the legitimate receiver as an AN against the eavesdropper.

10.5.8 Fountain coding

Fountain codes are rateless codes that are exploited for data transmission over erasure channels. A fountain encoder has the capability to generate an infinite number

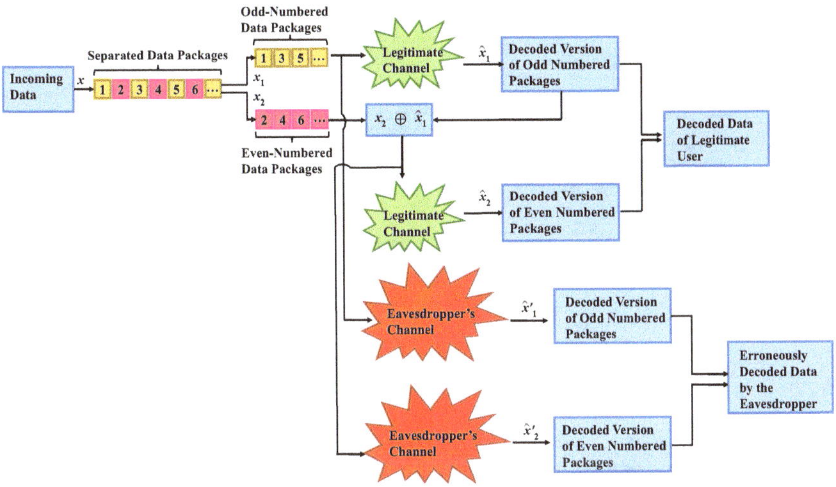

Figure 10.5 Illustration of noise aggregation

of encoded packages from a data sequence. This capability allows the encoded packages to be transmitted through erasure channels over and over again until the decoder at the reception side recovers all the transmitted data. In order to establish a communication between two nodes over an erasure channel by using fountain coding, first, the source file to be transmitted is divided into K small groups of equal length, which are termed the information packages. Then, the transmitter encodes the information packages to produce a potentially limitless number of fountain code packages and transmits the coded packages toward the receiver. At this point, the transmission order of each code package is randomly decided by being chosen among all the possible packages. As the decoding of the packages is completed at the reception side, new code packages are transmitted by the transmitter until the receiver sends a feedback signal to the transmitter that dictates the reconstruction of all the code packages is completed and it can stop producing new code packages [47].

Such a technique that has the ability to transmit data randomly in small portions can be exploited in security applications of communication networks. Thanks to the low encoding and decoding complexity of fountain codes, their usage in IoT networks becomes quite appealing to satisfy the unique requirements of IoT devices. In order to provide security between the transmitter and the legitimate receiver by using fountain codes, the required condition is to deliver all the code packages to the legitimate receiver before the eavesdropper obtains them. In such a security application, since the whole data is transmitted in pieces, even though the eavesdropper obtains some of the code packages, if the legitimate receiver obtains all of the code packages before the eavesdropper does, it can send a feedback signal to the transmitter to stop the transmission and disrupt the eavesdropping of the adversary in the middle.

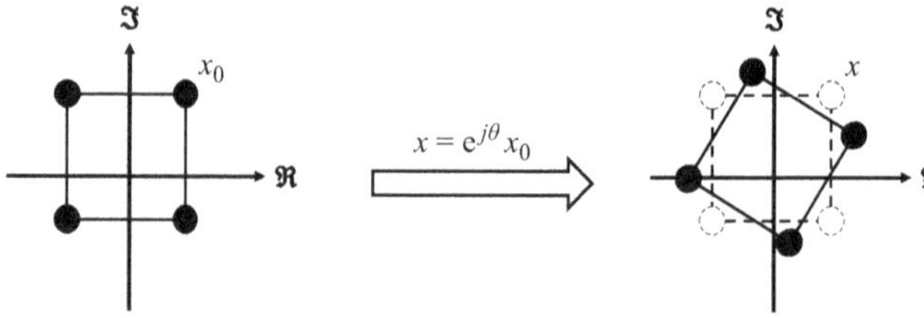

Figure 10.6 Constellation rotation

10.5.9 Constellation rotation

The key idea behind the constellation rotation is to apply an appropriate rotation angle to a complex constellation plane prior to data transmission as shown in Figure 10.6. In general, the constellation rotation technique is exploited in device-to-device (D2D) communication, which makes it a great candidate to be deployed in IoT networks. In D2D communication, while one device is assigned to use the real component, another device is assigned to use the imaginary component of the constellation [5] to exploit the advantages of constellation rotation. By following such a strategy, the orthogonality between the two dimensions of the constellation plane is exploited to avoid the data of two different devices to leak into each other.

In the literature, it is possible to encounter the different variations of constellation rotation techniques that are proposed to improve the security level of the communication systems at the physical layer. For example, in [48], the constellation rotation technique is applied by assigning one dimension of the signal for data transmission and assigning the other dimension of the signal for sending an AN. However, this technique comes with two main disadvantages. These disadvantages are the low power efficiency, which is caused by the usage of AN, and the reduced constellation size. Another PLS technique that exploits constellation rotation is given in [49]. In this work, the proposed PLS scheme exploits the random, independent, and unique channel characteristics between two communicators to guarantee the extraction of an uncorrelated key. In this regard, while the modulation order used at each node is decided based on the magnitude of the channel, the corresponding constellation diagram is rotated by an angle equal to the phase of the estimated channel. By following such a strategy, the proposed scheme shows immunity against intelligent eavesdroppers and robustness against channel estimation errors.

10.5.10 Machine learning

The rapid increase of the number of IoT devices that are connected to networks leads to huge data traffic to occur. Because of the large amount of real-time generated data, its processing, computation, and security is becoming a challenging task in the IoT environments. In this regard, machine learning (ML) is an emerging technology to reduce communication overhead, and signal processing complexity, and enhance security. ML tools are exploited to enhance the accuracy of RF fingerprint identification for IoT devices to improve PLS efficiency against different attacks.

In this regard, a physical layer continuous authentication and spoofing detection scheme based on RF fingerprinting for an actual wireless sensor network based on two ML approaches was proposed in [50]. Results show that the proposed algorithm can achieve a high level of accuracy in the classification that corresponds to the correct identification of the legitimate user. Another study considers deep neural network (DNN), convolutional neural network (CNN), and recurrent neural network (RNN) to identify RF fingerprints for IoT devices and distinguish among RF fingerprints for other IoT devices from the same manufacture to improve IoT security [51]. In [52], another RF fingerprint identification method that combined the differential constellation trace figure (DCTF) feature extraction, and the CNN-based classification without requiring any synchronization and compensation. An intrusion detection proposed to protect smart devices in a home environment by leveraging machine learning to detect illegitimate activities [53].

In [54], a proposed scheme used a smart home setting as a representative case study of the IoT. Regarding the proposed scheme, device classification methods proposed by applying ML algorithms on the data stored in the blockchain network which in turn helps to enhance the security of IoT network by detecting illegitimate devices. To resist spoofing attacks in industrial wireless sensor networks deep learning (DL)-based PLS was proposed in [55]. The proposed method can enhance the security of the industrial wireless network without sacrificing communication resources. Three different algorithms are adopted to implement the proposed scheme including the DNN-based sensor nodes' authentication method, the CNN-based sensor nodes' authentication method, and the conventional prepossessing neural network (CPNN)-based sensor nodes' authentication method.

Even though ML techniques have been used to enhance the accuracy of RF fingerprints for IoT devices as discussed before, further investigations to exploit ML to acquire a perfect CSI are needed nowadays. The existing channel estimation schemes that are based on channel modeling are insufficient for obtaining perfect and timely CSI according to that the PLS schemes based on these channel estimation models will not effectively secure the communication. The use of ML techniques in channel estimation has been proposed as a remedy. Hence, the performance of existing channel estimation techniques can be improved, with low complexity in practical applications.

10.5.11 Reconfigurable intelligent surfaces

As mentioned before, because of the compact sizes of IoT devices, it is unrealistic to propose attaching multiple antennas on the user side to exploit the benefits of MIMO. Therefore, the concept of reconfigurable intelligent surface (RIS) is proposed in the literature as a remedy to compensate for the effects of MIMO incapability in IoT networks. RIS can be defined as a large-scale array composed of a large number of passive low-cost elements, which can reflect or transmit the incident electromagnetic waves to the desired directions by appropriately adjusting their phase shifts [56].

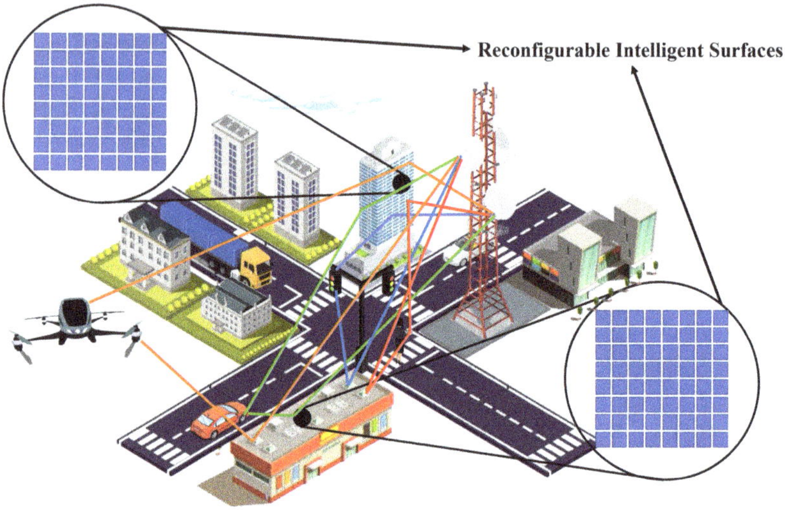

Figure 10.7 Reconfigurable intelligent surfaces in IoT networks

This process is illustrated in Figure 10.7. Even though RIS can provide many other advantages in wireless networks, in this section, the authors only considered its merits from the security perspective in IoT networks. If the possible use cases of RIS are considered from the security perspective, it can be inferred that the ability to control the channel by using RIS is an important enabler to develop advanced PLS techniques. Since the channel of the transmitted signal can be easily changed by adjusting the phase angle of RIS, in those cases where malicious attackers obtain the CSI of the transmitted data, the effective channel can be easily changed by switching the phase angle of RIS and the security at the physical layer can be provided.

In this regard, a PLS scheme based on RIS called user-specific RIS was proposed in [57] as a remedy for space-limited characteristics of compact devices. The proposed RIS scheme can partially control the signal amplitude whereas the existing RIS schemes can adjust only the phase of a signal. The proposed scheme provides an additional DoF for RIS beamforming design, which is an important notion in IoT networks since their capabilities of beamforming is highly limited.

However, although the possible incomes of RIS are appearing to be quite promising from the security perspective and therefore the research based on the RIS is a hot trend nowadays, RIS in wireless networks is relatively a new area to discover and its practical implementations are still not at an advanced level. For example, in today's conditions, it is only possible to practically realize the discrete phase shifts in RIS. However, the academic approach of PLS based on RIS in IoT networks assumes that the RIS phase shifts can be controlled continuously. This is only one example of RIS that requires more research and there are many more shady areas like this to light up. Thus, the usage of RIS to provide PLS requires a vigorous effort to explore its possible security applications in future IoT networks.

10.6 Recommendation and future directions

Since the vast majority of existing PLS schemes are not specifically designed for IoT networks, PLS is still in an early phase to be effectively used in this area. Therefore, the need to develop and implement novel PLS schemes to meet the unique features of IoT devices is crucial. In this regard, this section discusses the future directions of the PLS in IoT networks along with the authors' recommendations.

10.6.1 Multi-antenna systems in IoT devices

The unique features of 5G technologies, the growing internet applications, and the unstoppable data traffic among the connected devices along with the proliferation of IoT technologies have pushed the researchers to discover the unexplored parts of the frequency spectrum to eliminate the spectrum scarcity in the wireless networks. Throughout the passing years, these explorations at the higher portions of the frequency spectrum showed that the usage of high frequency bands for the data transmission brings many advantages to building more efficient communication networks.

One of the main advantages of the high frequency band signals is their low wavelength. As it is known, the frequency and wavelength of a signal are in an inverse proportion, which means that as the frequency of a signal increases, the wavelength of this signal decreases accordingly. Therefore, in the high frequency bands, such as the millimeter wave band and terahertz band, the wavelength of the signal is in millimeter or less. This allows the transceiver structures of the high frequency band devices to be modified in a more comfortable way from the hardware perspective. Since the wavelength of the transmitted signal is lower, the electronic devices that are used for communication can be equipped with more antennas by decreasing the spacing between each antenna. Moreover, the usage of higher frequencies along with the novel developments in the material science allows these antennas to be built in more compact sizes.

As it is mentioned earlier, since the application requirements of IoT networks require small size devices, the devices deployed in these networks do not provide too much freedom to their producers in terms of the hardware capabilities. However, since within the exploration of higher frequency bands, the antenna sizes and the spacing between these antennas are decreased, their installment on IoT devices can be much easier. By allowing IoT devices to be equipped with more antennas the barrier in front of the MIMO capabilities of IoT networks can be removed.

By removing these barriers in IoT networks, the applications of the PLS algorithms can be performed more easily. The reason behind this is the fact that the applicability of PLS techniques depends on the unique characteristics of the channel between IoT devices and nodes. Since the multiple antennas on both IoT devices and nodes allow the MIMO to provide a rich scattering environment, the diversity of the channel can be exploited to build a more robust PLS protection. Therefore, the usage of higher frequency bands is a considerable enabler to improve the applicability of PLS techniques in future IoT networks.

10.6.2 Energy harvesting

Energy harvesting is considered as a possible remedy for increasing the lifetime of the battery-powered IoT devices. In this concept, the required power of IoT devices is supplied from environmental resources such as solar, thermal, and RF resources. The aim of energy harvesting in IoT networks is to prolong the battery life of the devices until the next charge time or provide a sustainable and consistent energy source for IoT devices and push the batteries outside of the picture. By achieving such a powering technique, it is obvious that the security attacks that focus on the destruction of the sleep cycle of IoT devices can be eliminated at the physical layer because in such a case, these IoT devices no longer need to sleep to prolong their battery lives since they already have unlimited power supplied from the environment.

However, there are many concerns about the usage of energy harvesting techniques in IoT networks. One of the main concerns is the continuity of the energy source. In an IoT device that is fully dependent on the environmental energy sources, the absence of a battery may cause the device to be suddenly shut down when the continuity of the energy source is corrupted. For example, an IoT device that is equipped with solar energy harvesters may not function when the device is moved to a shadowy area. The other drawback of the usage of energy harvesting techniques in IoT networks is their occupation areas. Since IoT devices are produced in compact sizes, energy harvesters located in those devices must be designed wisely. To exemplify, if energy harvesters are built too big, IoT devices may lose their mobility. On the other hand, if energy harvesters are built too small, then the harvested power may not be enough to supply the required power to IoT devices to perform. Consequently, even though energy harvesting is a promising technique to provide longer lifetimes in IoT devices, which can be exploited to increase the security of IoT networks at the physical layer, it is an open area to investigate and bring about critical enhancements. However, once these enhancements are achieved, energy harvesting can be an enviable candidate to be used in IoT networks as a PLS solution.

10.7 Conclusion

Within the growing 5G technologies, the proliferation of IoT networks has accelerated more than ever. Due to this fact, the number of connected IoT devices has reached billions and is expected to grow even more in the future. This uncontrolled growth causes some concerns in the wireless communication society in terms of the security of these devices. The main reason for these concerns is originated from the fact that the nature of IoT networks is not similar to the conventional wireless networks, and the conventional security techniques fail to satisfy the unique requirements of IoT security. Since IoT devices are desired to be deployed in every possible object around the humanity, their sizes should be small, so that they can be mounted on the objects easily. This causes IoT devices to be equipped with low-end hardware that cannot support high processing operations. In addition to this, because of the mobility feature of IoT devices, their power requirements are provided by batteries rather than continuous grid power, which requires the batteries to be changed or recharged in certain time periods. Consequently, the usage of conventional security techniques may cause IoT devices not to function properly since their high computational requirements do not match with the hardware capabilities of IoT devices. This causes these devices to be overwhelmed by the security processing and not perform their actual task. Moreover, since the processing of the conventional security techniques puts an extra burden on the power usage, batteries of IoT devices can die earlier than the expected time and shut down the device without the knowledge of the user. Last but not least, even though conventional security techniques can provide a certain amount of security, within the advanced hacking techniques developed by malicious attackers, the usage of conventional security techniques becomes less desirable because of their high vulnerabilities. Therefore, current and future IoT networks require more advanced and robust security techniques to earn the trust of the users and be used without hesitation. In this regard, this chapter proposes the usage of PLS in IoT networks. Since PLS techniques exploit the unique characteristics of the wireless channel and noise between the devices and nodes, the security of IoT networks can be provided without increasing the burden of low-capacity IoT devices. Even though PLS has its own kind of challenges to be used in IoT networks, it is the authors' humble opinion that its usage in future IoT networks is a necessity to provide a better level of security.

Acknowledgment

This work was supported by the Scientific and Technological Research Council of Turkey (TUBITAK) under Grant 5200107, with the cooperation of Turkcell Technology and Istanbul Medipol University.

References

[1] Chettri L, Bera R. A comprehensive survey on Internet of Things (IoT) toward 5G wireless systems. *IEEE Internet of Things Journal*. 2019;7(1):16–32.

[2] Hamamreh JM, Guvenkaya E, Baykas T, *et al.* A practical physical-layer security method for precoded OSTBC-based systems. In: *2016 IEEE Wireless Communications and Networking Conference*. New York, NY: IEEE; 2016. p. 1–6.

[3] Hamamreh JM, Basar E, Arslan H. OFDM-subcarrier index selection for enhancing security and reliability of 5G URLLC services. *IEEE Access*. 2017;5:25863–25875.

[4] Zhang Y, Shen Y, Wang H, *et al.* On secure wireless communications for IoT under eavesdropper collusion. *IEEE Transactions on Automation Science and Engineering*. 2015;13(3):1281–1293.

[5] Sun L, Du Q. Physical layer security with its applications in 5G networks: a review. *China Communications*. 2017;14(12):1–14.

[6] Mukherjee A, Fakoorian SAA, Huang J, *et al.* Principles of physical layer security in multiuser wireless networks: a survey. *IEEE Communications Surveys I& Tutorials*. 2014;16(3):1550–1573.
[7] Yang Z, Yue Y, Yang Y, *et al.* Study and application on the architecture and key technologies for IOT. In: *2011 International Conference on Multimedia Technology*. New York, NY: IEEE; 2011. p. 747–751.
[8] Hassija V, Chamola V, Saxena V, *et al.* A survey on IoT security: application areas, security threats, and solution architectures. *IEEE Access*. 2019;7:82721–82743.
[9] Hu YC, Perrig A, Johnson DB. Packet leashes: a defense against wormhole attacks in wireless networks. In: *IEEE INFOCOM 2003. Twenty-second Annual Joint Conference of the IEEE Computer and Communications Societies* (IEEE Cat. No. 03CH37428). vol. 3. New York, NY: IEEE; 2003. p. 1976–1986.
[10] Zhang K, Liang X, Lu R, *et al.* Sybil attacks and their defenses in the internet of things. *IEEE Internet of Things Journal*. 2014;1(5):372–383.
[11] Cekerevac Z, Dvorak Z, Prigoda L, *et al.* Internet of things and the man-in-the-middle attacks – security and economic risks. *MEST Journal*. 2017;5(2): 15–25.
[12] Nahrstedt K, Li H, Nguyen P, *et al.* Internet of mobile things: Mobility-driven challenges, designs and implementations. In: *2016 IEEE First International Conference on Internet-of-Things Design and Implementation* (ioTDI). New York, NY: IEEE; 2016. p. 25–36.
[13] Angove P, O'Grady M, Hayes J, *et al.* A mobile gateway for remote interaction with wireless sensor networks. *IEEE Sensors Journal*. 2011; 11(12):3309–3310.
[14] Xu L, Zhou X, Li Y, *et al.* Intelligent power allocation algorithm for energy-efficient mobile internet of things (IoT) networks. *IEEE Transactions on Green Communications and Networking*. 2022;6(2):766–775.
[15] Giordano S. Mobile ad hoc networks. In: *Handbook of Wireless Networks and Mobile Computing*. New York, NY: John Wiley & Sons, Inc.; 2002;1:325–346.
[16] Zeng K. Physical layer key generation in wireless networks: challenges and opportunities. *IEEE Communications Magazine*. 2015;53(6):33–39.
[17] Wang N, Wang P, Alipour-Fanid A, *et al.* Physical-layer security of 5G wireless networks for IoT: challenges and opportunities. *IEEE Internet of Things Journal*. 2019;6(5):8169–8181.
[18] Mukherjee A. Physical-layer security in the Internet of Things: sensing and communication confidentiality under resource constraints. *Proceedings of the IEEE*. 2015;103(10):1747–1761.
[19] Yang Y, Wu L, Yin G, *et al.* A survey on security and privacy issues in Internet-of-Things. *IEEE Internet of Things Journal*. 2017;4(5):1250–1258.
[20] Hamamreh JM, Furqan HM, Arslan H. Classifications and applications of physical layer security techniques for confidentiality: a comprehensive survey. *IEEE Communications Surveys I& Tutorials*. 2019;21(2):1773–1828.

[21] Chen X, Ng DWK, Chen HH. Secrecy wireless information and power transfer: challenges and opportunities. *IEEE Wireless Communications*. 2016;23(2):54–61.

[22] Qi Q, Chen X, Zhong C, *et al.* Physical layer security for massive access in cellular Internet of Things. *Science China Information Sciences*. 2020;63(2):1–12.

[23] Gupta A, Christie R, Manjula R. Scalability in internet of things: features, techniques and research challenges. *International Journal of Computational Intelligence Research*. 2017;13(7):1617–1627.

[24] Sun L, Du Q. A review of physical layer security techniques for Internet of Things: challenges and solutions. *Entropy*. 2018;20(10). Available from: https://www.mdpi.com/1099-4300/20/10/730.

[25] Rojas P, Alahmadi S, Bayoumi M. Physical layer security for IoT communications—a survey. In: *2021 IEEE 7th World Forum on Internet of Things* (WF-IoT). New York, NY: IEEE; 2021. p. 95–100.

[26] Zhang Y, Xiang Y, Zhang LY, *et al.* Secure wireless communications based on compressive sensing: a survey. *IEEE Communications Surveys I& Tutorials*. 2019;21(2):1093–1111.

[27] Djelouat H, Amira A, Bensaali F. Compressive sensing-based IoT applications: a review. *Journal of Sensor and Actuator Networks*. 2018;7(4):45.

[28] Fragkiadakis A, Charalampidis P, Tragos E. Adaptive compressive sensing for energy efficient smart objects in IoT applications. In: *2014 4th International Conference on Wireless Communications, Vehicular Technology, Information Theory and Aerospace I& Electronic Systems* (VITAE). New York, NY: IEEE; 2014. p. 1–5.

[29] Wang N, Jiang T, Li W, *et al.* Physical-layer security in Internet of Things based on compressed sensing and frequency selection. *IET Communications*. 2017;11(9):1431–1437. Available from: https://ietresearch.onlinelibrary.wiley.com/doi/abs/10.1049/iet-com.2016.1088.

[30] Liu J, Hu Q, Suny R, *et al.* A physical layer security scheme with compressed sensing in OFDM-based IoT systems. In: *ICC 2020 – 2020 IEEE International Conference on Communications* (ICC); 2020. p. 1–6.

[31] Tian Q, Lin Y, Guo X, *et al.* New security mechanisms of high-reliability IoT communication based on radio frequency fingerprint. *IEEE Internet of Things Journal*. 2019;6(5):7980–7987.

[32] Soltanieh N, Norouzi Y, Yang Y, *et al.* A review of radio frequency fingerprinting techniques. *IEEE Journal of Radio Frequency Identification*. 2020;4(3):222–233.

[33] Nouichi D, Abdelsalam M, Nasir Q, *et al.* IoT devices security using RF fingerprinting. In: *2019 Advances in Science and Engineering Technology International Conferences* (ASET); 2019. p. 1–7.

[34] Wu Y, Huo Y. A survey of cooperative jamming-based secure transmission for energy-limited systems. *Wireless Communications and Mobile Computing*. 2021;2021:1–11.

[35] Hu L, Wen H, Wu B, *et al.* Cooperative jamming for physical layer security enhancement in Internet of Things. *IEEE Internet of Things Journal.* 2018;5(1):219–228.

[36] Gao Q, Huo Y, Jing T, *et al.* Cross-layer based intermittent jamming schemes for securing energy-constraint networks. arXiv; 2021. Available from: https://arxiv.org/abs/2103.13217.

[37] Bouabdellah M, El Bouanani F, Alouini MS. A PHY layer security analysis of uplink cooperative jamming-based underlay CRNs with multi-eavesdroppers. *IEEE Transactions on Cognitive Communications and Networking*. 2019;6(2):704–717.

[38] Guo W, Zhao H, Ma W, *et al.* Effect of frequency offset on cooperative jamming cancellation in physical layer security. In: *2018 IEEE Globecom Workshops* (GC Wkshps). New York, NY: IEEE; 2018. p. 1–5.

[39] Wang Q. Using secret spreading codes to enhance physical layer security in wireless communication. In: *2017 IEEE International Conference on Communications Workshops* (ICC Workshops); 2017. p. 447–450.

[40] Alsadi A, Mohan S. A new frequency hopping scheme to secure the physical layer in The Internet of Things (IoT). In: *2020 Wireless Telecommunications Symposium* (WTS). New York, NY: IEEE; 2020. p. 1–8.

[41] Letafati M, Kuhestani A, Behroozi H, *et al.* Jamming-resilient frequency hopping-aided secure communication for Internet-of-Things in the presence of an untrusted relay. *IEEE Transactions on Wireless Communications.* 2020;19(10):6771–6785.

[42] Liu Y, Zeng Q, Zhao Y, *et al.* Novel channel-hopping pattern-based wireless IoT networks in smart cities for reducing multi-access interference and jamming attacks. *EURASIP Journal on Wireless Communications and Networking*. 2021;2021(1):1–19.

[43] Taha FA, Althunibat S. Improving data confidentiality in chirp spread spectrum modulation. In: *2021 IEEE 26th International Workshop on Computer Aided Modeling and Design of Communication Links and Networks* (CAMAD); 2021. p. 1–6.

[44] Jeon H, Hwang D, Choi J, *et al.* Secure type-based multiple access. *IEEE Transactions on Information Forensics and Security*. 2011;6(3):763–774.

[45] Yaacoub E, Chehab A, Al-Husseini M, *et al.* Joint security and energy efficiency in IoT networks through clustering and bit flipping. In: *2019 15th International Wireless Communications I& Mobile Computing Conference* (IWCMC); 2019. p. 1385–1390.

[46] Sun L, Du Q. A review of physical layer security techniques for Internet of Things: challenges and solutions. *Entropy*. 2018;20(10):730.

[47] Sun L, Du Q. Physical layer security with its applications in 5G networks: a review. *China Communications*. 2017;14(12):1–14.

[48] Xu H, Sun L, Ren P, *et al.* Securing two-way cooperative systems with an untrusted relay: a constellation-rotation aided approach. *IEEE Communications Letters*. 2015;19(12):2270–2273.

[49] Alhasanat M, Althunibat S, Darabkh KA, *et al.* A physical-layer key distribution mechanism for IoT networks. *Mobile Networks and Applications.* 2020;25(1):173–178.

[50] Marabissi D, Mucchi L, Stomaci A. IoT nodes authentication and ID spoofing detection based on joint use of physical layer security and machine learning. *Future Internet.* 2022;14(2):61.

[51] Jafari H, Omotere O, Adesina D, *et al.* IoT devices fingerprinting using deep learning. In: *MILCOM 2018-2018 IEEE Military Communications Conference* (MILCOM). New York, NY: IEEE; 2018. p. 1–9.

[52] Peng L, Zhang J, Liu M, *et al.* Deep learning based RF fingerprint identification using differential constellation trace figure. *IEEE Transactions on Vehicular Technology.* 2020;69(1):1091–1095.

[53] Nobakht M, Sivaraman V, Boreli R. A host-based intrusion detection and mitigation framework for smart home IoT using OpenFlow. In: *2016 11th International Conference on Availability, Reliability and Security* (ARES). New York, NY: IEEE; 2016. p. 147–156.

[54] Dorri A, Roulin C, Jurdak R, *et al.* On the activity privacy of blockchain for IoT. In: *2019 IEEE 44th Conference on Local Computer Networks* (LCN); 2019. p. 258–261.

[55] Liao RF, Wen H, Wu J, *et al.* Deep-learning-based physical layer authentication for industrial wireless sensor networks. *Sensors.* 2019;19(11):2440.

[56] Basar E, Di Renzo M, De Rosny J, *et al.* Wireless communications through reconfigurable intelligent surfaces. *IEEE Access.* 2019;7:116753–116773.

[57] Liu K, Zhang Z, Dai L, *et al.* Compact user-specific reconfigurable intelligent surfaces for uplink transmission. *IEEE Transactions on Communications.* 2022;70(1):680–692.

Chapter 11

Physical layer security for wireless sensing and joint radar and communications

Saira Rafique[1], Ahmed Naeem[1], and Hüseyin Arslan[1]

This chapter focuses on providing physical layer security (PLS) to the joint radar and communication (JRC) systems. The integration of sensing and communication via shared spectral, hardware, and signal processing framework provides energy and cost efficiency. However, the inclusion of information into radar probing signals makes the communication susceptible to eavesdropping from the potential radar targets thereby generating a tradeoff, where the JRC transmitter needs to direct sufficient power toward the target to illuminate it while simultaneously limiting useful signal power to prevent eavesdropping. In addition to that, critical sensing information gathered via various sensing techniques including radar systems must also be protected. Therefore, in this chapter, various attacks such as exploratory, manipulation and disruption attacks on wireless sensing are discussed along with the techniques to counter them. Moreover, certain algorithms and PLS techniques are described for the realization of secure joint radar and communication systems.

11.1 Physical layer security for wireless sensing

11.1.1 Introduction to wireless sensing

Wireless sensing is the phenomenon of acquiring information about the environment through radio frequency (RF) signals. This information consists of certain critical parameters that can be further processed to achieve environmental awareness. RF signals transmitted by sensing-capable devices are reflected by the physical environment which can provide useful information regarding geographical and terrain maps, interference, signal strength, and propagation models. Furthermore, sensing is enabled through various devices such as cameras, wireless sensors, radars, and lidars. The next generation of wireless networks supports a diverse set of services such as remote sensing, smart home, environmental monitoring, and vehicle to everything (V2X) communication that require efficient sensing capabilities as described in Figure 11.1.

[1]Department of Electrical and Electronics Engineering, Istanbul Medipol University, Turkey

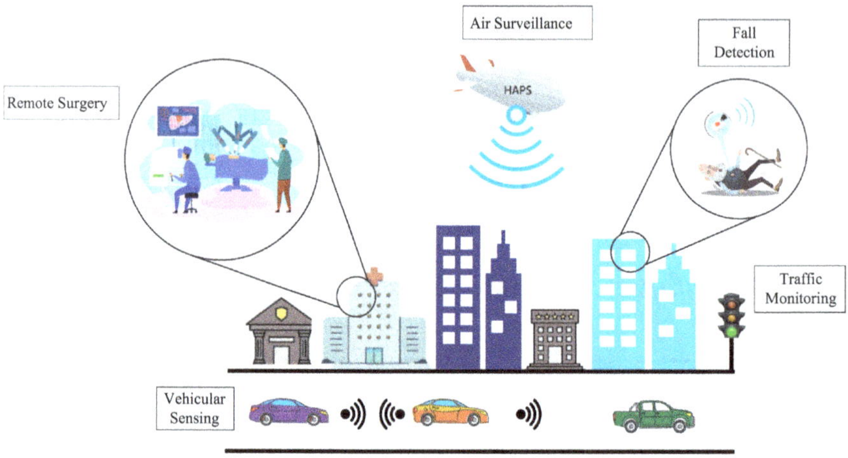

Figure 11.1 Wireless sensing applications

Wireless sensing can be further categorized as active and passive sensing. In active sensing, a dedicated sensing signal is transmitted which interacts with the scattered targets/objects present in the environment and its reflections are processed by the receiver to gain insights into the surroundings. Certain design parameters such as transmit power and high-correlation property must be taken into account for these dedicated sensing signals. On the contrary, passive sensing exploits the reflected signals that are extracted from various non-cooperative communication systems. These sources can be digital audio/video broadcast signals, television signals, and cellular base stations, etc. Passive sensing works first by transmitting the reference signal directly toward the passive sensing receiver via a LOS link. Meanwhile, the receiver also receives scattered replicas of the same reference signal due to its interaction with the target(s). Consequently, related sensing parameters are extracted by correlating signals obtained through two channels.

WiFi sensing which is also referred as wireless LAN (WLAN) sensing, exploits the existing WiFi signals to sense its environment. Common indoor WiFi sensing applications include motion detection, gesture recognition, fall detection, vital signs recognition, etc. The process of acquiring information about the surrounding environment through WiFi sensing is done through the detection and processing of IEEE 802.11 packets. Conventional 802.11 packets include training sequences that are used for channel estimation and synchronization which helps in the demodulation of the received data. Furthermore, the same training sequences can be used to sense the surrounding environment through the reflected echoes. WiFi sensing uses the same 802.11 waveforms to sense the environment and is backward compatible with the 802.11 standards. Moreover, for better and enhanced sensing capabilities and performance, new WiFi standards are being standardized such as IEEE 802.11bf. The main aim of 802.11bf is to provide standardization support for sensing applications at frequencies 1–7.125 GHz and above 45 GHz through modifications to the medium access control (MAC) layer, directional multi-gigabit (DMG), and enhanced-DMG (EDMG) physical layer designs.

11.1.2 Exploratory attacks on wireless sensing

In the literature, eavesdropping is referred to as an attack where the attacker attempts to gain access to or intercept content of the communication. Likewise, exploratory attacks can also be treated as an advanced form of eavesdropping; where, in addition to gaining access to the communication content, the attacker is concerned with the characteristics of the communication as well. Therefore, in exploratory attacks, the attacker is a sophisticated entity which not only listens to the ongoing communication but also aims to get sensing information regarding the objects constructing a particular environment.

From the sensing perspective, the exploratory attacks target sensing/monitoring nodes, processes and the environment. Here, the exploratory attacker may intend to gain insight into the sensing methods and modes, mapping techniques, control information, sensing node location, analog front-end characteristics, environment fingerprints, and the physical layer properties of the radio signals. The undesired exposure of this sensing information paves the way for manipulation and/or disruptive attacks [1]. A demonstration of exploratory attacks from the perspective of sensing and communication systems can be visualized in Figure 11.2, where the eavesdropper is not only intercepting the communication signal but also listening to the critical environmental information including pedestrian and vehicular movement.

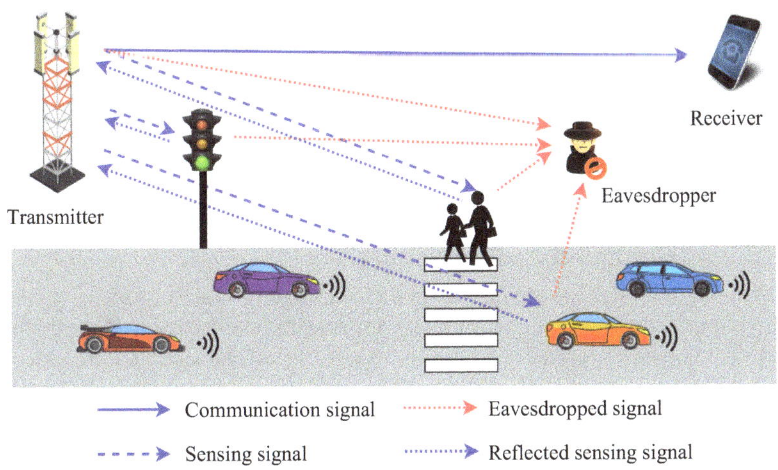

Figure 11.2 Exploratory attacks in communication and sensing systems

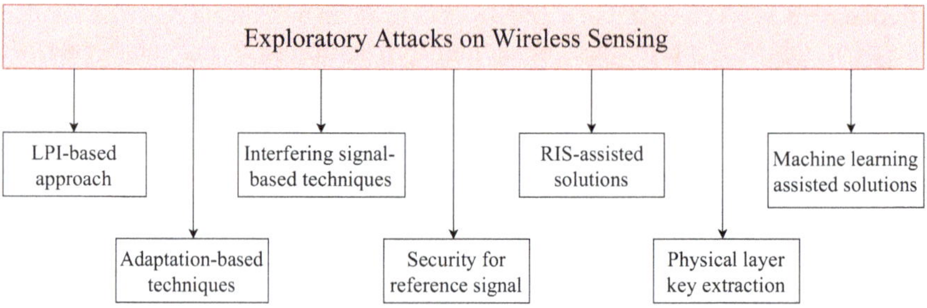

Figure 11.3 Exploratory attacks on wireless sensing

Some realizations of exploratory attacks include learning user behavior and data, control information, traffic dynamics, and preferences. This sensing information is critical and can lead to severe security breaches even if manipulation or disruptive attacks are not launched. Furthermore, learning a node's operational parameters such as bandwidth, carrier frequency, and pilots can lead to more intelligent and effective attacks, which undermines the efficacy of existing security techniques. Various techniques to counter exploratory attacks are listed in Figure 11.3.

11.1.2.1 LPI-based approach

The content and properties of RF signals can be protected against eavesdropping through transmission based on low probability of intercept (LPI). An LPI-based method secures the transmission of a legitimate user from potential attackers by hiding the useful data below the noise floor. This is done by employing a unique sequence to spread the signal's energy in time, frequency, and space. Wider beams, high-duty cycle waveforms, and wide bandwidths are used to achieve this spreading across the three domains. Some practical examples of these techniques include spread spectrum (SS) methods, frequency agility (fast switching between the operating frequencies takes place), and erratic scan patterns to circumvent interception [2]. Furthermore, waveform design in conjunction with antenna array design and optimization can also be used to guard critical information against exploratory attacks.

11.1.2.2 Adaptation-based techniques

The core idea behind these techniques is the adaptation of the transmission parameters according to the channel conditions, requirements, and location of the legitimate nodes [3]. Beamforming-based solutions are also one of the spatial anti-exploratory techniques where the signal power is exclusively enhanced toward legitimate receiver's direction while being suppressed in all other directions. Therefore, the attacker is deprived of receiving the transmitted signal or its replica [4,5]. Additionally, another technique known as directional modulation is also used to provide shield against exploratory attacks by maintaining the constellation points in their standard form towards legitimate receiver's direction while scrambling them in other directions to make their correct detection hard for the attacker [6]. Apart from these, other adaptation-based techniques include precoding, adaptive resource allocation, antenna selection, relay selection, signal constellation rotation, adaptive power control, and interference alignment. Although these adaptation-based techniques are fruitful in a diverse set of scenarios yet they cease to perform well in the presence of multiple collaborative scenarios. Therefore, interfering signal-based techniques are considered which are explained in the next section.

11.1.2.3 Interfering signal-based techniques

In these techniques, an interfering signal such as an artificial noise or a jamming signal is added to the transmission by a trusted node which can either be transmitter, receiver or a relay. The aim of this method is to degrade attacker's performance without affecting reception at the legitimate receiver. The null space of the legitimate receiver is exploited to add the interfering signal to the transmission [7].

These approaches are effective for both communication and sensing systems because they can simultaneously protect content and characteristics of transmission. One demonstration of these techniques is artificial noise-based approach with multiple antenna beamforming where an interfering signal is transmitted along with communication or sensing signal by leveraging legitimate channel's null space to reduce attacker's performance. Likewise, another example is cooperative jamming in which the external nodes are exploited to generate the interference signal [8]. Figure 11.4 shows the advantages and disadvantages of interfering signal-based techniques.

11.1.2.4 Security for reference signal

Specially tailored known reference signals are transmitted to gain insights into the environment. Unobstructed availability of these reference signals makes the critical sensing information prone to security threats. Therefore, securing these reference signals is of paramount importance for the integrity of a communication/sensing

Advantages	**Disadvantages**
Secure transmission irrespective of the better channel conditions at the attacker	Power consumption
Applicable to both TDD and FDD systems	Requires prior knowledge (CSI, receiver/object's location)
	Performance degradation due to improper noise design

Figure 11.4 Advantages and disadvantages of interference signal-based techniques

systems. Some critical techniques that ensure the security of reference signals include artificial noise insertion, eavesdropping resilient pilot design, and pilot tone manipulation.

- In ***artificial noise insertion*** techniques, an artificial noise is added to the pilot/reference signal depending on the uplink channel state information (CSI) to reduce the attacker's ability to correctly estimate the channel during the downlink pilot transmission [9].
- In ***eavesdropping resilient pilot design*** technique, unique anti-eavesdropping pilot signals with full duplex capabilities are designed for legitimate nodes. The pilots are specifically tailored such that the composite pilot matrix exhibits full rank for legitimate nodes, whereas it is rank deficit for the attacker. This approach limits the attacker's ability to exploit legitimate pilots for observing the subspace of its CSI [10].
- The ***pilot tone manipulation*** approach exploits the subcarriers' preceding instantaneous channel information to rotate the phases of pilot signals at the transmitter. By doing so the eavesdropper's performance is degraded because, during that time, the accurate channel estimation can only be performed by the intended receiver [11].

11.1.2.5 RIS-aided solutions

A reconfigurable intelligent surface (RIS) consists of various low-cost passive elements that can be oriented in different structures such as uniform planar array in which each element can be configured to modify the phase and/or amplitude of the incident signal. Since, RISs have the ability to control the propagation environment, they can prevent unauthorized receivers from intercepting wireless transmissions. This can be done by directing constructive and destructive interference toward legitimate receiver and the attacker respectively, which results in the reception of a strong signal at the legitimate receiver and a weaker signal at the attacker. However, this requires prior knowledge of CSI of both legitimate receiver and the attacker. In case when the attacker's CSI is unavailable, then the RIS-aided security solutions are based on channel, location, and requirements of the legitimate receiver only [12].

To ensure immunity against exploratory attacks, the protected zone concept is used to provide data confidentiality when the location of the attacker is unknown. This is achieved by forming a region that surrounds the legitimate nodes where the leaking signals that are emitted by the cooperating entities are masked with artificial noise [13]. The artificial noise transmission with various beamwidths and directions is based on the placement of the nodes, AoA/AoD, and the desired protection level.

Additionally, RISs can ensure sensing/communication security in specific scenarios when conventional PLS techniques exhibit certain limitations. For instance, when both legitimate receiver and the attacker are located in the same direction, then beamforming and directional modulation-based PLS techniques become ineffective. In these cases, RISs can provide security by providing alternative paths toward the receiver. Moreover, RISs and interfering signal-based techniques can be used together to direct the signals toward the attacker resulting in enhanced transmission security.

11.1.2.6 Physical layer key extraction

This approach exploits the features of a random wireless channel such as CSI, AoA, received signal strength (RSS) indicator (RSSI), feedback mechanisms, and sub-carrier indices to generate a secret key. The generated key can serve as a specific spreading sequence to protect the content and characteristics of a wireless transmission in LPI-based security solutions. The key-based security solutions depend on three basic assumptions, which are given as follows:

1. **Channel reciprocity:** It states that in a TDD system the two communicating nodes observe similar channel response (uplink channel response is similar to the downlink channel response).
2. **Channel de-correlation:** It states that in a rich scattering environment, an independent channel is observed at the attacker if it is at least half a wavelength away from the legitimate node.
3. **Channel randomness:** It assists to generate random key bits in temporal, spatial, and spectral domains.

The fundamental steps for secret key generation are summarized in Figure 11.5. Initially, channel probing is performed between the communicating nodes via

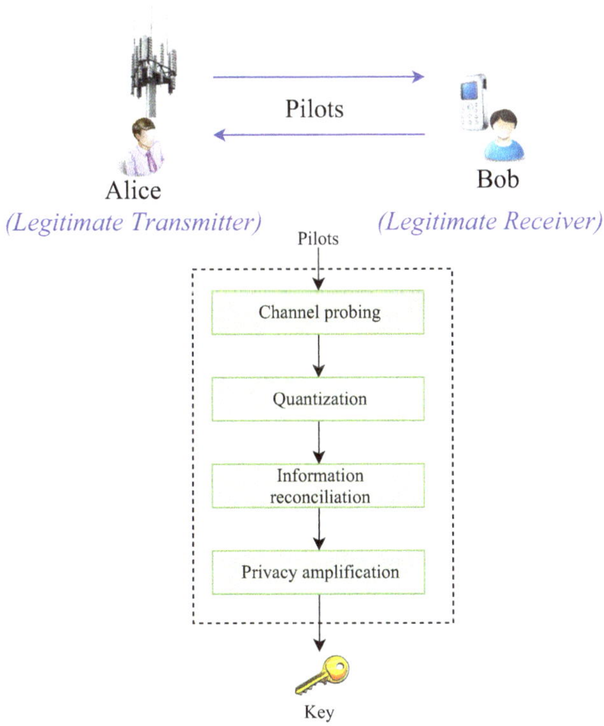

Figure 11.5 Process explaining key generation algorithm

<table>
<tr><th>Advantages</th><th>Disadvantages</th></tr>
<tr><td>Efficient key sharing and management</td><td>Requires additional signalling and processing at legitimate nodes</td></tr>
<tr><td rowspan="2">Performs well when an attacker experiences better channel conditions compared to the legitimate receiver</td><td>Sensitive to channel mismatch errors between legitimate nodes</td></tr>
<tr><td>Not effective in poor scattering environments</td></tr>
</table>

Figure 11.6 Advantages and disadvantages of physical layer key extraction techniques

sounding techniques. Channel quantization is then used to generate the secret key bits using specific channel features. In the next phase, information reconciliation is used to reduce the mismatch between the key bits generated at legitimate nodes. Lastly, privacy amplification is used to optimize and enhance the randomness of the resulting key bits. Several advantages and disadvantages of secret key generation-based techniques using the channel characteristics are summarized in Figure 11.6.

11.1.2.7 Machine-learning and deep learning-assisted solutions

Artificial intelligence (AI) has emerged as a promising tool for learning and adapting according to the wireless environment. PLS techniques including beamforming and precoding, subcarrier allocation, and interfering signal-based methods are complex optimization problems. Moreover, their complexity further increases with futuristic wireless network enhancements, RIS-aided channel manipulation, multi-user scenarios, and mmWave mMIMO systems. ML and DL algorithms can be used to simplify complex modeling and optimization problems. They can also be exploited to achieve better CSI which is a critical part of PLS techniques.

11.1.3 Manipulation attacks on wireless sensing

Manipulation attack in wireless communication and sensing is a conventional generalization of a spoofing attack. The spoofer in a spoofing attack can detect and modify the information exchanged between the legitimate transmitter and receiver. However, a manipulation attack has an additional ability to compromise the integrity of wireless transmission's characteristics. The information gathered through an exploratory attack such as participating nodes, sensing processes, and the radio environment can be exploited to launch these attacks [1]. Different types and aspects of manipulation attacks are summarized in Figure 11.7 and discussed in the following sections.

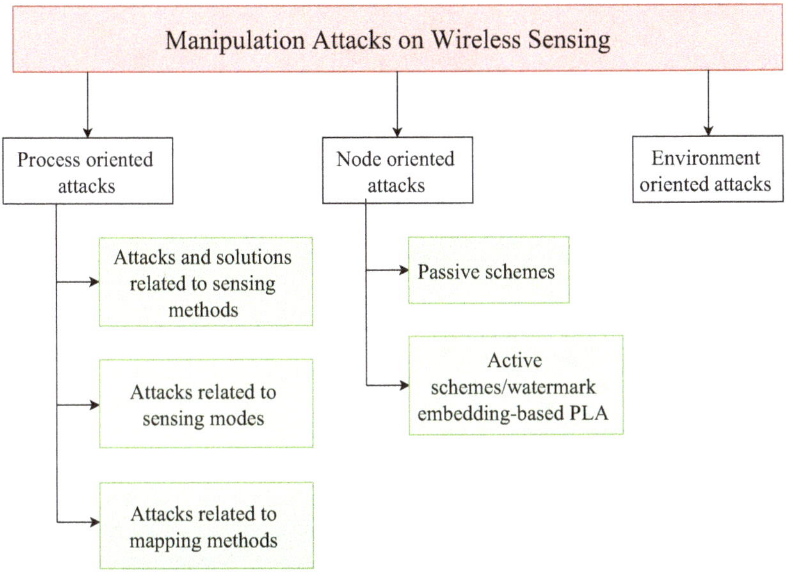

Figure 11.7 Manipulation attacks on wireless sensing

11.1.3.1 Process-oriented attacks

In this section, different types of attacks regarding manipulation of the wireless sensing process by either tampering or degrading the sensing modes, methods, or mapping approaches are explained.

- ***Attacks and solutions related to sensing methods:*** Various approaches are found in the literature to perform the sensing operation which include radar-based sensing, communication, and reference signal-based sensing. The spoofing attack on radar-based sensing requires the spoofer to add, remove, and/or change the location of the actual target(s) [14]. In Figure 11.8, the spoofer/attacker spoofs the actual car position therefore, the intended vehicle gets manipulated position of the same car, causing a collision or false alarms. The spoofer, in a spoofing attack, alters the original location and velocity of an object by adding various spoofed signals with different delays and velocities that correspond to fake objects. In short, it imitates the presence of an object but with compromised characteristics. The spoofer can achieve this by either re-transmitting previously recorded legitimate signals or by sending a signal with the same characteristics as the legitimate signal.

  The manipulation attacks can be defended by properly detecting and filtering off the spoofed radar signals. These illegitimate signals need to be detected timely and consistently while not disrupting other radar-based functionalities of the system. Techniques such as introducing a randomized probing in time to

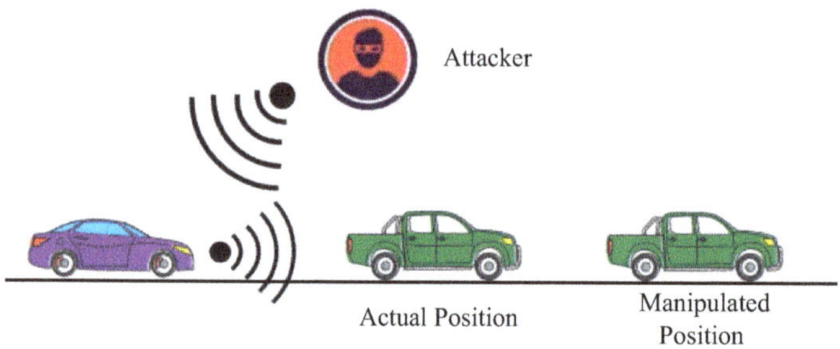

Figure 11.8 Manipulation attacks on wireless sensing

detect the manipulated attacks, in which the radar randomly stops its transmission and listens to any incoming illegitimate signals. This technique will protect the radar system from any spoofing attack but at a cost of causing latency in sensing the environment because of the transmission, discontinuity [15]. Another way to secure the radar system from spoofing attacks is by exploiting the spatio-temporal domain in which multiple narrow beams are transmitted in random directions and their echoes are observed. When the legitimate node receives any echo from an unprobed direction it will indicate that there is a possibility of a manipulating node. Moreover, techniques such as frequency hopping-based sensing is also a method to overcome spoofing attacks. Network-based solutions are also possible in which the radar traffic can be monitored through machine learning which can detect spoofing signals.

The attacker may also manipulate the communication and reference-based sensing signals. If the attacker has accurate knowledge about the reference signals, pilots, and the control signals, it can launch replay, pilot contamination, and conventional spoofing attacks. Similarly, in pilot contamination, the pilots similar to the legitimate node can be transmitted by the attacker which can cause incorrect channel estimation, causing incorrect CSI-based mapping and sensing. For such attacks to be countered, the legitimate node needs solutions such as secret key generation [16], pilot superposition methods with random signal, employment of random pilots that can be selected from a set of known constellation symbols, energy ratio test [17], etc.

- ***Attacks related to sensing modes:*** The sensing modes here are referred to as the signal parameters, which vary according to various applications. Such as the sensing rate f, period T, update threshold Th, etc. In this section, attacks on these signal parameters and their detrimental effect on overall performance are explained. In case the attacker knows the sensing session duration, it can hide its presence using the resources outside that specific time duration. Alternatively, the spoofer can attack intelligently during the sensing duration to degrade the sensing performance. The sensing duration can be randomized for the spoofer so that it is unable for him to know the exact sensing duration [18]. Another sensing

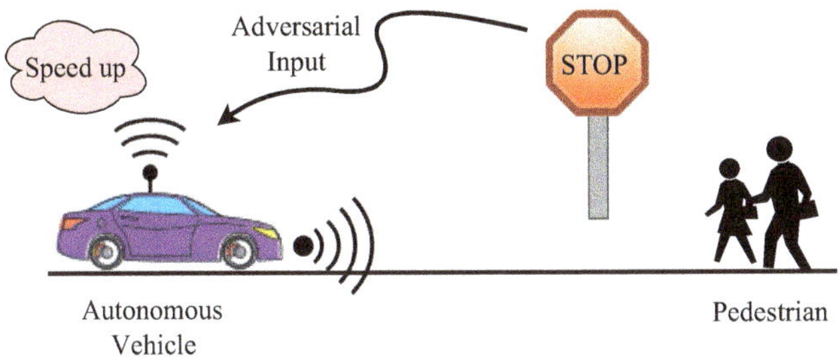

Figure 11.9 Adversarial attack in vehicular communication and sensing

parameter that is critical and can be manipulated by the spoofer is the threshold, which decides when and if sensing transmission needs to be initiated and with which nodes in the network it needs to share sensing information. In this case, unnecessary sensing updates are monitored so there is less probability that the spoofer can spoof and gain knowledge about the threshold value.

- ***Attacks related to mapping methods:*** The mapping knowledge if accessed by the manipulator can be used to derive extra information from the observable data that can be used for attacking and spoofing. Attacks such as adversarial and spatial interpolation-based attacks are a few ways in which the spoofer can manipulate the system related to mapping methods. Adversarial threats are classified into *evasion, poisoning, model inversion* and *extraction*. All these attacks involve the malicious corruption of training data or the attacker adding certain features to damage the real data. Figure 11.9 shows an example of an adversarial attack where the attacker adds some features to mislead the vehicle, instead of stopping for the pedestrians the vehicle is commanded/suggested to speed up. To combat such attacks, solutions such as *distillation* are used to enhance the robustness of the system against adversarial attacks by exploiting knowledge from deep neural networks [19]. Moreover, solutions based on *region-based classification* techniques are also used where samples around the test data are also processed through the trained model making sure that all the samples are the desired ones [20]. The main aim of these solutions is to make sure that authentic and required data is fed to the mapping algorithm. Besides, these solutions, interpolation-based methods are also extensively used to combat these attacks. Furthermore, some classical methods including inverse distance weighting [21] and Kriging algorithm [22,23] are used against these attacks.

11.1.3.2 Node-oriented attacks

Attacks that are node-driven manipulates different nodes through compromising the integrity or authenticity of the wireless signal characteristics. A spoofer is able to manipulate the position, data, identity, synchronization, patterns, etc. The attacker can manipulate and violate the authentication process in a way that it becomes a trustworthy participant of the system without no one knowing it [24]. Moreover, the integrity of sensing system violation is also a critical aspect, such as breaching the data integrity by launching malicious attacks. These attacks include threats like tempering of the data, spoofing attacks on the global positioning system (GPS), false reporting about the location of targets, etc. To resist such node-oriented manipulation attacks, different techniques are proposed which are divided into passive and active schemes.

- ***Passive schemes:*** This scheme provides solutions that leverage the physical layer characteristics and properties of the received signal such as channel and/or the analog front-end for authentication purposes. Passive physical layer authentication (PLA) includes two stages: training and message transmission [25]. For a reliable fingerprint database, the features associated with the legitimate nodes are gathered during the training stage. While the signal in the transmission phase, the receiver extracts out valuable information from a noisy signal that is received from an unknown transmitter. Then, in order to confirm the transmitter's identity, it compares the extracted features with the database fingerprints. This process is explained in detail in Figure 11.10 as a flowchart diagram. Furthermore, channel-based PLA techniques are based on exploiting different features of the propagation environment between the communicating nodes which includes; angle of departure (AoD), angle of arrival (AoA), round trip time (RTT), RSSI, and CSI [26]. PLA approaches based on the channel perform well in a slow-changing environment. However, when considering a fast-fading or time-variant environment, modifications are needed in the basic algorithms. The modification approaches to overcome the channel dynamics include channel tracking methods based on time-varying multipath correlation models [27–29] or clustering the observations leveraging ML and DL approaches [30].
- ***Active schemes/watermark embedding-based PLA:*** In the active node-oriented attack schemes, there is intentional incorporation of some identity information at the physical layer in the communication instead of only depending on the front-end of the device or the channel. *Tag* is the identity information mentioned above, it is added based on a pre-shared key and a complicated tag generation. This solution and approach provides better authentication and ensures the integrity of the message/data because any manipulation of the message will also change and alter the tag as well. Moreover, time-varying tags also provide robustness against replay attacks, which might otherwise pass the conventional authenticity checks. These authentication tags can be added to the training pilots [31] or to the data frame [32]. Some other solutions for active schemes for node oriented attacks includes; pre-shared key including the transmission parameters modification [33,34], tag insertion by replacing source data bits [35], and frame structure modification [36].

11.1.3.3 Environment-oriented attacks

Attacks that are environment-oriented typically target and affect the propagation environment, changes the physical objects, and alter the environmental conditions. The main aim of these attacks is to fool sub-optimal parameters selection with

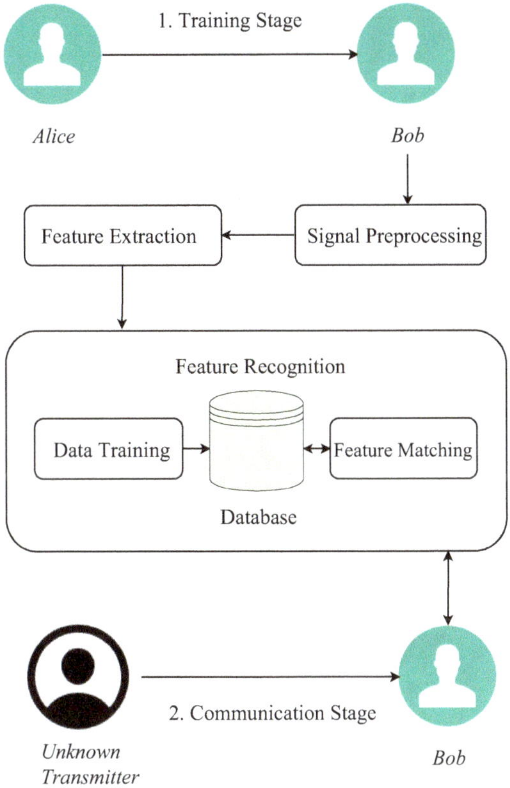

Figure 11.10 Basic steps for passive PLA algorithms

improper resource allocation and optimization algorithms. Moreover, devices that are sensitive to heat, humidity, rain, pressure, electromagnetic waves and, other physical and environmental conditions can be manipulated by artificially changing these conditions. From the perspective of wireless sensing, the solutions against artificially manipulating the environment is still an open issue. External sensors, including cameras, sensor fusion methods, or sensor networks, can, however, be used to identify such attacks. Crowds-sourcing and collaborative sensing are two such examples.

11.1.4 Disruption attacks on wireless sensing

Disruption attacks are launched by the attacker to impose disturbance in the ongoing sensing/communication processes. The most common demonstration of this attack is the intentional interference directed at sensing and/or communication devices which disrupts the characteristics and content of wireless signals. Various types of disruption attacks are summarized in Figure 11.11. In wireless sensing or communication systems, the received signal must have more power than the noise and interference.

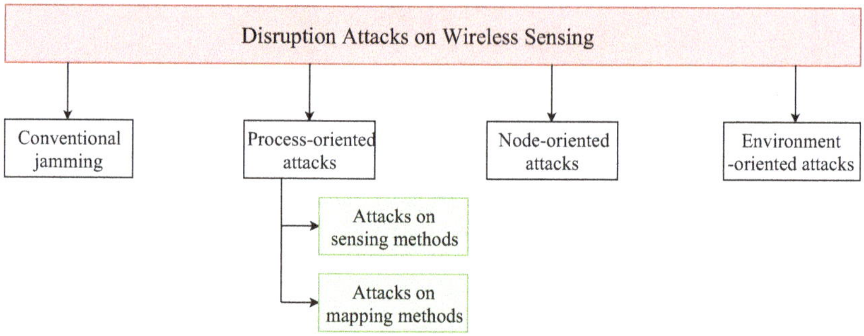

Figure 11.11 Disruption attacks on wireless sensing

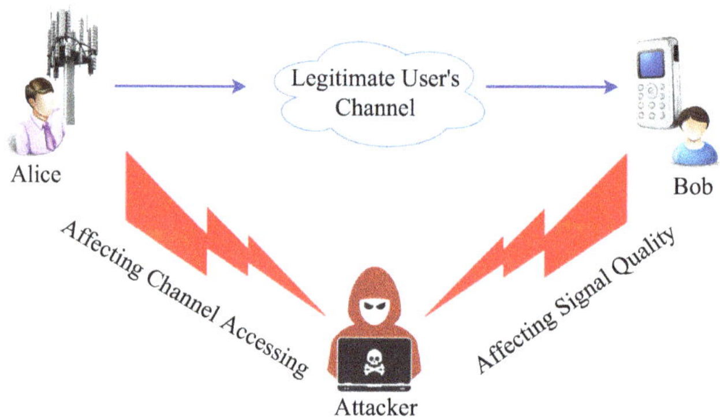

Figure 11.12 Disruption attack on transmitter and receiver

However, the attacker enhances the interference level in the transmission channel which results in transmission disruption for legitimate nodes. This results in wastage of resources and performance degradation for both sensing and communication systems. Figure 11.12 shows a disruption attack on Alice and Bob, where Alice experiences a false channel occupancy alarm while Bob experiences signal quality degradation.

11.1.4.1 Conventional jamming types and detection

The goal of a jammer or disruptive attacker is to cause discontinuity in the ongoing sensing/communication process. **Constant** jamming is one of the basic forms of jamming attack in which the illegitimate entity continuously transmits an interfering signals across the same spectral resources used by the legitimate nodes. The interfering signal can be a modulated waveform or a pseudorandom noise. Such an attack results in degraded SINR while also making the channel occupied at the legitimate node. The detection of jamming attacks can be done via certain observation parameters such as packet error rate (PER), carrier sensing time and RSS. One of the advanced variants of jamming attack is the **intermittent** attack in which the attacker randomly switches between transmitting an interfering signal and the sleeping mode when it transmits nothing. As a result, there is a trade-off between energy consumption and jamming efficiency. The techniques provided for a constant attacker can also be used to identify this kind of attacker.

A more intelligent attacking paradigm is provided by **reactive** jamming. In this case, the attacker identifies the channel and only launches an attack when it discovers an active communication between legitimate nodes. These attacks are therefore more energy-efficient than continuous and intermittent ones. However, it also implies that the legitimate nodes' channel access mechanism is unaffected by this attack. Consequently, the carrier sensing time approach is not applicable for the detection of reactive jamming. However, PER and RSS approaches work for the detection of reactive jamming. An **adaptive** attacker has the enhanced ability to change its transmit power in response to the channel conditions experienced by the legitimate user. In line with reactive jamming, an adaptive attacker detects the legitimate transmission and then adjusts the strength of the jamming signal based on the RSS observed at the legitimate node to disturb its transmission. Therefore, this is the most energy-efficient jamming technique. However, this requires prior knowledge about the legitimate channel which makes the reactive jamming approach impractical. These attacks can be identified by RSS and PER statistics.

11.1.4.2 Process-oriented attacks

- **Attacks on sensing methods:** Disruption attacks can compromise the reliability of critical sensing information that is gathered through various sensing techniques by transmitting jamming signals. These types of attacks are averted through the provision of diversity or by deploying mechanisms to avoid interfering signals. Diversity can be achieved through multi-antenna systems and cooperative relaying techniques. Whereas, protocol hopping and SS techniques are used to avoid interfering signals.
    1. **Cooperative relaying schemes:** Cooperative communication with trusted relay(s) is one of the simplest anti-jamming strategies for both communication and sensing. In particular, using cooperative beamforming and relay selection methods provides an alternative route for signal propagation between the legitimate nodes, minimizing the impact of the attacker's disruption.
    2. **RIS-aided anti-jamming solutions:** The ability of RIS to control the channel can be exploited to prevent jamming by adjusting the amplitude and/or phase of the reflected jamming signals to reduce their effect at the legitimate nodes. For instance, in the presence of a disruption attack, Bob receives signals from Alice (legitimate Tx), the attacker, and the reflected jamming signal from RIS. Here, by modifying phases of its elements, an RIS can ensure destructive interference at Bob between the original and reflected jamming signals. Additionally, RIS can also provide additional path(s) for the legitimate signal.

3. **Spread spectrum techniques:** The spread spectrum-based security solutions explained earlier for exploratory attacks are also applicable to counter disruptive attacks. Spreading a signal's energy in spatial, temporal, or frequency domain makes it robust against disruptive attacks. Some dominant SS techniques include direct-sequence spread spectrum (DSSS), frequency hopping spread spectrum (FHSS), parallel sequence spread spectrum (PSSS), chirp spread spectrum (CSS), and time-hopping spread spectrum (THSS). However, among these techniques, FHSS and DSSS are widely used. The carrier frequency is constantly shifted in the FHSS modulation technique to prevent narrowband jamming. The frequency shifting across the entire spectrum is realized through PLS key generation or pre-shared secret sequence. Likewise, DSSS is another approach to tackle narrowband jamming through the expansion of the transmitted signal into a wide frequency band by multiplying it with a secret sequence. Consequently, only a small part of the transmitted signal's frequency spectrum can be affected by the narrowband jamming signal. However, these schemes are spectrally inefficient because they require a wide spectrum.
4. **Channel surfing and protocol hopping:** Channel surfing is an adaptive variation of FHSS in which there is no continuous hopping of the carrier frequency as in FHSS. Instead, after learning that the current band is being jammed, the frequency is changed to a different frequency. The high PER and high RSS measurements can be used to identify jamming. Similarly, protocol hopping is another potential anti-jamming technique. In these approaches, reliable communication is maintained even in the presence of attackers by having legitimate nodes hop between several available protocol parameters (or even technologies).
5. **Multi-antenna-based techniques:** The deployment of multiple antennas paves the way for adaptive antenna arrays and digital beamforming which enables directional communication to prevent interference from undesired sources. In particular, it provides adaptive beam nulling in the direction of jamming sources and permits the node to receive signals from a specific direction. As a result, it provides immunity against disruption attacks to both radar and communication systems and improves their performance.
6. **Machine learning and AI-based techniques:** Recently, ML-based solutions have been used to counter disruption attacks. For example, in some cases, deep reinforcement learning is used to understand the attacker's strategy, and then based on that legitimate nodes take appropriate counter steps [37]. As potential mitigating measures, techniques such as transmission rate adaptation, energy harvesting from the received interfering signal, or exploiting the interfering signal to broadcast the legitimate signal by utilizing ambient backscatter communication (ABC) techniques are also being explored. Reactive attacks, however, cannot be tackled with this strategy since the attacker first performs channel sensing before launching the attack. One approach to counter this limitation is through the use of "fake" transmission to provoke the reactive attacks [38]. As a result, interference signals are

sent, which are subsequently either employed for ABC or energy harvesting. This strategy not only drains the attacker's resources but also increases the transmission power of the legitimate nodes (using the harvested energy). The IoT and other use cases with limited power sources may find this strategy particularly interesting. In RIS-assisted mitigation approaches, reinforcement learning has also been employed to jointly optimize anti-jamming power allocation and reflecting beamforming.

- **Attacks on mapping methods:** When there is more data to be processed in a particular time interval, it is referred to as a flooding attack and it can disrupt mapping methods. This may take place when a mapping model is being created during the model learning procedure. Here, the attacker's goal is to impair or disrupt the mapping mechanism by flooding the responding or initiating node with data. For instance, an illegitimate user may send more repeated measurements in a packet than required. This can greatly prolong the learning period during the learning stage, hence disrupting the process. This can fill the processor's buffer during the mapping stage thereby, delaying access to the actual measurement and producing outputs or choices that are out of date.

  Here, the solution's objective should be to quickly and easily identify and eliminate incoming false data or sensing signals. The attacker may send random data or many replications of the true measurement in the packet when feedback-based periodic sensing is enabled. By forecasting the anticipated measurement and eliminating any data that deviates from the prediction, fraudulent data may be found in the first scenario. The final measurement may then be obtained by averaging the remaining readings.

11.1.4.3 Node-oriented attacks

In node-oriented attacks, a responding node might bombard the initiating node with unnecessary data/messages; henceforth limiting its ability to perform sensing or communication with other nodes. This process is somehow similar to disruption attacks related to mapping methods. However, node-oriented disruption attacks deal with the manipulation of the responding node. Additionally, the solutions provided in Sections 11.1.4.2 and 11.1.3.2 are also applicable in this case.

11.1.4.4 Environment-oriented attacks

The main idea behind environment-based sensing attacks includes creating interference in specific areas to deteriorate the sensing or communication processes. For instance, the responding or initiating node may be unable to extract the sensing or communication signals and collect measurements if smart environment technologies such as RIS are utilized to reflect any incoming signals to those places [39]. Alternately, numerous and repetitive artificial environment manipulations can be performed, which may cause the REM process to be initiated repeatedly. The solutions provided for *attacks related to sensing modes (threshold-based manipulation)* and *attacks related to mapping methods* are also applicable to these types of attacks.

11.2 Physical layer security for joint radar and communication systems

Futuristic mobile communication networks are envisioned as multi-functional entities that can deliver critical sensing information including accurate position, velocity, angular orientation, and material properties of targets in addition to conventional high-rate communication. Moreover, the enhanced features equipped with centimeter-level positioning accuracy enable the network to achieve improved context awareness. Henceforth, radar sensing is considered an integral component of next-generation wireless networks. However, these advanced network capabilities yield certain security concerns; since the attackers may tamper with critical sensing information[40]. Therefore, ensuring communication secrecy and JRC covertness is of paramount significance.

11.2.1 Physical layer security for dual-functional radar communication systems

A dual function radar communication (DFRC) system focuses on the co-design of radar and communication systems. In such systems, a single emission (waveform/signal) is transmitted through a jointly designed hardware thereby guaranteeing cost and spectral efficiency. To realize a DFRC system, two approaches are dominating the JRC literature. These are radar-centric and communication-centric DFRC systems. In the former method, communication data is embedded on top of a radar waveform, whereas in the later approach, a communication waveform is optimized to perform radar sensing. A DFRC system based on *IEEE 802.11 a/g* is implemented in [41] to detect forward collisions in autonomous vehicles. Likewise, a dedicated short-range communication (DSRC) standard *IEEE 802.11p* is used as a DFRC system which enables the vehicles to realize vehicle-to-vehicle and vehicle-to-roadside communications. Additionally, specific frame structures are also designed to realize DFRC systems such as in [42].

11.2.1.1 Radar target(s) as potential eavesdropper

A DFRC system incurs security challenges owing to the shared spectrum and broadcast nature of transmission [43]. The incorporation of communication information into radar probing signal makes it vulnerable to eavesdropping from the potential targets. The targets that are being sensed can also extract the information embedded in the DFRC signal that is intended for communication users. This situation frequently happens in defense-related applications where the radar targets can potentially be combat platforms of the adversaries. Therefore, the emission of a common DFRC waveform might risk the leakage of critical information towards the target(eavesdropper). Hence, the provision of physical layer security is obligatory for DFRC systems. Such a system is demonstrated in Figure 11.13:

1. A solution for this scenario where the radar target is considered as a malicious entity and can eavesdrop the transmitted information from base station (BS)

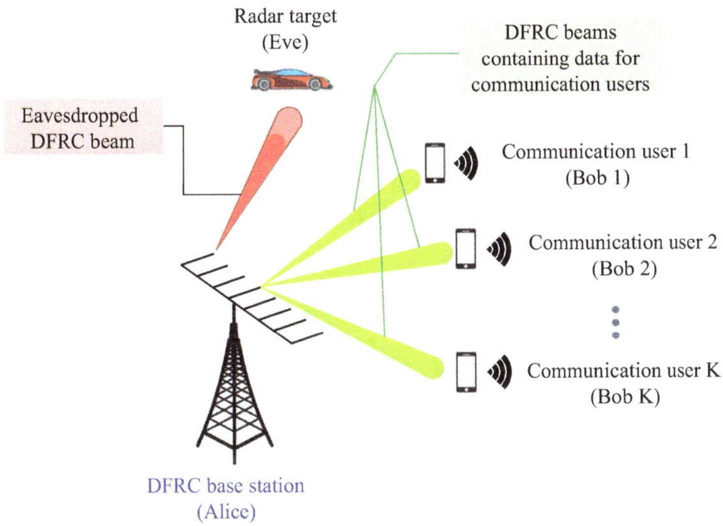

Figure 11.13 DFRC system imposed eavesdropping

to communication users (CUs) is by using the constructive interference approach. Moreover, designing a transmit waveform and receiver beamformer is considered to maximize the receive radar signal-to-interference-plus-noise ratio (SINR) in order to have a secure DFRC system, where the undesired multi-user interference (MUI) is converted into useful power. In this method, the phases of received signals at CUs are rotated into the relaxed decision zone, and the undesired MUI is designed to contribute in useful power. In the problem formulation, the transmit waveform and the receive beamformer are jointly designed to maximize the radar receive SINR, where the MUI is designed to be constructive at the CUs.

2. One another solution to the above mentioned vulnerability is the exploitation of artificial noise to enhance communication secrecy. For example, directional modulation in combination with artificial noise can be used to improve communication performance in a particular direction while degrading it in other directions [44] as shown in Figure 11.14. Likewise, artificial noise-based linear precoding design can be used to guarantee secrecy performance in a multiple-input multiple-output (MIMO) system [45]. Additionally, optimization problems can also be formulated to ensure communication secrecy through SINR minimization at the target by exploiting artificial noise [46]. Meanwhile, by enforcing SINR criteria, the communication secrecy rate for the legitimate users is ensured. Hence, a tradeoff is inherent to a DFRC system where the joint transmitter aims to illuminate the potential targets by directing the sufficient power towards them while also limiting the useful signal power toward sensing targets to prevent eavesdropping.

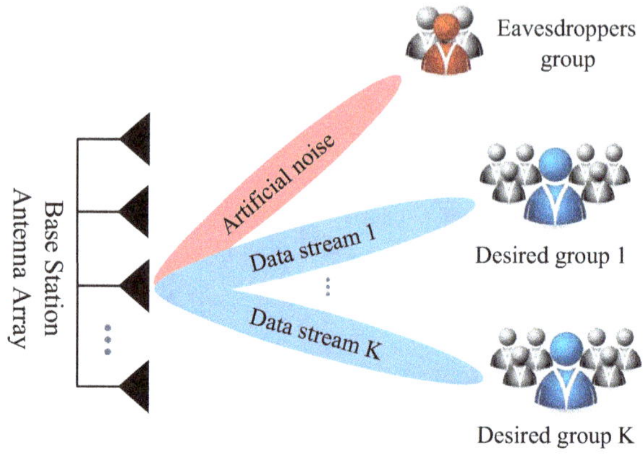

Figure 11.14 Directional modulation with artificial noise in a DFRC system

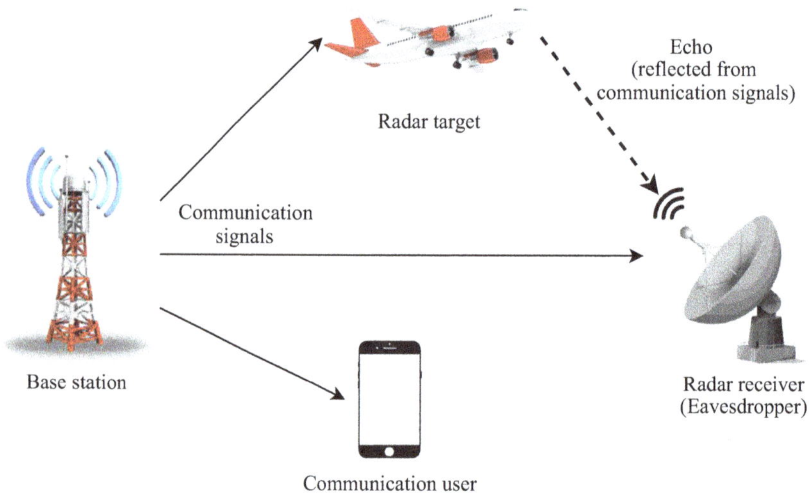

Figure 11.15 Radar receiver as an eavesdropper

11.2.1.2 Radar receiver as an eavesdropper

Another security vulnerability in radar-communication coexistence systems arises when the radar receiver acts as a potential attacker. The scenario is explained in Figure 11.15 where the radar receiver detects the communication signal reflected from the target as well as from the communication transmitter directly. Here, the radar receiver launches an eavesdropping attack on the direct signal coming from the communication source. Therefore, in the presence of eavesdropping, it is required to increase the radar SINR while ensuring the secrecy rate. Interested readers are referred to [47,48] for detailed description.

11.2.1.3 Spread spectrum-based JRC security

The convergence of radar and communication has been well studied and focused by now in literature in terms of DFRC systems. Another paradigm of such systems are to have effective solutions to secure them. One such solution is exploiting SS techniques such as JRC-based frequency hopping. Conventionally, all SS techniques are used for secure wireless communication by increasing resistance to noise, interference, eavesdropping and jamming. Frequency hopping transmits radio signals by changing the carrier frequency rapidly while occupying a large spectrum. The frequencies are varied and changed on basis of a hopping sequence which is known to both the transmitter and the receiver.

This technique of conventional secure wireless communication is translated into JRC domain to secure the DFRC system and the sensed information that is being communicated to the receiver. Frequency hopping multiple-input and multiple-output (FH-MIMO) DFRC-based system is used to combine the benefits of both a secure signalling and simultaneous sensing and communication. The hopping of frequencies are divided in each pulse of a pulse radar in a way that each sub-pulse is transmitted at a different frequency [49]. To ensure that each waveform pulse is orthogonal, it is taken care of that no two antennas transmit at same sub-band that is used as a hopping frequency. Moreover, the communication data is embedded, by multiplying communication signal onto the radar pulses that are transmitted in frequency hopped manner, resulting in continuous sensing and communication in a secure manner.

Another technique to overcome and mitigate both inter-radar interference and spoofing is by using blue frequency-modulated continuous-wave (BlueFMCW) which randomly hops frequencies [50]. In contrary, frequency-modulated continuous-wave (FMCW) is susceptible to spoofing and jamming attacks which causes *ghost* objects to appear and hinders sensing performance. Due to the spoofing attack the false alarm probability increases because of the *ghost* objects. In conventional FMCW, the receivers know the system parameters such as the f_c, bandwidth, chirp transmission period, and modulation type, this is why the attacker can synchronize its self with these parameters and easily spoof the radar system. The detail working and mechanism of BlueFMCW is explained in Figure 11.16. The example of a ghost object in the conventional FMCW is shown in Figure 11.16 (a) and in (c) the beat frequency after the correlation of transmitted and received signal is shown in which the ghost object's beat frequency is also appearing and will make a false alarm for the radar system. To overcome this issue, the BlueFMCW is used as shown in Figure 11.16(b) and (c), in which the ghost object's beat frequency will be randomly disappeared over various fast Fourier transform (FFT) bins Which will result in significantly smaller peaks in the spectrum and will be undetectable.

11.2.1.4 Clutter-masked waveform-based JRC security

One technique to improve communication data secrecy in a JRC system is via clutter-masked transmissions. Generally, clutter is considered as having a negative impact on radar surveillance that reduces sensing capabilities and leads to performance degradation. Therefore, clutter suppression techniques have been investigated in the

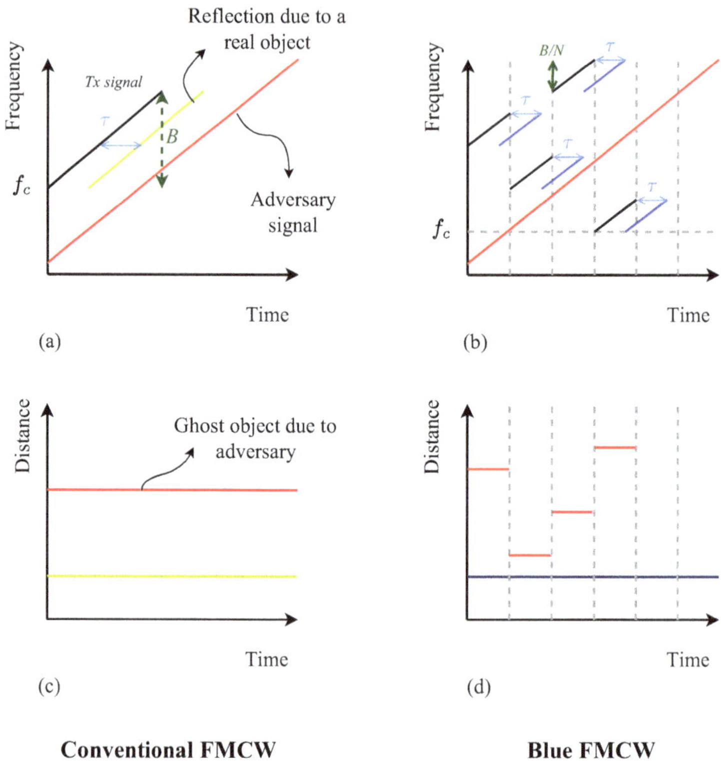

Figure 11.16 Transmitted signal and beat signal of (a) and (c) conventional FMCW and (b) and (d) BlueFMCW

literature such as space–time adaptive processing (STAP) [51]. Moreover, clutter modeling and characterization is a pre-requisite for efficient clutter removal. For instance, deconvolution can be employed to eliminate sea clutter from the radar data based on theoretical model of the sea clutter [52]. However, after the separation of noise from the clutter, the later can be exploited to realize covert sensing and communication. The JRC signal generated by using the local clutter model is highly likely to be interpreted as a clutter return by the unauthorized interceptor [53]. An example of this process is illustrated in Figure 11.17 in which initially the probability density function (PDF) of a tree is modelled, afterwards communication and radar signals are merged in a way such that the resultant signal follows the same distribution as that of a clutter (tree). The attacker receives this signal and treat it as a clutter and thus does not make any attempt to decode it thereby guaranteeing LPI-capable joint radar and communication transmission. In a nutshell, the waveform is designed as a *smart noise* which fools the attacker towards making ambiguous assumptions. Therefore, such a JRC signal/waveform design method has a strong likelihood of achieving secure communications under critical adversarial circumstances while concurrently meeting radar sensing requirements.

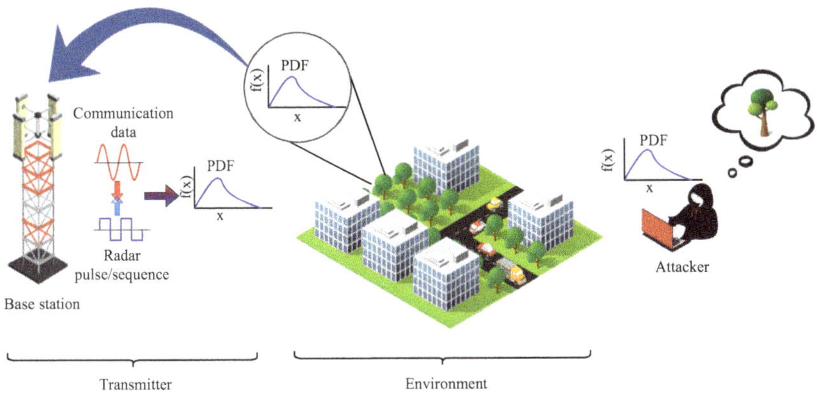

Figure 11.17 Clutter-masked waveform-based JRC security scenario

11.2.2 Physical layer security for radar–communication coexistence

Due to the increase of wireless devices and requirements the amount of bandwidth required for all these high data-rate services is causing spectral congestion. To cope up with this problem, various spectrum sharing techniques are developed, one of which is the spectrum sharing of radar and communications bands. There is a large portion of spectrum still available in the mid-band which is already being used by the radar technology, in order for communication to coexist in these bands in harmony and in a secure manner, different techniques are proposed.

MIMO architecture is widely deployed in both radar and communication systems because of its benefit of providing additional spatial degrees of freedom for better performance gains. Conventionally radar transmits at a significantly high power compared to that of communication systems, this causes interference for the communication receivers. Existing techniques for these scenarios are mostly related to interference mitigation but from security perspective this radar interference can be utilized to disrupt eavesdropper and to safeguard confidential information [54]. One such solution to have PLS is to exploit the coexisted radar signals as inherent jamming signals to disrupt the eavesdropper's reception as illustrated in Figure 11.18. The transmit beamformers of communication and radar systems are jointly designed to ensure the secure transmission as well as achieve satisfactory performance of radar target detection. Specifically, the aim is to minimize the maximum eavesdropping signal-to-interference-plus-noise ratio (SINR) on multiple legitimate users, while satisfying the legitimate communication SINR constraints to guarantee the quality-of-service (QoS) of legitimate transmission, the radar target detection probability constraint for guaranteeing the radar performance, and the transmit power constraints of radar and communication systems.

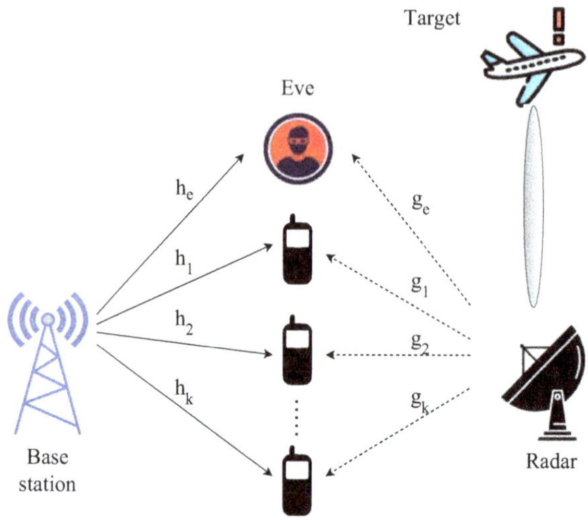

Figure 11.18 A radar communication coexistence system

11.3 Conclusion

The security solutions provided by fifth generation (5G) and beyond wireless communication systems demonstrate significant enhancements over the previous standards. However, the emergence of novel technologies namely integrated sensing and communication (ISAC) brings about unique use-cases in which conventional security methods such as classical cryptography are unable to comply. Moreover, critical sensing information such as traffic dynamics, a user's location, and speed and environmental information including the number of scatterers, reflectors, mobility detection, can be accessed and exploited by adversaries. All these factors have galvanized the scientific research community to bring about novel security solutions including physical layer security for sensing and communication. This chapter elaborates on several security challenges incurred by sensing/communication systems and provides insights into the techniques to counter them. Furthermore, security vulnerabilities and their solutions for JRC systems have also been discussed.

Acknowledgment

This work was supported by the Scientific and Technological Research Council of Turkey (TUBITAK) under Grant no. 120C142.

References

[1] Furqan HM, Solaija MSJ, Türkmen H, *et al.* Wireless communication, sensing, and REM: a security perspective. *IEEE Open Journal of the Communications Society*. 2021;2:287–321.

[2] Lynch D. *Introduction to RF Stealth.* Raleigh, NC: Scitech Publishing Inc, 2004, p. 573.

[3] Khodakarami H, Lahouti F. Link adaptation for physical layer security over wireless fading channels. *IET Communications*. 2012;6(3):353–362.

[4] Khisti A, Wornell GW. Secure transmission with multiple antennas—Part II: The MIMOME wiretap channel. *IEEE Transactions on Information Theory*. 2010;56(11):5515–5532.

[5] Sheng Z, Tuan HD, Duong TQ, *et al.* Beamforming optimization for physical layer security in MISO wireless networks. *IEEE Transactions on Signal Processing*. 2018;66(14):3710–3723.

[6] Hafez M, Yusuf M, Khattab T, *et al.* Secure spatial multiple access using directional modulation. *IEEE Transactions on Wireless Communications*. 2017;17(1):563–573.

[7] Cumanan K, Xing H, Xu P, *et al.* Physical layer security jamming: Theoretical limits and practical designs in wireless networks. *IEEE Access*. 2016;5: 3603–3611.

[8] Dong L, Han Z, Petropulu AP, *et al.* Improving wireless physical layer security via cooperating relays. *IEEE Transactions on Signal Processing*. 2009;58(3):1875–1888.

[9] Liu TY, Lin SC, Hong YWP. On the role of artificial noise in training and data transmission for secret communications. *IEEE Transactions on Information Forensics and Security*. 2016;12(3):516–531.

[10] Zhu Q, Wu S, Hua Y. Optimal pilots for anti-eavesdropping channel estimation. *IEEE Transactions on Signal Processing*. 2020;68:2629–2644.

[11] Soltani M, Baykaş T, Arslan H. Achieving secure communication through pilot manipulation. In: *2015 IEEE 26th Annual International Symposium on Personal, Indoor, and Mobile Radio Communications* (PIMRC). New York, NY: IEEE; 2015. p. 527–531.

[12] Almohamad A, Tahir AM, Al-Kababji A, *et al.* Smart and secure wireless communications via reflecting intelligent surfaces: a short survey. *IEEE Open Journal of the Communications Society*. 2020;1:1442–1456.

[13] Wu Q, Mei W, Zhang R. Safeguarding wireless network with UAVs: a physical layer security perspective. *IEEE Wireless Communications*. 2019; 26(5):12–18.

[14] Sun Z, Balakrishnan S, Su L, *et al.* Who is in control? Practical physical layer attack and defense for mmWave-based sensing in autonomous vehicles. *IEEE Transactions on Information Forensics and Security*. 2021;16:3199–3214.

[15] Shoukry Y, Martin P, Yona Y, *et al.* Pycra: Physical challenge-response authentication for active sensors under spoofing attacks. In: *Proceedings of the 22nd ACM SIGSAC Conference on Computer and Communications Security*; 2015. p. 1004–1015.

[16] Zhou X, Maham B, Hjorungnes A. Pilot contamination for active eavesdropping. *IEEE Transactions on Wireless Communications*. 2012;11(3): 903–907.

[17] Xiong Q, Liang YC, Li KH, *et al.* An energy-ratio-based approach for detecting pilot spoofing attack in multiple-antenna systems. *IEEE Transactions on Information Forensics and Security*. 2015;10(5):932–940.

[18] Dutta RG, Guo X, Zhang T, *et al.* Estimation of safe sensor measurements of autonomous system under attack. In: *Proceedings of the 54th Annual Design Automation Conference 2017*; 2017. p. 1–6.

[19] Papernot N, McDaniel P, Wu X, *et al.* Distillation as a defense to adversarial perturbations against deep neural networks. In: *2016 IEEE Symposium on Security and Privacy* (SP). New York, NY: IEEE; 2016. p. 582–597.

[20] Cao X, Gong NZ. Mitigating evasion attacks to deep neural networks via region-based classification. In: *Proceedings of the 33rd Annual Computer Security Applications Conference*; 2017. p. 278–287.

[21] Denkovski D, Atanasovski V, Gavrilovska L, *et al.* Reliability of a radio environment map: Case of spatial interpolation techniques. In: *2012 7th International ICST Conference on Cognitive Radio Oriented Wireless Networks and Communications* (CROWNCOM). New York, NY: IEEE; 2012. p. 248–253.

[22] Han Z, Liao J, Qi Q, *et al.* Radio environment map construction by kriging algorithm based on mobile crowd sensing. *Wireless Communications and Mobile Computing*. 2019;2019:1–12.

[23] Xia H, Zha S, Huang J, *et al.* Radio environment map construction by adaptive ordinary Kriging algorithm based on affinity propagation clustering. *International Journal of Distributed Sensor Networks*. 2020; 16(5):1550147720922484.

[24] Fragkiadakis AG, Tragos EZ, Askoxylakis IG. A survey on security threats and detection techniques in cognitive radio networks. *IEEE Communications Surveys & Tutorials*. 2012;15(1):428–445.

[25] Bai L, Zhu L, Liu J, et al. Physical layer authentication in wireless communication networks: a survey. *Journal of Communications and Information Networks*. 2020;5(3):237–264.

[26] Wang X, Hao P, Hanzo L. Physical-layer authentication for wireless security enhancement: current challenges and future developments. *IEEE Communications Magazine*. 2016;54(6):152–158.

[27] Xiao L, Greenstein L, Mandayam N, *et al.* A physical-layer technique to enhance authentication for mobile terminals. In: *2008 IEEE International Conference on Communications*. New York, NY: IEEE; 2008. p. 1520–1524.

[28] Xiao L, Greenstein LJ, Mandayam NB, *et al.* Using the physical layer for wireless authentication in time-variant channels. *IEEE Transactions on Wireless Communications*. 2008;7(7):2571–2579.

[29] Liu Y, Chen HH, Wang L. Physical layer security for next generation wireless networks: theories, technologies, and challenges. *IEEE Communications Surveys & Tutorials*. 2016;19(1):347–376.

[30] Weinand A, Karrenbauer M, Lianghai J, *et al.* Physical layer authentication for mission critical machine type communication using Gaussian mixture

model based clustering. In: *2017 IEEE 85th Vehicular Technology Conference* (VTC Spring). New York, NY: IEEE; 2017. p. 1–5.

[31] Xie N, Zhang S. Blind authentication at the physical layer under time-varying fading channels. *IEEE Journal on Selected Areas in Communications*. 2018;36(7):1465–1479.

[32] Paul LY, Verma G, Sadler BM. Wireless physical layer authentication via fingerprint embedding. *IEEE Communications Magazine*. 2015;53(6):48–53.

[33] Wang X, Liu FJ, Fan D, *et al.* Continuous physical layer authentication using a novel adaptive OFDM system. In: *2011 IEEE International Conference on Communications* (ICC). New York, NY: IEEE; 2011. p. 1–5.

[34] Kumar V, Park JM, Bian K. Blind transmitter authentication for spectrum security and enforcement. In: *Proceedings of the 2014 ACM SIGSAC Conference on Computer and Communications Security*; 2014. p. 787–798.

[35] Xie N, Chen C, Ming Z. Security model of authentication at the physical layer and performance analysis over fading channels. *IEEE Transactions on Dependable and Secure Computing*. 2018;18(1):253–268.

[36] Supangkat S, Eric T, Pamuji A. A public key signature for authentication in telephone. In: *Asia-Pacific Conference on Circuits and Systems*. vol. 2. New York, NY: IEEE; 2002. p. 495–498.

[37] Van Huynh N, Nguyen DN, Hoang DT, *et al.* "Jam Me If You Can:" defeating jammer with deep dueling neural network architecture and ambient backscattering augmented communications. *IEEE Journal on Selected Areas in Communications*. 2019;37(11):2603–2620.

[38] Hoang DT, Nguyen DN, Alsheikh MA, *et al.* "Borrowing Arrows with Thatched Boats": the art of defeating reactive jammers in IoT networks. *IEEE Wireless Communications*. 2020;27(3):79–87.

[39] Lyu B, Hoang DT, Gong S, *et al.* IRS-based wireless jamming attacks: When jammers can attack without power. *IEEE Wireless Communications Letters*. 2020;9(10):1663–1667.

[40] Chorti A, Barreto AN, Köpsell S, *et al.* Context-aware security for 6G wireless: the role of physical layer security. *IEEE Communications Standards Magazine*. 2022;6(1):102–108.

[41] Daniels RC, Yeh ER, Heath RW. Forward collision vehicular radar with IEEE 802.11: feasibility demonstration through measurements. *IEEE Transactions on Vehicular Technology*. 2017;67(2):1404–1416.

[42] Rafique S, Arslan H. A novel frame design for integrated communication and sensing based on position modulation. In: *2021 IEEE 94th Vehicular Technology Conference* (VTC2021-Fall); 2021. p. 1–5.

[43] Wei Z, Liu F, Masouros C, *et al.* Toward multi-functional 6G wireless networks: integrating sensing, communication, and security. *IEEE Communications Magazine*. 2022;60(4):65–71.

[44] Shu F, Xu L, Wang J, *et al.* Artificial-noise-aided secure multicast precoding for directional modulation systems. *IEEE Transactions on Vehicular Technology*. 2018;67(7):6658–6662.

[45] Zhu J, Schober R, Bhargava VK. Linear precoding of data and artificial noise in secure massive MIMO systems. *IEEE Transactions on Wireless Communications*. 2015;15(3):2245–2261.

[46] Su N, Liu F, Masouros C. Enhancing the physical layer security of dual-functional radar communication systems. In: *2019 IEEE Global Communications Conference* (GLOBECOM). New York, NY: IEEE; 2019. p. 1–6.

[47] Chalise BK, Amin MG. Performance tradeoff in a unified system of communications and passive radar: a secrecy capacity approach. *Digital Signal Processing*. 2018;82:282–293.

[48] Ma WKK. Semidefinite relaxation of quadratic optimization problems and applications. *IEEE Signal Processing Magazine*. 2010;1053(5888/10).

[49] Wu K, Zhang JA, Huang X, *et al.* Frequency-hopping MIMO radar-based communications: an overview. *IEEE Aerospace and Electronic Systems Magazine*. 2021:42–54.

[50] Moon T, Park J, Kim S. BlueFMCW: random frequency hopping radar for mitigation of interference and spoofing. *EURASIP Journal on Advances in Signal Processing*. 2022;2022(1):1–17.

[51] Guerci JR. *Space–Time Adaptive Processing for Radar*. Norwood, MA: Artech House; 2014.

[52] Wang Y, Zha Y, Huang Y, *et al.* A deconvolution method for ship detection in sea clutter environment. In: *2015 IEEE International Geoscience and Remote Sensing Symposium* (IGARSS). New York, NY: IEEE; 2015. p. 1590–1593.

[53] Washington R, Bischof B, Garmatyuk D, *et al.* Clutter-masked waveform design for LPI/LPD radarcom signal encoding. *Sensors*. 2021;21(2):631.

[54] Chu J, Liu R, Liu Y, *et al.* Joint transmit beamforming design for secure communication and radar coexistence systems. In: *2022 IEEE Wireless Communications and Networking Conference* (WCNC); 2022. p. 205–209.

Chapter 12
Physical layer security in non-terrestrial networks

Olfa Ben Yahia[1], Eylem Erdogan[2], and Gunes Karabulut Kurt[1]

12.1 Introduction

Since the establishment of the first analog communication network in the 1980s, the mobile communication network has undergone phenomenal development. This development is not a one-step process, but rather consists of several generations with different standards, potentials, and techniques. On average, a new generation is introduced every ten years. So far, as the fifth-generation (5G) began to be commercialized, research institutes around the world turned their attention to sixth-generation (6G) networks, which are expected to be rolled out around 2030.

6G networks are anticipated to outperform 5G networks in terms of capacity and support for developing services and applications, making them a crucial enabler for the intelligent information society of the 2030s while 5G wireless networks are the primary enabler for the information society of the 2020s. In order to meet the needs of the smart data society of 2030, 6G wireless network research has been added to the agenda. The smart data society of the 2030s will be fully digitized, data-driven on a global scale, and intelligent due to near-instant and unrestricted broad wireless connectivity. 6G will be crucial for the execution of this plan; it will connect everything, enable full coverage, and combine all features, including communication, sensing, caching, computation, control, location, radar, navigation, and imaging, to enable full-vertical applications. Consequently, a multidimensional network integrating non-terrestrial and terrestrial networks is required to support a variety of applications, including aerial flight, maritime travel, and land vehicles.

The evolution of telecommunication technologies, the ever-increasing demand for new services, and the rapid evolution of smart devices are the driving forces behind the development of non-terrestrial networks (NTN) as an effective solution to complement terrestrial networks in providing services over uncovered or under-served geographic areas [1]. An NTN is a network in which spaceborne [i.e., geostationary Earth orbit (GEO), medium Earth orbit (MEO), and low Earth

[1]Department of Electrical Engineering, Polytechnique Montréal, Canada
[2]Department of Electrical and Electronics Engineering, Istanbul Medeniyet University, Turkey

orbit (LEO)] or airborne [i.e., unmanned aircraft system (UAS) and high altitude platform station (HAPS)] vehicles function as relay nodes or base stations, distinguishing transparent and regenerative satellite systems [2]. The different types of satellites (or UAS platforms) are given in Table 12.1 according to the 3rd Generation Partnership Project (3GPP) [3]. NTNs distinguishing characteristic is the capability to provide wide-area connectivity by delivering connections across regions that are costly or difficult to serve with terrestrial networks (e.g., rural areas, vessels, airplanes). As a result, NTN is expanding the terrestrial network's coverage in a global market where customer requirements are changing rapidly. In addition, NTN guarantees scalability, service continuity, and availability, thereby revolutionizing the existing cellular communications infrastructure. Industrial and academic interest in NTN systems over the past decade have provided the initial motivation for this investigation.

One of the most promising architectural features of 6G networks is the seamless integration of space and terrestrial networks. This architecture, known as a vertical heterogeneous network (VHetNet), consists of space, aerial, and terrestrial networks. The envisioned space network layer is comprised of three different types of satellite constellations, which can be classified based on their altitudes and orbit types and are connected by high-speed inter-satellite links.

Specifically, GEO satellites are orbiting at about 36,000 km above the Earth. MEO satellites are between 2,000 and 20,000 km, and LEO satellites are deployed on a circular orbit at 500–2,000 km of altitude. By offering distance-insensitive point-to-multipoint communications, one key component of satellite communication (SatCom) is to simultaneously and efficiently transmit the signal across the Earth. Satellites have been used in the past for a variety of purposes, including television broadcasting, data transmission, navigation, weather prediction, surveillance systems, planetary exploration, and so on. However, high orbit satellite communications are unsuitable for delay-sensitive traffic due to their limited resolution and low latency. It is expected that LEO satellites will play a crucial role in the future of

Table 12.1 Types of NTN platforms

Platforms	Altitude range	Orbit	Typical beam footprint size
LEO satellite	300–1,500 km	Circular around	100–1,000 km
MEO satellite	7,000–25,000 km	the earth	100–1,000 km
GEO satellite	35,786 km	Notional station	200–3,500 km
UAS platform (including HAPS)	8–50 km (20 km for HAPS)	keeping position fixed in terms of elevation azimuth with respect to a given earth point	5–200 km
High Elliptical Orbit (HEO)	400–50,000 km	Elliptical around the earth	200–3,500 km

wireless communications by providing a high data rate, low latency, and wide coverage that includes rural areas where wired communication is difficult or impossible to implement. For this reason, thousands of LEO satellites are expected to be launched and deployed in the next decade as a variety of private and public organizations seek to create a flexible network with ubiquitous global coverage and high capacity.

Different from SatCom systems, aerial networks can be categorized into three groups: low-altitude platform stations, unmanned aerial vehicles (UAV), and HAPS systems. According to the International Telecommunication Union's (ITU) definition, a HAPS system is a fixed object between 20 and 50 km in altitude. However, the majority of recent deployments have focused on an altitude of 18–20 km [4]. In particular, HAPS systems positioned in the stratosphere can combine the characteristics of terrestrial networks (low cost and latency) and satellites (wide coverage area) [5]. Moreover, HAPS systems are quasi-stationary aerial vehicles that are less susceptible to atmospheric turbulence because they are positioned above cloud formations [6]. Nonetheless, HAPS systems are susceptible to stratospheric attenuation, which may be caused by sulfuric acid components resulting from volcanic activity, gases, and polar clouds [7]. Alongside the HAPS systems, UAV platforms employed near ground level play an important role by providing a low-cost solution for temporary wireless delivery services to remote areas that cannot be reached by satellites or HAPS systems.

For the majority of applications, the current architecture of SatCom relies on the microwave radio frequency (RF) band [8]. In RF communication, 100 MHz–50 GHz frequencies are considered depending on the types and applications of satellites. Furthermore, the main advantage of RF communication is that is available on both line-of-sight (LOS) and non-LOS (NLOS). However, RF communication suffers from limited capacity, spectrum congestion, low bandwidth, and regulatory constraints. Additionally, RF links are susceptible to interception or jamming, which creates security issues. To meet the high data rates, free-space optical (FSO) communication links are regarded as a crucial component for NTNs.

On the other hand, FSO communication ensures high data rate transmission with a low probability of interception and offers LOS connectivity over unlicensed spectrum [9,10]. Despite their unique characteristics, FSO links are susceptible to significant variations in the phase and intensity of the received signal due to fluctuations in the index of refraction caused by changes in temperature and pressure, particularly for long-distance communications [11]. In addition, FSO communication is highly affected by turbulence-induced signal degradation and attenuation. More precisely, fog or snow can significantly degrade FSO links, whereas rain attenuation has less of an effect [12]. In addition, optical SatCom is susceptible to beam scintillation and beam wandering, which are primarily caused by large-scale atmospheric inhomogeneities. In uplink communications, beam wander can be a significant factor as the beam is affected heavily by turbulent eddies which may cause beam displacement and link failure [13]. However, it may be negligible for downlink transmissions. In addition, acquisition and pointing are critical in FSO SatCom due to device vibration, platform jitter, or any type of strain in electronic or mechanical equipment. In order to prevent link failure, the transmitter and receiver must maintain LOS connectivity by using proper alignment techniques [13].

In the existing studies, different fading models are proposed for RF communication. However, it has been demonstrated that shadowed-Rician fading models the RF channels in SatCom very well. Shadowed-Rician was initially proposed by [14] in 2003. It has applications in a variety of frequency bands. In addition, it is assumed that the amplitude of the LOS that follows the Nakagami fading is random due to LOS path obstructions by buildings, trees, and mountains, making it more realistic. This model's primary advantage is that it yields closed-form and mathematically tractable expressions [14].

In the current literature, FSO communication performance has been examined utilizing fading channels such as Log-normal, Gamma–Gamma, and Malága. Likewise, exponentiated Weibull (EW) fading can be used in optical NTNs as is the best fit for various aperture sizes across a wide range of atmospheric conditions, particularly when aperture averaging is employed. In the aperture averaging technique, scintillation is spatially averaged in the aperture averaging approach to reduce the influence of turbulence and to increase overall performance [15,16]. Spatial diversity can be built between satellites and HAPS systems, similar to aperture averaging, to provide stable communication and decrease the impact of atmospheric turbulence conditions. Due to masking effects and barriers, LOS communication cannot be maintained between the satellites and the ground receivers due to the optical beam's limitations. Therefore, in SatCom, using a HAPS as an intermediary relay node can be considered a promising approach to mitigate channel impairments and to benefit from both communication systems' advantages, including greater coverage and reduced delay. The complementing features of RF and FSO communications can also be used in combination to reduce the impacts of turbulence and improve the overall performance of the proposed systems [17]. There are two possible configurations for this combination. The first configuration is the mixed RF-FSO model, which uses RF communication at one hop and FSO communication at the other in a dual-hop architecture [18]. The second one is a hybrid RF/FSO configuration in which, FSO is used in parallel with RF communication [19,20]. Using these configurations, it is possible to reap the benefits of both RF and FSO communications while limiting the effects of weather-dependent consequences.

The ability to communicate private data in NTNs is a serious concern. Traditionally, communication security is achieved by implementing encryption techniques that rely on the upper layers of the open systems interconnection (OSI) model. These methods are classified according to whether the shared key is public or private. Eavesdroppers will not be able to decrypt the message's secret key without the use of sophisticated encryption techniques and powerful computers. For applications that require a high level of security, these techniques are no longer appropriate due to the rapid development of computer processing power and ways to interrupt encryption schemes. Instead of cryptographic models, physical layer security (PLS) which has been proven to boost information security, has recently received a lot of attention. As compared to traditional cryptographic algorithms, PLS uses signal and channel properties, such as difference and reciprocity, to provide secure transmission between users without respect to the computational capability of illegitimate users. PLS is a frequent term for information-theoretic security. When it comes to cipher systems, Shannon was the first to develop information-theoretic security in 1949 by not relying on computational limits [21]. Latter, Wyner introduced the wiretap model, in which he suppose that the main channel should be superior to the eavesdropper to ensure perfect secrecy [22].

In the existing literature, the majority of research emphasizes RF eavesdropping, in which the eavesdropper intercepts information from RF channels. PLS has recently been implemented for FSO communication. In optical communication, the eavesdropper must be placed in close proximity to the communicating users in order to collect the beam reflected from aerosols or gases in the environment. Nevertheless, if the attacker blocks the laser beam while intercepting the signal, the legitimate receiver can detect the power loss and therefore the transmission is ended due to security concerns. Another possibility of eavesdropping can occur when severe pointing errors or diffraction cause misalignment between the legitimate users. In this case, the beam disperses and the eavesdropper may get refracted beam [23]. Even though optical communication can be intercepted, PLS performance has received little attention. The authors in [24] investigated the PLS for three realistic use-cases while assuming different eavesdropper locations. It was observed that when the attacker is located close to the transmitter, atmospheric conditions have less effect on the secrecy performance. The situations in which the attacker is close to either the receiver or the transmitter are also investigated in [25]. The authors revealed that when the eavesdropper is close to the main transmitter, it is easier to capture the signal due to the huge attenuation experienced by the signal when propagating through the atmosphere. The closed-form expressions of average secrecy capacity (ASC), secrecy outage probability (SOP), and strictly positive secrecy capacity (SPSC) over Málaga atmospheric turbulence fading in the presence of pointing errors are derived in [26]. The authors of [27] examined how well FSO communication worked in terms of a new metric called "effective secrecy throughput." Also, the authors of [28] investigated the ASC for the Log-normal fading model under conditions of generalized misalignments and atmospheric turbulence. Recently, the authors of [29] suggested adding artificial noise to FSO channels over Gamma–Gamma fading when there are pointing errors. This would improve the secrecy performance of FSO channels. Security issues for FSO communication systems are getting more and more attention, according to the most recent studies. But these works concentrate on optical eavesdropping on the ground level. So far, there is a big gap in what we know about PLS for FSO in NTNs.

In what follows, we present interesting system models to discuss optical eavesdropping in NTN systems.

12.2 Eavesdropping in RF communication

In the current literature, several studies have been presented to provide secure communication in SatCom [30–33]. The authors in [30] studied SOP and ASC performance of colluding and non-colluding eavesdroppers for a multiuser SatCom architecture. To improve system security, they proposed a scheduling strategy based on the predetermined threshold. In [31], the authors investigated the impact of hardware impairments on the PLS performance for colluding and non-colluding eavesdropping while considering a NOMA-based integrated satellite with partial relay selection in cooperative networks. In [32], the authors evaluated SOP and ASC performance for terrestrial cooperative SatCom considering amplify-and-forward (AF) and decode-and-forward (DF) relaying techniques. In addition, the SOP performance of a multi-antenna satellite-terrestrial setup was investigated for the random location of the ground receivers and eavesdroppers [33]. Different from the current literature that considers terrestrial relay nodes, the authors in [34] investigate the security performance of SatCom using a non-terrestrial platform relay scheme. The system model in [34] is presented in the following subsections.

12.2.1 System model

In this system model, to examine RF eavesdropping in downlink SatCom, the authors proposed a new HAPS-assisted dual-hop downlink communication [34]. As depicted in Figure 12.1, they consider an LEO satellite (S) that is circularly orbiting at 500 km

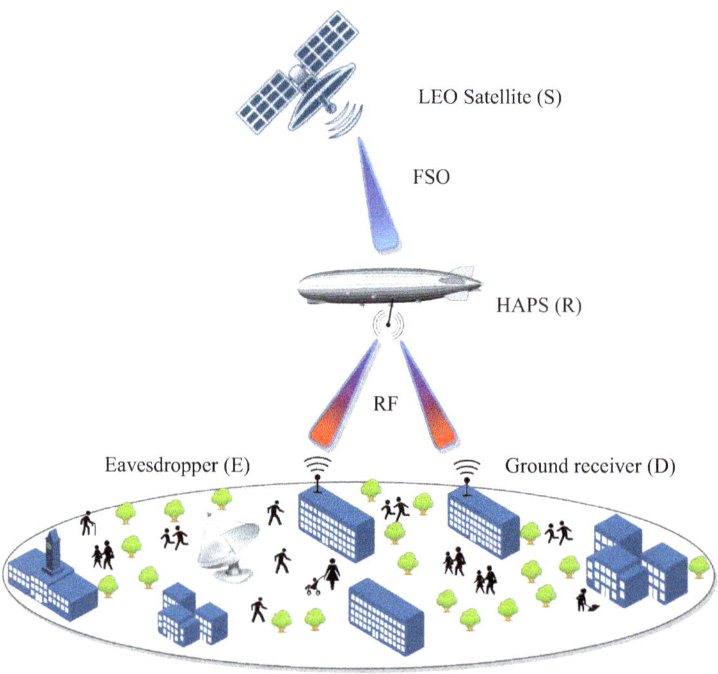

Figure 12.1 Illustration of HAPS-assisted FSO-RF SatCom

communicating with a GS (D) via a quasi-stationary HAPS node at an altitude of 14 km while an unauthorized attacker (E) located on the ground tries to capture the transmitted signal. FSO communication is utilized between the LEO satellite and the HAPS node, while RF communication is considered between HAPS and ground receivers. For this setup, they derived closed-form expressions for SOP and probability of positive secrecy capacity (PPSC) by analyzing the implications of pointing errors and various shadowing levels as shown in the following subsection.

12.2.2 Secrecy performance analysis

In this scenario, the FSO communication follows the Gamma–Gamma fading and the RF channels follow the shadowed-Rician fading model. Considering the DF relaying scheme, the overall signal-to-noise ratio (SNR) of the system can be expressed as

$$\gamma_0 = \min(\gamma_{SR}^{FSO}, \gamma_{RD}^{RF}), \tag{12.1}$$

where $\gamma_{SR}^{FSO} = \frac{P_s}{N_0} I_{SR}^2$ defines the instantaneous SNR of the FSO channel and $\gamma_{RD}^{RF} = \frac{P_R}{N_0}|h_{RD}|^2$ presents the instantaneous SNR of the intended receiver.

For the RF link, the probability density function (PDF) and the cumulative distribution function (CDF) of the received SNR are given by [32]

$$f_{\gamma_z}^{RF}(\gamma) = \sum_{k=0}^{m_z-1} \frac{\alpha_z(1-m_z)_k \cdot (-\delta_z)^k \gamma^k}{\overline{\gamma}_z^{k+1} \cdot (k!)^2} \exp(-\lambda_z \gamma), \tag{12.2}$$

$$F_{\gamma_z}^{RF}(\gamma) = 1 - \sum_{k=0}^{m_z-1} \sum_{i=0}^{k} \frac{\alpha_z(1-m_z)_k \cdot (-\delta_z)^k \gamma^i}{i! \lambda_z^{k-i+1} \overline{\gamma}_z^{k+1} . k!} \exp(-\lambda_z \gamma), \tag{12.3}$$

where $\lambda_z = \frac{\beta_z - \delta_z}{\overline{\gamma}_z}$, $\overline{\gamma}_z$ indicates the average SNR of h_z. $\alpha_z = \frac{1}{2b_z}(\frac{2b_z m_z}{2b_z m_z + \Omega_z})^{m_z}$, $\beta_z = \frac{1}{2b_z}$, $\delta_z = \frac{\Omega_z}{2b_z(2b_z m_z + \Omega_z}$ with m_z being the Nakagami parameter of the corresponding link, and Ω_z, $2b_z$ are the average power of the LoS component and multipath component. Finally, $(\cdot)_k$ denotes the Pochhammer symbol.

For the FSO communication, the PDF and CDF of the instantaneous SNR at R can be given as [32]

$$f_{\gamma_{SR}}^{FSO}(\gamma) = \frac{\xi^2}{r\Gamma(\alpha)\Gamma(\beta)\gamma} G_{1,3}^{3,0}\left(h\alpha\beta(\tfrac{\gamma}{\mu_r})^{\frac{1}{r}} \middle| \begin{matrix} \xi^{2+1} \\ \xi^2, \alpha, \beta \end{matrix} \right), \tag{12.4}$$

$$F_{\gamma_{SR}}^{FSO}(\gamma) = \frac{\xi^2 r^{\alpha+\beta-2}}{2\pi^{r-1}\Gamma(\alpha)\Gamma(\beta)} G_{r+1,3r+1}^{3r,1}\left(\tfrac{(h\alpha\beta)^r}{\mu_r r^{2r}}\gamma \middle| \begin{matrix} 1, k_1 \\ k_2, 0 \end{matrix} \right), \tag{12.5}$$

where the parameter r represents the used type of detection at the HAPS node (i.e., $r = 1$ for HD and $r = 2$ denotes intensity modulation with direct detection (IM/DD)), the coefficient ξ indicates the pointing error displacement at the receiver, α and β imply the severity of fading and scintillation produced by the atmospheric turbulence conditions, $h = \frac{\xi^2}{\xi^2+1}$, μ_r is the average SNR of the corresponding FSO link within the specific type of detection, and $G_{p,q}^{m,n}\left(x\middle|\begin{matrix}a_1,\ldots a_p\\ b_1,\ldots b_q\end{matrix}\right)$ is the Meijer G-function [35]. Furthermore, $k_1 = \Delta(r, \xi^2 + 1)$, $k_2 = \Delta(r, \xi^2), \Delta(r, \alpha), \Delta(r, \beta)$ where the notation $\Delta(k, a)$ denotes $\Delta(k, a) = \frac{a}{k}, \frac{a+1}{k}, \ldots, \frac{a+k-1}{k}$ including k values. According to [11, Section 12], α and β can be written as

$$\alpha = \left\{\exp\left[\frac{0.49\sigma_R^2}{(1 + 1.11\sigma_R^{12/5})^{7/6}}\right] - 1\right\}^{-1}, \tag{12.6}$$

$$\beta = \left\{\exp\left[\frac{0.51\sigma_R^2}{(1 + 0.69\sigma_R^{12/5})^{5/6}}\right] - 1\right\}^{-1}, \tag{12.7}$$

where σ_R is the Rytov variance given as [11, Section 12]

$$\sigma_R^2 = 2.25K^{7/6}sec^{11/6}(\zeta)\int_{h_0}^{H} C_n^2(h)(h - h_0)^{5/6}dh, \tag{12.8}$$

where $K = \frac{2\pi}{\lambda}$ represents the wave number, λ is the wavelength, ζ indicates the zenith angle, and $C_n^2(h)$ is the refractive-index structure parameter expressed as [36]

$$\begin{aligned} C_n^2 = {} & 0.00594(w/27)^2(10^{-5}h)^{10}\exp(-h/1{,}000) + 2.7\\ & \times 10^{-16}\exp(-h/1500) + A\exp(-h/100), \end{aligned} \tag{12.9}$$

where A is a nominal value of $C_n^2(0)$ at the ground in $\mathrm{m}^{-2/3}$, w is the RMS wind spread [m/s], h is the altitude, H is the altitude of S, and h_0 represents the height of R node above ground level.

As mentioned before, the SOP expression can be written as

$$\begin{aligned} P_{SO} &= \Pr\left[C_s < R_s\right],\\ &= \Pr\left[\log_2(1 + \gamma_D) - \log_2(1 + \gamma_E) < R_s\right]\\ &= \int_0^{\infty} F_{\gamma_0}(\gamma\gamma_{th} + \gamma_{th} - 1) f_{\gamma_E}(\gamma)\, d\gamma,\\ &\simeq \int_0^{\infty} F_{\gamma_0}(\gamma\gamma_{th}) f_{\gamma_E}(\gamma)\, d\gamma, \end{aligned} \tag{12.10}$$

where $\gamma_{th} = 2^{2R_s}$ with R_s presenting the secrecy rate and γ_E is the instantaneous SNR of the eavesdropper.

Therefore, by using [35, Eqn. 07.34.21.0011.01] and with the aid of $\exp(-bx) = G_{0,1}^{1,0}\left(bx \middle| \begin{matrix} - \\ 0 \end{matrix}\right)$, the closed-form expression of SOP can be derived as

$$\begin{aligned} P_{SO} = & 1 - \sum_{p=0}^{m_{RD}-1} \sum_{t=0}^{p} \frac{\alpha_{RD}(1-m_{RD})_p \cdot (-\delta_{RD})^p}{t! \lambda_{RD}^{p-t+1} \overline{\gamma}_{RD}^{p+1} . p!} \\ & \times \sum_{q=0}^{m_{RE}-1} \frac{\alpha_{RE}(1-m_{RE})_q \cdot (-\delta_{RE})^q}{\overline{\gamma}_{RE}^{q+1} \cdot (q!)^2} \gamma_{th}^t \\ & \times \left[(\lambda_{SD}\gamma_{th})^{-(q+t+1)} G_{1,1}^{1,1}\left(\frac{\lambda_{SE}}{\lambda_{SD}\gamma_{th}} \middle| \begin{matrix} -(q+t) \\ 0 \end{matrix} \right) \right. \\ & - G_{r+2,3r+1}^{3r,2}\left(\frac{D\gamma_{th}}{(\lambda_{RE}+\lambda_{RD}\gamma_{th})} \middle| \begin{matrix} 1, -(q+t), k_1 \\ k_2, 0 \end{matrix} \right) \\ & \left. \times B(\lambda_{RE}+\lambda_{RD}\gamma_{th})^{-(q+t+1)} \right]. \end{aligned} \tag{12.11}$$

In addition, mathematically speaking, the PPSC expression can be obtained using the following formula:

$$P_{PPSC} = \Pr\left[\log_2(1+\gamma_0) - \log_2(1+\gamma_E) > 0\right], \tag{12.12}$$

and after some manipulations, it can be written as

$$P_{PPSC} = 1 - \int_0^{\infty} F_{\gamma_D}(\gamma) f_{\gamma_E}(\gamma) d\gamma. \tag{12.13}$$

Thus, the final expression of PPSC for this model can be obtained as

$$\begin{aligned} P_{PPSC} = & \sum_{p=0}^{m_{RD}-1} \sum_{t=0}^{p} \frac{\alpha_{RD}(1-m_{RD})_p \cdot (-\delta_{RD})^p}{t! \lambda_{RD}^{p-t+1} \overline{\gamma}_{RD}^{p+1} \cdot p!} \\ & \times \sum_{q=0}^{m_{RE}-1} \frac{\alpha_{RE}(1-m_{RE})_q \cdot (-\delta_{RE})^q}{\overline{\gamma}_{RE}^{q+1} \cdot (q!)^2} \gamma_{th}^t \\ & \times \left[(\lambda_{SD}\gamma_{th})^{-(q+t+1)} \times G_{1,1}^{1,1}\left(\frac{\lambda_{SE}}{\lambda_{SD}\gamma_{th}} \middle| \begin{matrix} -(q+t) \\ 0 \end{matrix} \right) \right. \\ & - G_{r+2,3r+1}^{3r,2}\left(\frac{D\gamma_{th}}{(\lambda_{RE}+\lambda_{RD}\gamma_{th})} \middle| \begin{matrix} 1, -(q+t), k_1 \\ k_2, 0 \end{matrix} \right) \\ & \left. \times B(\lambda_{RE}+\lambda_{RD}\gamma_{th})^{-(q+t+1)} \right]. \end{aligned} \tag{12.14}$$

In the following part, numerical results for the proposed system model are provided under different conditions. For the FSO channel, they considered a heterodyne detection technique for an optical wavelength $\lambda = 1550$ nm. Windy weather is assumed with a wind speed level of $w = 65$ m/s and a zenith angle of $\zeta = 75°$. For the RF links different shadowing levels are considered: frequent heavy shadowing ($m = 1.0$, $b = 0.063$, $\Omega = 8.94 \times 10^{-4}$), average shadowing ($m = 10$, $b = 0.126$, $\Omega = 0.835$), and infrequent light shadowing ($m = 19$, $b = 0.158$, $\Omega = 1.29$). The secrecy threshold is set to $R_s = 0.01$ nats/s/Hz.

Figure 12.2 depicts the SOP as a function of the average SNR per-hop $\overline{\gamma}$ for various eavesdropper SNRs assuming frequent heavy shadowing. The theoretical curves, which are depicted by solid lines, are validated by Monte Carlo (MC) simulations depicted by circles. In addition, we can observe the simulations' good agreement with the obtained derivations, which validates them. In addition, it is evident from this graph that greater $\overline{\gamma}_E$ values indicate lower SOP performance. Furthermore, the figure shows that the proposed scenario outperforms the direct RF SatCom scenario due to the high path loss caused by the great distance between the satellite and the GS.

Figure 12.3 investigates the pointing error effect for frequent heavy shadowing and average shadowing. As shown in the figure, when $\xi = 1.1$, indicating a lower level of pointing, the SOP under average shadowing performs better than the SOP under frequent heavy shadowing. However, when $\xi = 0.8$, the overall performance of the SOP deteriorates under average shadowing. As a result of the severe

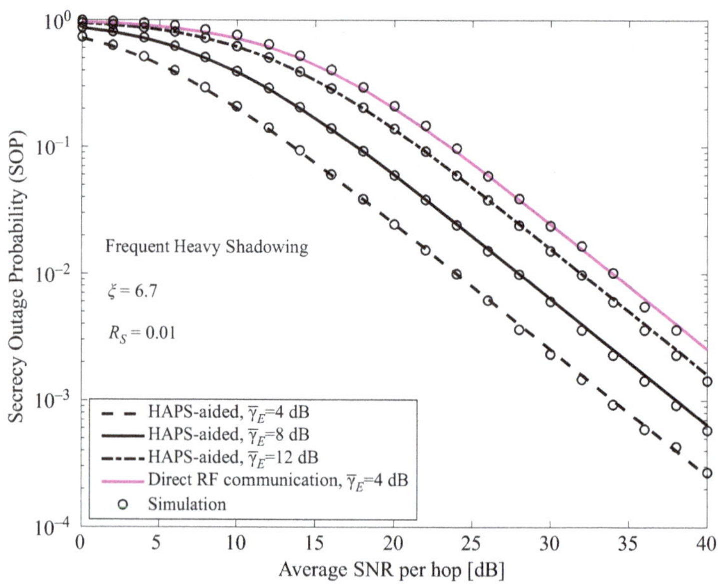

Figure 12.2 SOP performance of the proposed model for different $\overline{\gamma}_E$

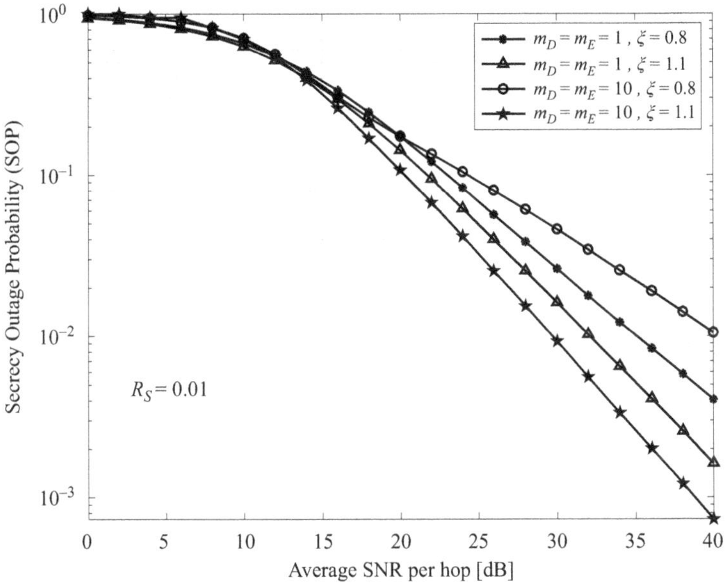

Figure 12.3 Impact of pointing error on the SOP performance for different shadowing levels, $\overline{\gamma}_E = 12$ dB

misalignment caused by a high level of pointing error, the link between the satellite and HAPS dominates the communication, despite the fact that the channel between HAPS and the ground has been improved.

Figure 12.4 illustrates the PPSC performance of the RF eavesdropper attack with varying γ_E values when both channels of D and E are frequently heavily shadowed when $\xi = 6.7$. According to the obtained results, the PPSC increases for lower γ_E. This demonstrates the significance of γ_E in ensuring secure communication.

To conclude the findings of this work can be summarised as follows:

- Extreme shadowing conditions can have a negative impact on the overall secrecy performance of HAPS-aided FSO-RF SatCom.
- Given that the HAPS node is positioned in the stratosphere, the FSO channel between the satellite and the HAPS node cannot be significantly degraded by adverse weather. Consequently, HAPS can ensure secure communication even in the presence of strong wind levels, and the overall secrecy performance can be improved.
- The obtained results have demonstrated that larger values of pointing inaccuracy improve overall performance. As ξ is associated with the misalignment of the communicating nodes, it can be viewed as a crucial parameter in the design of FSO communication.
- The average SNR of the eavesdropper can be viewed as a significant metric for securing HAPS-assisted SatCom.

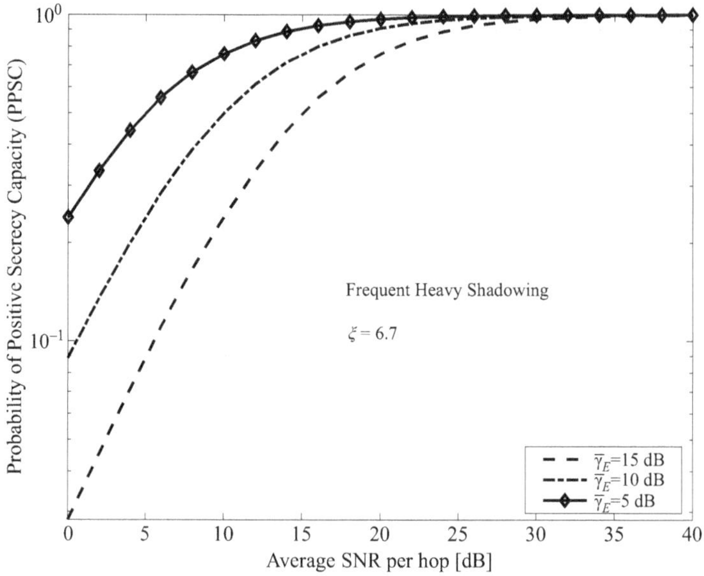

Figure 12.4 PPSC performance for different $\overline{\gamma}_E$ levels under frequent heavy shadowing

In this subsection, we presented a system model that studies the RF eavesdropping for NTNs. In the following subsection, we will present FSO eavesdropping in space and air.

12.3 Eavesdropping in FSO communication

It is true that recent research has focused on security issues in optical networks for both visible light and FSO communications. However, the majority of these works are limited to terrestrial communications. As a result, there is a gap in the body of knowledge on eavesdropping on FSO communications in space and air. Additionally, EW fading has not been taken into account in any of the previous works on FSO eavesdropping. To the best of the authors' knowledge, the works in [37,38] introduced the PLS performance in optical communication for NTN systems. In the following subsections, the works in [37,38] are presented along with the main results and the main observations.

12.3.1 Eavesdropping in space/air

12.3.1.1 System model

In this model [37], the authors consider two use-cases of FSO eavesdropping for aerial networks' downlink communication, as illustrated in Figure 12.5. In both scenarios, a single-antenna transmitter is in communication with a flying receiver node

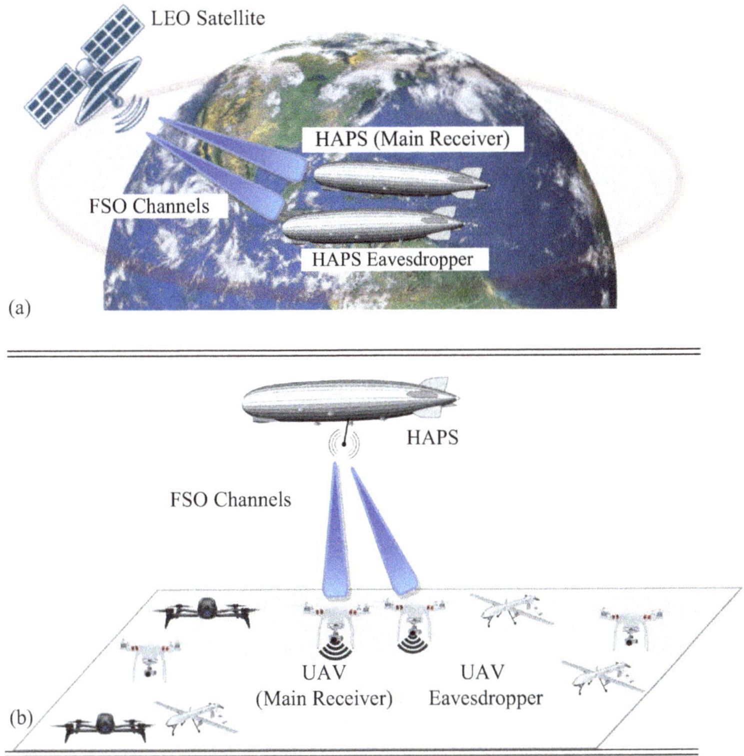

Figure 12.5 (a) HAPS eavesdropping and (b) UAV eavesdropping

while being overheard by an external attacker (E). In the first scenario, a HAPS eavesdropper tries to intercept the signal sent when an LEO satellite is in communication with a legitimate quasi-stationary HAPS node located in the stratosphere. In the second scenario, the authors assume that a HAPS node is transmitting sensitive data to a UAV system hovering near ground level while a nearby UAV is trying to capture the transmission. Both scenarios involve EW fading model for FSO channels.

For this model, the expressions of SOP and PPSC for FSO eavesdropping are derived over EW fading channels.

12.3.1.2 Secrecy performance analysis

For both models, the instantaneous received SNR at node j where $j \in \{D, E\}$ can be given as

$$\gamma_j = \frac{P_j}{N_0} f_j^2 I_j^2 = \overline{\gamma}_j I_j^2, \tag{12.15}$$

where f_j indicates the attenuation depending on the altitude of the flying platform, $\overline{\gamma}_j$ indicates the average SNR at node j, and $E[I_j^2] = 1$.

Considering the EW distribution for all links, the PDF and the CDF can be expressed, respectively, as [39]

$$f_{I_j}(I) = \frac{\alpha_j \beta_j}{\eta_j} \left(\frac{I}{\eta_j}\right)^{\beta_j - 1} \exp\left[-\left(\frac{I}{\eta_j}\right)^{\beta_j}\right] \times \left(1 - \exp\left[-\left(\frac{I}{\eta_j}\right)^{\beta_j}\right]\right)^{\alpha_j - 1}, \tag{12.16}$$

$$F_{\gamma_j}(\gamma) = \sum_{\rho=0}^{\infty} \binom{\alpha_j}{\rho} (-1)^{\rho} \exp\left[-\rho\left(\frac{\gamma}{\eta_j^2 \overline{\gamma}_j}\right)^{\frac{\beta_j}{2}}\right], \tag{12.17}$$

where η_j is the scale parameter. α_j, β_j are the shape parameters that can be given as

$$\alpha_j = \frac{7.220 \times \sigma_{I_j}^{2/3}}{\Gamma\left(2.487\sigma_{I_j}^{2/6} - 0.104\right)},$$
$$\beta_j = 1.012\left(\alpha_j \sigma_{I_j}^2\right)^{-13/25} + 0.142,$$
$$\eta_j = \frac{1}{\alpha_j \Gamma\left(1 + 1/\beta_j\right) g_1(\alpha_j, \beta_j)}, \tag{12.18}$$

where $\Gamma(\cdot)$ represents the Gamma function and $g_1(\alpha_j, \beta_j)$ is α_j and β_j-dependent constant variable given by [16].

Additionally, the PDF of γ_E can be derived from (12.17) with respect to γ as

$$f_{\gamma_E}(\gamma) = \frac{\alpha_E \beta_E \gamma^{\frac{\beta_E}{2} - 1}}{2(\eta_E^2 \overline{\gamma}_E)^{\frac{\beta_E}{2}}} \sum_{k=0}^{\infty} \binom{\alpha_E - 1}{k} (-1)^k \times \exp\left[-(k+1)\left(\frac{\gamma}{\eta_E^2 \overline{\gamma}_E}\right)^{\frac{\beta_E}{2}}\right]. \tag{12.19}$$

Thereafter, by using [40, 3.478.1], SOP expression which is defined as given in (12.10), and after a few manipulations, the final expression of SOP for both models can be obtained as

$$P_{SO} = \frac{\alpha}{(\eta^2 \overline{\gamma}_E)^{\frac{\beta}{2}}} \sum_{\rho=0}^{\infty} \binom{\alpha}{\rho} (-1)^{\rho} \sum_{k=0}^{\infty} \binom{\alpha - 1}{k} (-1)^k \times \left((k+1)\left(\frac{1}{\eta^2 \overline{\gamma}_E}\right) + \rho\left(\frac{\gamma_{th}}{\eta^2 \overline{\gamma}_D}\right)\right)^{-\frac{\beta}{2}}. \tag{12.20}$$

Furthermore, the final expression of PPSC can be obtained by using (12.13) as

$$P_{PPSC} = 1 - \frac{\alpha}{(\eta^2\overline{\gamma}_E)^{\frac{\beta}{2}}} \sum_{\rho=0}^{\infty} \binom{\alpha}{\rho} (- 1)^{\rho} \sum_{k=0}^{\infty} \binom{\alpha - 1}{k} \\ \times (- 1)^k \left((k+1) \left(\frac{1}{\eta^2 \overline{\gamma}_E} \right) + \rho \left(\frac{\gamma_{th}}{\eta^2 \overline{\gamma}_D} \right) \right)^{-\frac{\beta}{2}}. \quad (12.21)$$

The parameters γ_D and γ_E define the average SNR of the main receiver and the average SNR of the illegitimate receiver, respectively. The other parameters can be found in [37].

In this part, the analytical and simulation results using MC simulations are provided to analyze the secrecy performance of the proposed models. For the HAPS eavesdropping scenario, the LEO satellite is orbiting 500 km above the Earth's surface, while the HAPS receiver and eavesdropper are located 18 km above the Earth's surface. The wind speed level is set to $w = 65$ m/s and they consider a different size for the aperture D_G. For UAV eavesdropping, the HAPS node is located 20 km, while the UAV receiver and eavesdropper are hovering at 200 m above the ground. For this scenario, the authors consider the presence of Mie scattering and thin cirrus clouds. In addition, different wind speed levels are considered. For both models, the optical wavelength is λ =1,550 nm, and the secrecy rate is set to $Rs = 0.01$ nats/s/Hz.

Figure 12.6 investigates the effect of the zenith angle on the secrecy performance of both scenarios. The fact that theoretical findings and the MC simulations show good agreement. Furthermore, when comparing the two models, it is evident that HAPS eavesdropping outperforms UAV eavesdropping in terms of secrecy performance as the HAPS layer encounters less attenuation and atmospheric turbulence. Additionally, we can see that both scenarios' overall secrecy performance degrades at higher zenith angle values.

Figure 12.7 illustrates the SOP performance for the case of HAPS eavesdropping, taking into account various aperture sizes for the legitimate receiver. As depicted in this figure, increasing the aperture diameter of the main receiver improves the secrecy performance because a larger aperture allows the receiver to collect more signals. In addition, the aperture size has a significant effect on the scintillation index, allowing to reduce the negative effects of atmospheric turbulence.

The authors examine the PPSC performance of the suggested models while taking into account various $\overline{\gamma}_E$ values in Figure 12.8. For HAPS eavesdropping in this figure, the zenith angle is set to $\xi = 80°$, while for UAV eavesdropping, it is set to $\xi = 70°$. It is obvious from the figure that the PPSC performance improves as $\overline{\gamma}_E$ drops, showing the significant impact of $\overline{\gamma}_E$ on the secrecy performance. Additionally, the good agreement with the MC simulations demonstrates the correctness of the analytical study.

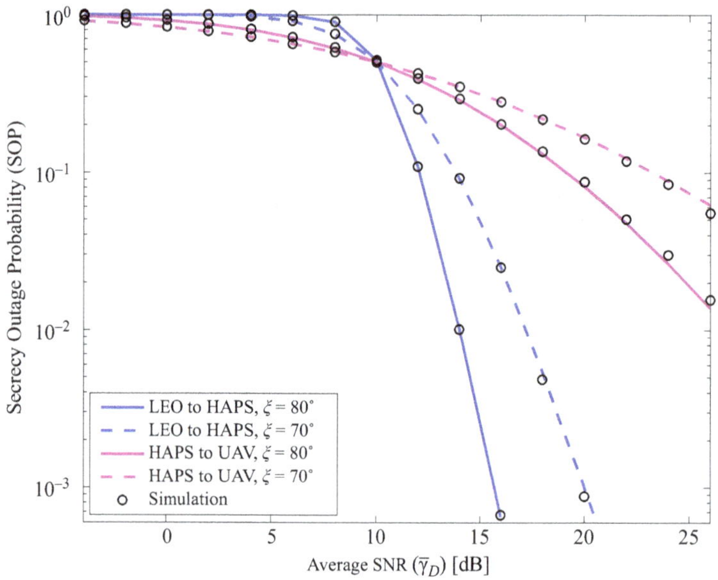

Figure 12.6 SOP performance of the proposed models vs. $\overline{\gamma}_D$ for different zenith angles, $\overline{\gamma}_E = 10$ dB

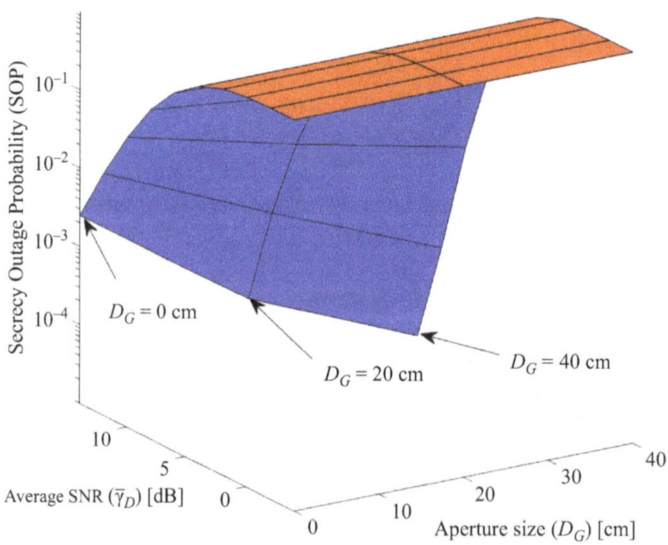

Figure 12.7 SOP performance of the HAPS eavesdropping vs. $\overline{\gamma}_D$ for different destination's aperture size, $\overline{\gamma}_E = 10$ dB

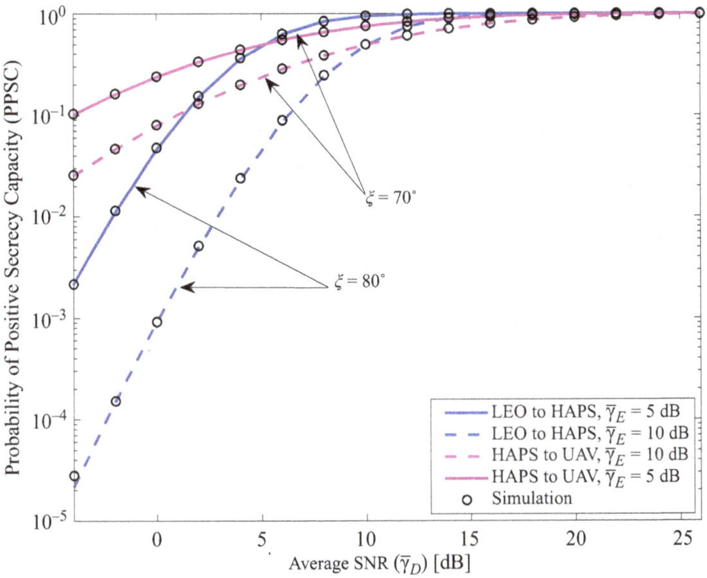

Figure 12.8 PPSC performance the proposed models vs. $\overline{\gamma}_D$ under different $\overline{\gamma}_E$

Finally, the main conclusions of this work can be summarised as follows:

- The altitude of the main receiver with respect to the ground level affect directly the atmospheric turbulence in downlink communication.
- According to the results, raising the zenith angle or the wind speed level degrades the overall secrecy performance as they increase atmospheric turbulence.
- Implementing larger aperture sizes for the legitimate receiver improves its ability to keep information more secure as it helps to reduce atmospheric turbulence impact. This makes the main channel better than the eavesdropper channel and reduces the leaked beam to the eavesdropper.

In the following subsection, we present optical satellite eavesdropping.

12.3.2 Satellite eavesdropping

12.3.2.1 System model

In this work, the authors proposed a novel scenario involving space-based eavesdropping attacks [38]. As illustrated in Figure 12.9, they assume an LEO satellite S that seeks to communicate with a HAPS $\mathscr{H}$ node in the presence of a powerful eavesdropping spacecraft E located in close proximity to S. First, they consider S to $\mathscr{H}$ downlink communication, where E is within the convergence area of S's optical beam and can therefore eavesdrop on the transmitted beam. Second, they assume $\mathscr{H}$ to S uplink communication, in which E attempts to obtain $\mathscr{H}$'s information. Due to the quasi-stationary position of the HAPS node, tracking and accuracy issues, as

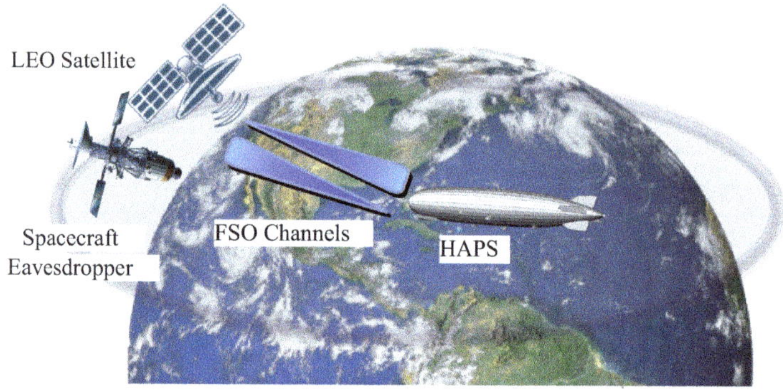

Figure 12.9 Illustration of satellite eavesdropping model

well as the Doppler shift, are tolerable for the communication between S and $\mathscr{H}$. In both cases, E can only capture a small fraction r_e of the transmitted signal, whereas the intended receiver collects a larger fraction r_b, where $r_e + r_b \leq 1$. Note that the parameters r_b and r_e depend on the aperture size and beam divergence angle of each device. In the analysis, the authors suppose that the positions of the satellite, HAPS, and eavesdropper spacecraft can be used to extract the physical model of the channel, which is representative of the channel state information (CSI).

12.3.2.2 Secrecy performance analysis

The instantaneous SNR at $\mathscr{H}$ can be given as

$$\gamma_{\mathscr{H}} = \frac{r_b P_S}{N_0} I_{\mathscr{H}}^2 = \overline{\gamma}_{\mathscr{H}} I_{\mathscr{H}}^2, \tag{12.22}$$

where $\overline{\gamma}_{\mathscr{H}} = \frac{r_b P_S}{N_0}$ defines the average SNR at $\mathscr{H}$ with $\mathbb{E}[I_{\mathscr{H}}^2] = 1$, and the instantaneous SNR at S and E can be expressed similarly after changing the subscripts as $\gamma_{\mathscr{S}} = \frac{r_b P_{\mathscr{H}}}{N_0} I_S^2 = \overline{\gamma}_S I_S^2$ and $\gamma_E = \frac{r_e P_j}{N_0} I_E^2 = \overline{\gamma}_E I_E^2$ with $\mathbb{E}[I_E^2] = \mathbb{E}[I_S^2] = 1$. In addition, the PDF and CDF of the FSO channels are given by (12.16) and (12.17) after changing the subscripts.

For these models, the authors derived new expressions for ASC, SOP, and secrecy throughput (ST).

In S to $\mathscr{H}$ downlink communication, E is assumed to be positioned very close to S. Therefore, the turbulence-induced fading can be ignored to provide realistic conditions [25]. Therefore, the ASC for the downlink eavesdropping can be obtained by averaging the secrecy capacity C_s as

$$\overline{C}_s = \frac{1}{\ln(2)} \mathbb{E}\big[\ln(1 + \gamma_{\mathscr{H}}) - \ln(1 + \gamma_E)\big]. \tag{12.23}$$

Thereafter, by using Jensen's inequality, for high SNR values, the ASC for the downlink communication can be obtained as

$$\begin{aligned}\overline{C}_s &\cong \frac{1}{\ln(2)}\left[\ln\left(1+\mathbb{E}\left[\gamma_{\mathcal{H}}\right]\right) - \ln\left(1+\mathbb{E}[\gamma_E]\right)\right] \\ &\cong \frac{1}{\ln(2)}\left[\ln\left(1+\frac{r_b P_S}{N_0}\right) - \ln\left(1+\frac{r_e P_S}{N_0}\right)\right]. \end{aligned} \tag{12.24}$$

For the uplink communication, the eavesdropper is considered to be close to the receiver, therefore E experiences the same turbulence induced-fading as the main receiver. Thus, the final expression of ASC can be given as

$$\begin{aligned}\overline{C}_s &= \frac{1}{\ln 2}\sum_{\rho=0}^{\infty}\binom{\alpha}{\rho}(-1)^{\rho}\frac{2}{\beta}\eta^2\overline{\gamma}_E\left(\frac{1}{\rho}\right)^{\frac{2}{\beta}}\int_0^{\infty}X^{\frac{2}{\beta}-1}G_{1,1}^{1,1}\left(\eta^2\overline{\gamma}_E\left(\frac{X}{\rho}\right)^{\frac{2}{\beta}}\middle|\begin{matrix}0\\0\end{matrix}\right) \\ &\quad\times G_{0,1}^{1,0}\left(X\middle|\begin{matrix}-\\0\end{matrix}\right)dX - \frac{1}{\ln 2}\sum_{\rho=0}^{\infty}\sum_{t=0}^{\infty}(-1)^{\rho+t}\binom{\alpha}{\rho}\binom{\alpha}{t}\frac{2}{\beta}\left(\frac{1}{\rho A+tB}\right)^{\frac{2}{\beta}} \\ &\quad\times\int_0^{\infty}Y^{\frac{2}{\beta}-1}G_{1,1}^{1,1}\left(\left(\frac{Y}{\rho A+tB}\right)^{\frac{2}{\beta}}\middle|\begin{matrix}0\\0\end{matrix}\right)\times G_{0,1}^{1,0}\left(Y\middle|\begin{matrix}-\\0\end{matrix}\right)dY \\ &= \frac{1}{\ln 2}\frac{2\times 2^{\frac{2}{\beta}-\frac{1}{2}}}{(2\pi)^{\beta-\frac{1}{2}}}\eta^2\overline{\gamma}_S\sum_{t=1}^{\infty}\binom{\alpha}{t}(-1)^{t+1}\left(\frac{1}{t}\right)^{\frac{2}{\beta}} \\ &\quad\times G_{\beta+2,\beta}^{\beta,\beta+2}\left(4\times\left(\frac{\eta^2\overline{\gamma}_S}{t^{\frac{2}{\beta}}}\right)^{\beta}\middle|\begin{matrix}\Delta(\beta,0),\frac{1-(2/\beta)}{2},\frac{2-(2/\beta)}{2}\\ \Delta(\beta,0)\end{matrix}\right) \\ &\quad-\frac{1}{\ln 2}\frac{2\times 2^{\frac{2}{\beta}-\frac{1}{2}}}{(2\pi)^{\beta-\frac{1}{2}}}\times\sum_{\rho=1}^{\infty}\sum_{t=1}^{\infty}\binom{\alpha}{\rho}\binom{\alpha}{t}(-1)^{\rho+t+2}\left(\frac{1}{\rho A+tB}\right)^{\frac{2}{\beta}} \\ &\quad\times G_{\beta+2,\beta}^{\beta,\beta+2}\left(4\times\left(\frac{1}{\rho A+tB}\right)^{2}\middle|\begin{matrix}\Delta(\beta,0),\frac{1-(2/\beta)}{2},\frac{2-(2/\beta)}{2}\\ \Delta(\beta,0)\end{matrix}\right).\end{aligned} \tag{12.25}$$

In what follows and by using (12.10), the final expressions of SOP for both scenarios can be obtained.

For the downlink communication, P_{SO} can be obtained as

$$P_{\mathrm{SO}} = \sum_{\rho=0}^{\infty}\binom{\alpha_{\mathcal{H}}}{\rho}(-1)^{\rho}\exp\left[-\rho\left(\frac{2^{R_s}+2^{R_s}\overline{\gamma}_E-1}{\eta_{\mathcal{H}}^2\overline{\gamma}_{\mathcal{H}}}\right)^{\frac{\beta_{\mathcal{H}}}{2}}\right]. \tag{12.26}$$

For the uplink eavesdropping, the final expression of SOP can be written as

$$P_{SO} = \frac{\alpha}{(\eta^2\overline{\gamma}_E)^{\frac{\beta}{2}}} \sum_{\rho=0}^{\infty} \binom{\alpha}{\rho} (-1)^\rho \sum_{q=0}^{\infty} \binom{\alpha-1}{q} (-1)^q \times \left(\rho \left(\frac{\gamma_{th}}{\eta^2\overline{\gamma}_S} \right) + (q+1) \left(\frac{1}{\eta^2\overline{\gamma}_E} \right) \right)^{-\frac{\beta}{2}}. \tag{12.27}$$

To evaluate and characterize the overall efficiency of achieving reliable and secure transmission, ST is proposed [41]. Mathematically, the expression of ST can be written as

$$ST = R_s(1 - P_{SO}). \tag{12.28}$$

The expressions of ST for uplink and downlink communication can be easily obtained by substituting (12.26) and (12.27) in (12.28).

This part evaluates the secrecy performance of satellite eavesdropping for the proposed scenarios. For both models, the authors consider the same input parameters. S is orbiting at an altitude of 500 km, while the HAPS node is positioned at 18 km above ground level. The zenith angles are set to $\xi_S = \xi_{\mathscr{H}} = 70°$, the wind speed level $w_S = w_{\mathscr{H}} = 65$ m/s, and the secrecy rate is set as $R_s = 0.01$ bit/s/Hz for SOP simulations.

The SOP performance for both scenarios is obtained as a function of the SNR of the main receiver in Figure 12.10. The average SNR of the spacecraft attacker is set to $\overline{\gamma}_E = 10$ dB. As can be seen from the figure, the secrecy performance of S to $\mathscr{H}$ communication outperforms the uplink $\mathscr{H}$ to S communication. This can be justified by the fact that as the malicious receiver approaches the legitimate receiver, it is able to capture more information due to the greater distance at which more beams can be reflected. Moreover, in both scenarios, the secrecy performance degrades as the fraction of the leaked power received by the eavesdropper, r_e, increases. In addition, as depicted in the figure, increasing the secrecy rate R_s degrades the SOP performance. Finally, there is an agreement between the theoretical expressions and the MC simulations, which validates the obtained theoretical expressions.

The effect of the zenith angle on the proposed satellite eavesdropping scenarios is depicted in Figure 12.11. The authors set the average SNR of the spacecraft attacker to $\overline{\gamma}_E = 4$ dB and $r_e = 0.4$ for the fraction of power received by E. As can be seen, increasing the zenith angle reduces the amount of information captured by the authorized receiver and increases the information transmitted to the attacker. In addition, the signal fluctuation becomes significant as the zenith angle increases. Consequently, we conclude that there is a significant performance loss. Moreover, as depicted in the figure, for a lower zenith angle, the SOP performance is nearly identical in both cases, and simulation results reveal that the scintillation indexes are identical. However, when the zenith angle is increased to 80°, there is a 2 dB

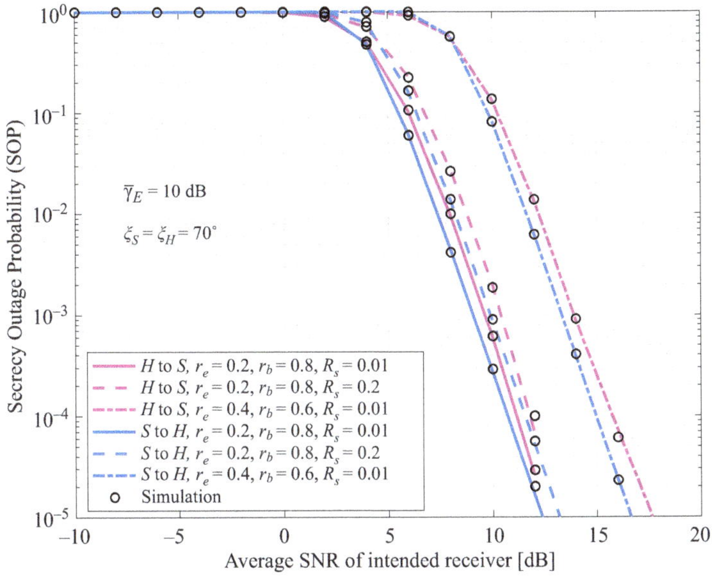

Figure 12.10 SOP performance of the proposed scenarios for different fractions of received power and different secrecy rate when $\overline{\gamma}_E = 10$ *dB*

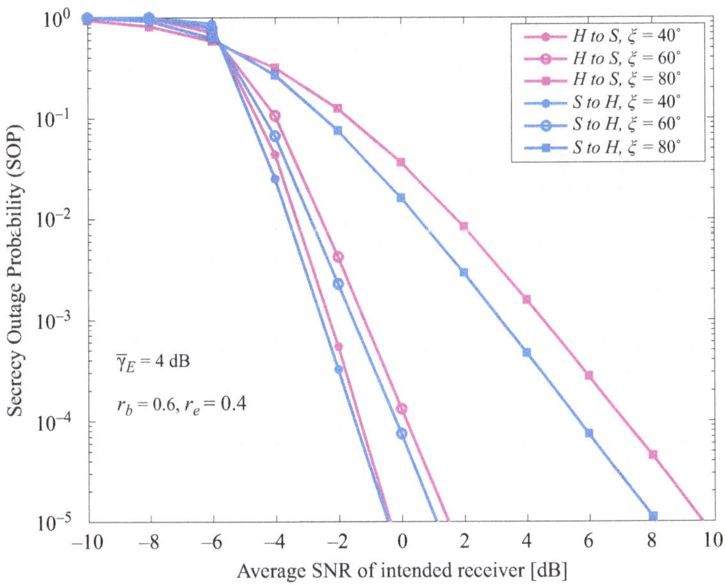

Figure 12.11 SOP performance of the proposed models under different zenith angles, $\overline{\gamma}_E = 4$ *dB*

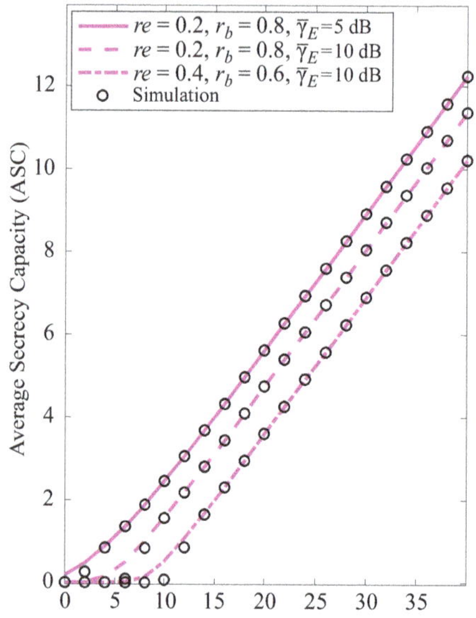

(a)

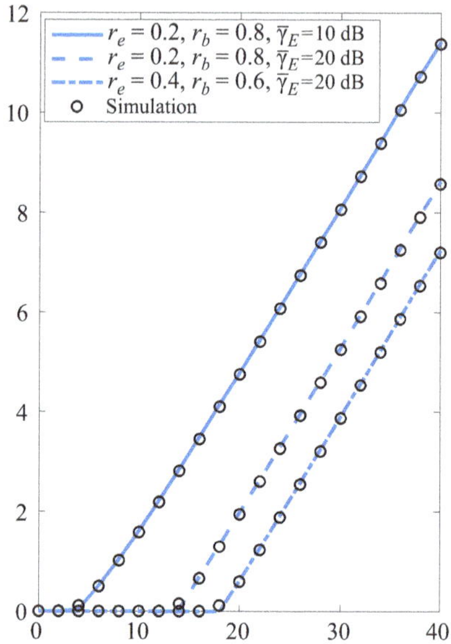

(b)

Figure 12.12 ASC performance of both scenarios for different $\overline{\gamma}_E$ and different fractions of received power. (a) Average SNR ($\overline{\gamma}_S$) [dB] H to S and (b) average SNR ($\overline{\gamma}_H$) [dB] S to H

difference between the two curves. Also, the impact of beam wander increases as the zenith angle increases.

Figure 12.12 depicts the performance of the ASC in both scenarios. As shown in the figure, the overall secrecy performance degrades as the average SNR of the eavesdropper increases or as the amount of power captured by the intended receiver decreases. In addition, for the *S* to $\mathcal{H}$ communication, the authors assume that the eavesdropper has a higher SNR due to its proximity to the LEO satellite and the absence of turbulence-induced fading. In addition, as shown in the figure, the analytical expressions match the MC simulations precisely. Finally, when the average SNR of the legitimate receiver is greater than that of the eavesdropper, perfect secrecy is achieved.

Figure 12.13 evaluates the ST in relation to the target secrecy rate R_s for the proposed models at various average SNR levels of the main receiver. In all situations, the spacecraft's average SNR is set to 2 dB. As can be seen, the ST increases as R_s increases to a certain level and then decreases. This is a result of ST's dependence on *Rs*. When the R_s is relatively low, the ST increases. However, if R_s exceeds a certain threshold, the system cannot sustain reliability and security. Moreover, it is observed that the *S* to $\mathcal{H}$ downlink scenario performs slightly better than the $\mathcal{H}$ to *S* uplink scenario at lower R_s. Nevertheless, after a specific value of R_s, the uplink scenario outperforms the downlink communication. Lastly, as anticipated, increasing the average SNR of the intended receiver enhances ST performance.

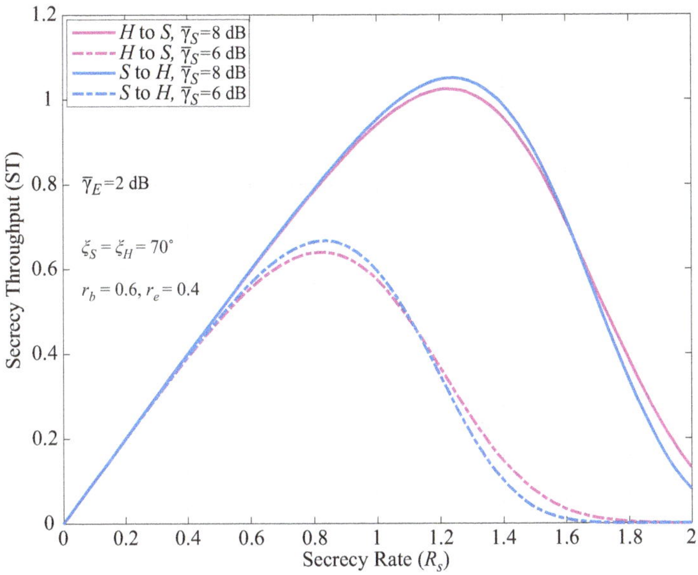

Figure 12.13 ST performance of the proposed models vs. R_s

Finally, the main observations from this work can be summarized as follows:

- Increasing the zenith angles between the communicating nodes deteriorates the overall secrecy performance.
- The spacecraft attacker can gather more data in the uplink scenario since more beams can be reflected because of the larger distance.
- The secrecy performance is directly impacted by the variations in the signal introduced by atmospheric turbulence.
- Critical parameters for ensuring secure communication are the eavesdropper's SNR γ_E and the quantity of power leaked to the eavesdropper r_e.
- From an ST perspective, it is seen that after a specific value of the secrecy rate Rs, the system's reliability and secrecy are compromised.

12.4 Conclusion

In this chapter, we have presented the secrecy performance of NTN systems from a physical layer aspect. First, RF eavesdropping on the ground for dual-hop HAPS-aided downlink SatCom was covered. After that, we presented the PLS for NTN systems by covering three distinct FSO eavesdropping scenarios; HAPS eavesdropping, UAV eavesdropping, and satellite eavesdropping. The obtained results showed that the signal fluctuations caused by atmospheric conditions have a direct impact on the secrecy performance.

To summarize, significant consideration must be given to channel modeling and the radio propagation environment in order to meet the anticipated quality of service and security requirements of wireless networks.

The future work directions can be summarized as follows:

- Secrecy performance analysis for multicast communication can be carried out while considering the presence of multiple eavesdroppers.
- Propose some techniques to enhance the secrecy performance of NTN systems and favor the communication to the legitimate user.
- Security can be improved by proposing a cross-layer framework while combining encryption and PLS techniques at the same time.
- It is essential to investigate massive MIMO's vulnerabilities in the face of potentially malicious attacks, such as jamming and eavesdropping, as it is one of the primary physical layer technologies that is anticipated to be employed in future networks. Additionally, it has been demonstrated that distributed jamming strategies can compromise massive MIMO technology. A promising research area could be the development of various beamforming methods as distributed jammer countermeasures.
- Intelligent Reflective Surface (IRS) is a cutting-edge transmission technique that offers many transmission channels by altering the amplitude, phase, and frequency of received signals. Therefore, by engaging with only a legitimate user, on the other hand, IRS can be used to deliver PLS. Therefore, IRS techniques can be used to provide security while only interacting with legitimate users on the alternate route.
- By learning the various behavioral patterns of malicious attackers and detecting active intrusions, integrating artificial intelligence/machine learning (AI/ML) approaches with PLS can be an efficient solution for future networks. PLS can thus be improved further by implementing AI/ML techniques.

References

[1] Rinaldi F, Maattanen HL, Torsner J, *et al.* Non-terrestrial networks in 5G & beyond: a survey. *IEEE Access*. 2020;8:165178–165200.

[2] 3GPP. Solutions for NR to support Non-Terrestrial Networks (NTN), document TR 38.821, Release 16; Jan. 2020. Available from: [Online]. Available:https://www.3gpp.org.

[3] Solutions for NR to Support non-terrestrial networks (NTN); June 2021.

[4] Alam MS, Karabulut Kurt G, Yanikomeroglu H, *et al.* High altitude platform station based super macro base station constellations. *IEEE Communications Magazine*. 2021;59(1):103–109.

[5] Karabulut Kurt G, Khoshkholgh MG, Alfattani S, *et al.* A vision and framework for the high altitude platform station (HAPS) networks of the future. *IEEE Communications Surveys Tutorials*. 2021;23(2):729–779.

[6] Fidler F, Knapek M, Horwath J, *et al.* Optical communications for high-altitude platforms. *IEEE Journal of Selected Topics in Quantum Electronics.* 2010;16(5):1058–1070.

[7] Giggenbach D, Purvinskis R, Werner M, *et al.* Stratospheric optical inter-platform links for high altitude platforms. In: *Int. Commun. Satellite Systems Conf. and Exhibit*; 2002. p. 1910.

[8] Vishwakarma N, R S. Capacity analysis of adaptive combining for hybrid FSO/RF satellite communication system. In: *National Conference on Communications* (NCC); 2021. p. 1–6.

[9] Michailidis ET, Nomikos N, Bithas P, *et al.* Outage probability of triple-hop mixed RF/FSO/RF stratospheric communication systems. In: *IEEE Int. Conf. on Adv. in Satellite and Space Commun.* (SPACOMM); 2018. p. 1–6.

[10] Erdogan E. On the performance of cognitive underlay RF/FSO communication systems with limited feedback. *Optics Communications.* 2019;444: 87–92.

[11] Andrews LC, Phillips RL. *Laser Beam Propagation Through Random Media* (SPIE Press Monograph). Bellingham, WA: SPIE; 2005.

[12] Nadeem F, Kvicera V, Awan MS, *et al.* Weather effects on hybrid FSO/RF communication link. *IEEE Journal on Selected Areas in Communications.* 2009;27(9):1687–1697.

[13] Kaushal H, Kaddoum G. Optical communication in space: challenges and mitigation techniques. *IEEE Commun Surveys Tuts.* 2016;19(1):57–96.

[14] Abdi A, Lau WC, Alouini MS, *et al.* A new simple model for land mobile satellite channels: first- and second-order statistics. *IEEE Transactions on Wireless Communications.* 2003;2(3):519–528.

[15] Barrios R, Dios F. Exponentiated Weibull distribution family under aperture averaging for Gaussian beam waves. *Optics Express.* 2012;20(12):13055–13064.

[16] Erdogan E, Altunbas I, Kurt GK, *et al.* Site diversity in downlink optical satellite networks through ground station selection. *IEEE Access.* 2021;9:31179–31190.

[17] Arezumand H, Zamiri-Jafarian H, Soleimani-Nasab E. Outage and diversity analysis of underlay cognitive mixed RF-FSO cooperative systems. *Journal of Optical Communications and Networking.* 2017;9(10):909–920.

[18] Ben Yahia O, Erdogan E, Kurt GK. HAPS-assisted hybrid RF-FSO multicast communications: error and outage analysis. *IEEE Transactions on Aerospace and Electronic Systems* (early access). 2022;1–13.

[19] Ben Yahia O, Erdogan E, Karabulut Kurt G, *et al.* HAPS selection for hybrid RF/FSO satellite networks (early access). *IEEE Transactions on Aerospace and Electronic Systems.* 2022;58(4):2855–2867.

[20] Ben Yahia O, Erdogan E, Kurt GK, *et al.* A weather-dependent hybrid RF/FSO satellite communication for improved power efficiency. *IEEE Wireless Communications Letters.* 2022;11(3):573–577.

[21] Shannon CE. Communication theory of secrecy systems. *The Bell System Technical Journal.* 1949;28(4):656–715.

[22] Wyner AD. The wire-tap channel. *Bell System Technical Journal.* 1975;54(8):1355–1387.

[23] Boluda-Ruiz R, Qaraqe K. Effect of misalignment error on secrecy outage capacity of FSO communication links. In: *IEEE Wireless Commun. and Netw. Conf.* (WCNC); 2019. p. 1–7.

[24] Ai Y, Mathur A, Verma GD, *et al.* Comprehensive physical layer security analysis of FSO communications over Málaga channels. *IEEE Photonics Journal.* 2020;12(6):1–17.

[25] Lopez-Martinez FJ, Gomez G, Garrido-Balsells JM. Physical-layer security in free-space optical communications. *IEEE Photon Journal.* 2015;7(2):1–14.

[26] Saber MJ, Sadough SMS. On secure free-space optical communications over Málaga turbulence channels. *IEEE Wireless Communications Letters.* 2017;6(2):274–277.

[27] Monteiro MEP, Rebelatto JL, Souza RD, Brante G. Maximum secrecy throughput of MIMOME FSO communications with outage constraints. *IEEE Transactions on Wireless Communications.* 2018;17(5):3487–3497.

[28] Zhu J, Chen Y, Sasaki M. Average secrecy capacity of free-space optical communication systems with on-off keying modulation and threshold detection. In: *2016 International Symposium on Information Theory and Its Applications* (ISITA). New York, NY: IEEE; 2016. p. 616–620.

[29] Sikri A, Mathur A, Verma G. Secrecy performance enhancement of artificial noise injection scheme-based FSO systems. In: *2021 IEEE 94th Vehicular Technology Conference* (VTC2021-Fall). New York, NY: IEEE; 2021. p. 01–05.

[30] Guo K, An K, Zhang B, *et al.* Physical layer security for multiuser satellite communication systems with threshold-based scheduling scheme. *IEEE Transactions on Vehicular Technology.* 2020;69(5):5129–5141.

[31] Guo K, An K, Zhou F, *et al.* On the secrecy performance of NOMA-based integrated satellite multiple-terrestrial relay networks with hardware impairments. *IEEE Transactions on Vehicular Technology.* 2021;70(4):3661–3676.

[32] Ai Y, Mathur A, Cheffena M, *et al.* Physical layer security of hybrid satellite-FSO cooperative systems. *IEEE Photon Journal.* 2019;11(1):1–14.

[33] Zhang Y, Ye J, Pan G, *et al.* Secrecy outage analysis for satellite-terrestrial downlink transmissions. *IEEE Wireless Communications Letters.* 2020; 9(10):1643–1647.

[34] Ben Yahia O, Erdogan E, Karabulut Kurt G. On the use of HAPS to increase secrecy performance in satellite networks. In: *IEEE International Conference on Communications Workshops* (ICC Workshops); 2021. p. 1–6.

[35] The Wolfram Functions Site;. Available from: http://www.wolfram.com.

[36] ITU-R. Propagation data required for the design of Earth-space systems operating between 20 THz and 375 THz. International Telecommunication Union, Recommendation P.1622; 2003.

[37] Ben Yahia O, Erdogan E, Kurt GK, *et al.* Physical layer security framework for optical non-terrestrial networks. In: *2021 28th International Conference on Telecommunications* (ICT); 2021. p. 162–166.

[38] Ben Yahia O, Erdogan E, Kurt GK, *et al.* Optical satellite eavesdropping. *IEEE Transactions on Vehicular Technology*. 2022;71(9):10126–10131.

[39] Barrios Porras R. Exponentiated Weibull fading channel model in free-space optical communications under atmospheric turbulence. Ph.D. dissertation, Dept. Signal Theory Commun., Univ. Politècnica de Catalunya (UPC), Barcelona, Spain; May 2013.

[40] Gradshteyn IS, Ryzhik IM. *Table of Integrals, Series, and Products*. London: Academic Press; 2014.

[41] Pattanayak DR, Dwivedi VK, Karwal V, *et al.* On the physical layer security of a decode and forward based mixed FSO/RF co-operative system. *IEEE Wireless Communications Letters*. 2020;9(7):1031–1035.

Chapter 13

Security in physical layer of cognitive radio networks

Deemah H. Tashman[1] and Walaa Hamouda[1]

Fifth-generation (5G) networks and beyond are anticipated to support a vast number of connections and services. Due to the enormous amount of confidential data shared between devices in these networks, the risk of security vulnerabilities escalates proportionally. Cognitive radio networks (CRNs) are no exception since they are vulnerable to a variety of physical-layer threats; hence, a physical-layer approach is necessary to safeguard these networks. Consequently, physical-layer security (PLS) has recently been applied for examining and strengthening the security of wireless networks, including CRNs. In light of this, this chapter examines a brief overview of CRNs, as well as the main physical layer attacks and their respective primary countermeasures. In addition, the employment of energy harvesting (EH) techniques to improve CRNs security is discussed. The security threats on cognitive unmanned aerial vehicles (UAVs) and their primary defense mechanisms are presented. In addition, the consideration of cascaded fading channels and their effect on the security of CRNs are explored. Finally, this chapter includes conclusions and potential future directions.

13.1 Introduction

Physical-layer security (PLS) has emerged as a solid strategy for enhancing the privacy of wireless networks. The security of communications can be improved without the use of encryption or decryption through PLS approach. Wyner presented a three-node wiretap model, in which two channels should be examined when investigating a system's PLS; the main channel and the wiretap channel. The transmitter and legitimate receiver communicate on the main channel, whereas the transmitter and an eavesdropper communicate through the wiretap channel. From a PLS perspective, it is possible to improve the privacy of legitimate users by either improving the conditions of the main channel or worsening those of the wiretap channel. Given this, users of cognitive radio networks (CRNs) may employ PLS to protect their communications due to its effectiveness in safeguarding networks.

[1]Department of Electrical and Computer Engineering, Concordia University, Canada

13.1.1 Motivation of physical-layer security

Network security has been mainly reliant on the encryption strategy deployed in the network's upper layers. Nevertheless, there are various disadvantages to using encryption for security, particularly when assuming the fifth generation (5G) communications [1]. For example, the software and hardware complexity increases significantly in 5G networks and beyond and this requires huge processing power [2,3]. It is also challenging to exchange secret keys because of the heterogeneous networks featured in 5G networks. Additionally, mostly, encryption is performed using strong algorithms that assume the eavesdropper's receiver has limited computing capacity. In spite of this, recent advancements in devices' computing power have made it less difficult to crack encryption protocols, which facilitate the task of eavesdroppers. It is thus vital to take specific precautions in order to increase communication system security. As a result, PLS has emerged as an intriguing method for studying and enhancing the security of secret information exchanged between authorized parties in the 5G era. Moreover, PLS does not need the exchange of security keys since it does not rely on encryption and decryption procedures.

13.1.2 Wiretap channel

The three-node wiretap communication architecture is composed of the main channel and the wiretap channel as depicted in Figure 13.1. When analyzing a system's PLS, we need to take into account both channels. The main channel exists between the transmitter and the authorized receiver, whereas the wiretap channel exists between the transmitter and the eavesdropper. When the channel between authorized users (main channel) has better conditions than the channel between the transmitter and the eavesdroppers (wiretap channel), the data is assured to be secure, as Wyner discovered [4].

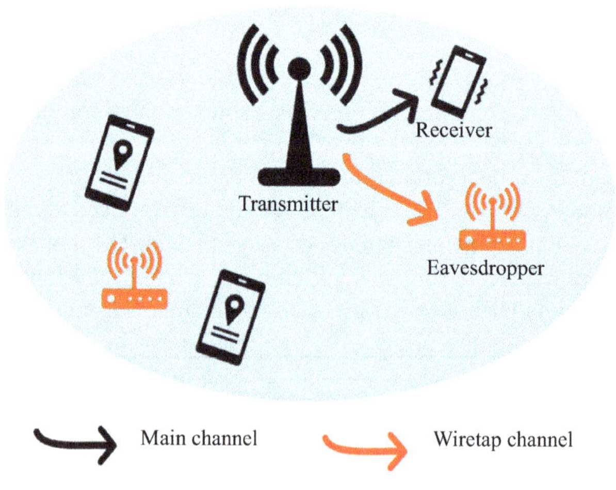

Figure 13.1 Three-node wiretap channel

13.1.3 Physical layer security metrics

In this section, the most commonly employed PLS metrics for assessing network security are provided:

- *Secrecy outage probability:* Secrecy outage probability (SOP) is defined as the probability that the secrecy capacity (C_s) falls below a certain secrecy rate threshold (C_{th}) as

$$SOP = P_r \left(C_s < C_{th} \right). \tag{13.1}$$

For the system model considered in Figure 13.1, the secrecy capacity is defined as [5]

$$C_s = \begin{cases} C_M - C_E, & \text{if } \gamma_M > \gamma_E \\ 0, & \text{if } \gamma_M \leq \gamma_E \end{cases}, \tag{13.2}$$

with C_M is the main channel capacity and C_E is the wiretap channel capacity. γ_M and γ_E are the received signal-to-noise ratios (SNRs) at the legitimate and the eavesdropper receivers, respectively. Passive eavesdroppers (eavesdroppers who monitor without performing transmissions) render it difficult to discern the channel state information (CSI) of the wiretap channel. Given this, the transmitter does not have access to CSI, thus it broadcasts at a constant rate (C_{th}). In this case, SOP is the most effective method for determining how secure a system model is since leakage of secret information across a wiretap connection is inevitable.

- *Probability of non-zero secrecy capacity:* The probability of a non-zero secrecy capacity (P_r^{nzc}) stands for the probability that the secrecy capacity is greater than zero. Alternatively, it is a chance that the capacity of the main channel is likely to be greater than the capacity of the wiretap channel. This occurrence takes place when the circumstances of the main channel are superior to those of the wiretap channel in terms of the received SNRs,

$$P_r^{nzc} = P_r \left(C_s > 0 \right) = P_r \left(\gamma_M > \gamma_E \right) \cdot \tag{13.3}$$

These metrics indicate that the main channel's conditions should be improved or the wiretap channel's conditions must be degraded in order to prevent the communications from eavesdropping. Indeed, wireless signals are subjected to multipath fading and path loss. Multi-path propagation causes fading, whereas path loss occurs as a consequence of increased distance. Particularly, the channel is more secure against attacks when the distance between legitimate users is less than the distance between the transmitter and the eavesdropper. Moreover, when it comes to PLS, the goal is to make use of fading and interference to provide a secure system by using the environment's impairments [3].

13.2 Cognitive radio networks

A possible strategy to address the issue of scarcity and the inefficient utilization of the allocated radio spectrum is the implementation of cognitive radio networks (CRNs). The reason for this issue in the spectrum is the federal communications commission's (FCC's) strict regulation regarding the allocation of radio frequencies. Moreover, there will be greater demand for CRNs as the number of connected devices and users grows with the introduction of 5G communications and beyond [6]. The two main categories of CRNs users are the primary users (PUs), i.e., licensed users, and the secondary users (SUs), which are also known as cognitive users or unlicensed users [7,8]. Within a CRN, SUs may use one of three distinct access methods: overlay, underlay, or interweave [8–14]. Using the overlay approach, SUs collaborate with PUs for mutual benefit. Given this, a common method for SUs to gain access to the licensed spectrum is to relay the broadcasts of PUs. Underlay CRNs allow SUs to use the radio spectrum simultaneously with PUs as long as the SUs' broadcast power will not interfere with the PUs' transmissions [15–21]. In this scenario, SUs must adjust their transmission power such that the interference they cause does not exceed the interference threshold tolerable at the PU receiver. Finally, in an interweave CRN, SUs are only allowed to access the PUs' bands if they are unoccupied. Here, SUs acquire radio environment characteristics and begin sensing the licensed bands in search of spectrum gaps. To sense the spectrum, one of the spectrum sensing techniques can be utilized, such as energy detection, wave-form-based sensing, cyclostationary sensing, and matched filter-based sensing [22].

There are essentially three steps to a cognitive radio cycle: spectrum sensing, spectrum processing, and spectrum decision. Given that subsequent steps are dependent on the precision and reliability of the sensing data, spectrum sensing is a crucial step in the process. After collecting data through spectrum sensing, the SU may begin performing spectrum analysis. Spectrum decision is the process by which an SU chooses a certain band to utilize based on that band's characteristics and the requirements of the user. Once SU has settled on a frequency range, it chooses the transmission characteristics, such as modulation type and data rate. Additionally, SU needs to keep monitoring the bands in case the PU reappears [22].

13.2.1 Securing cognitive radio networks

Attackers can target the physical layer of CRNs, regardless of the mode of transmission that is employed by SUs, i.e., underlay, overlay, or interweave. To illustrate, when SUs choose to communicate via an underlay paradigm, there should be a continual adjustment made to the amount of the transmitted power. Since the variation in the transmit power has the possibility to affect both the main channel capacity as well as the wiretap channel capacity, the circumstances of both of these channels are impacted. This suggests that the received SNRs acquired at both ends are affected, and, as a result, the secrecy of the transmission will also be compromised [22]. For instance, assuming that the circumstances of the wiretap channel remain the same, but the average received SNR at the legitimate receiver is the one that changes, decreasing the transmit power owing to the underlay mode would result in a degradation of the PLS for the legitimate users. In addition, when the overlay mode is utilized, a collaboration between the SUs and PUs networks is accomplished [23]. In the event that a secondary malicious node is present, the PU transmission may be put in jeopardy since the SU network may not be trusted to support the PUs network. Furthermore, in the interweave access mode, the SUs have to sense the bands searching for a vacant one. In this case, malicious users may attempt to inject false sensed data and prevent SUs from accessing the band. Given all these reasons, utilizing a security approach is essential for achieving the requisite degree of secrecy for CRNs. PLS has been shown to be a dependable and effective strategy to secure CRNs as will be discussed in the following subsections.

13.2.2 Differences in securing CRNs and other conventional networks

In comparison to non-cognitive networks, the application of PLS to CRNs has been deemed more difficult, and this is for several reasons. First, the three steps of cognition, namely spectral sensing, processing, and decision, are subjected to attacks. Second, SUs should be able to recognize the distinction between lawful and harmful PUs. It is indeed possible that while the band is empty, hostile selfish SUs will attempt to mimic the transmission characteristics of the PUs in order to prevent other SUs from accessing the band. Consequently, the secondary network is unable to take advantage of spectrum availability, which may result in reduced throughput. Third, SUs should be able to differentiate between lawful and harmful SUs, particularly SUs should be attentive to the integrity of the sensed data. That is, an SU attacker attempts to interfere with the fusion center (centralized sensing) so that it cannot accurately determine whether the bands are active or inactive [24]. Fourth, CRNs might be the target of attacks launched from outside the network. That is, due to the broadcasting nature of the communication, users within transmission distance of the transmitter may overhear the secret information. The SU or PU receiver may also be targeted by an adversary who intends to disrupt communication by sending destructive signals (jamming). Fifth, since an attack on the SUs or PUs network impacts the security and performance of the other one, they both need to be secured from various sorts of attacks. Lastly, there are particular attacks that can only be carried out against CRNs, and as a result, different countermeasures and procedures must be employed to resist them. Hence, comparing this to non-cognitive networks, there is an elevation in the complexity of the safeguarding process.

13.3 Attacks on the physical layer of cognitive radio networks and countermeasures

In this section, some of the most common security threats on the physical layer of CRNs are discussed, along with several countermeasures that have been proposed in the literature to mitigate them. The majority of these attacks are shown in Figure 13.2, including a primary user emulation attack (PUEA), jamming attack, and eavesdropping. The transmitter is denoted by Tx, while Rx is used to represent the receiver.

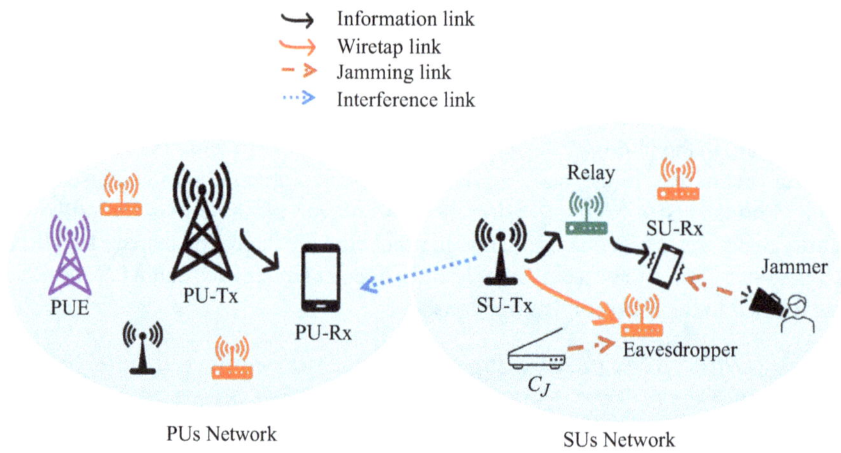

Figure 13.2 Illustration of the main attacks on the physical layer of CRNs

13.3.1 Primary user emulation attack

One of the most common and well-known attacks against the physical layer of CRNs is the primary user emulation attack (PUEA). An occurrence of PUEA begins whenever an SU acts as an emulator for a PU; this user is regarded as a primary user emulator (PUE). This attempt originates at the stage when the spectrum is being sensed by SUs. PUEA is carried out by transmitting a signal with the same characteristics as the PU's signal across the licensed band while the PUs are not present [25]. As a result, the possibility of making use of the band will be lost, and the throughput of SUs will be degraded. There are three primary categories of a PUEA, which are as follows:

- *Selfish PUE*: This is an SU that imitates a PU to prevent other SUs from utilizing the available bands. Consequently, this selfish user will use the band for its own broadcasts.
- *Malicious PUE*: This is an SU that intends to only obstruct the SUs' activities without adopting the bands for transmissions.
- *Mixed PUE*: This is a combination of the first and second categories.

Techniques to tackle PUEA:

In earlier studies, researchers have investigated a variety of ways to detect or resist this danger. The throughput of SUs under the PUEA has been the focus of previous research. For instance, in [26], the authors studied the throughput of an SU under the attack of a PUE. The impact of certain parameters on the throughput has been studied, such as the sensing time, attacker's presence probabilities, and attacker strength. In [27], a throughput maximization mechanism in the presence of a PUEA based on detecting the optimal spectrum sensing strategy through the Neyman–Pearson criterion is explored. Methods for combating PUEA often rely on distinguishing between actual and emulated PUs by exploiting one of the PUs' features. For instance, one of the approaches is the transmitter location verification scheme [28]. In this scheme, a PU signal is verified by checking the location of the transmitter through the received signal characteristics, such as its energy level. Then, the SU compares the detected level to the recorded one to identify any abnormalities. Moreover, in [29], a detection mechanism based on generalized likelihood ratio test (GLRT) is proposed to differentiate between a legitimate PU and an emulator. In [30], a PUEA detection approach that is based on the received power differentiation was adopted. In addition, in [31], a surveillance approach is utilized to combat the PUEA by determining the identification of an active user and a selfish PUE attacker. This is accomplished through a game-theory analysis and Nash equilibrium (NE) approach.

13.3.2 Jamming attack

The message broadcast begins when the SU decides which band to access. The CRN may be vulnerable to jamming attacks at this phase. Interfering signals (extremely high-power signals broadcast on the same frequency band) are emitted by a node known as a jammer. In CRNs, jamming has a wide variety of adverse effects on users. That is, during the jamming period, the ongoing communication may be terminated or a denial-of-service might occur. In CRNs, the most two common kinds of jamming are:

- *Continuous jamming*: This happens when a jammer continuously sends destructive signals on a single band or multiple bands of the spectrum without ceasing the transmission. The strength of the received signal may be used to identify this form of jamming as the power of the received signals in this instance should be rather high.
- *Discrete jamming*: This is a more hazardous sort of jamming since the jammer emits destructive signals periodically rather than continually, making it more difficult to detect.

The manner in which jammers select spectral bands varies, and thus they are divided into the following categories [32]:

- *Static jammer*: This jammer keeps transmitting jamming signals at the same frequency band.
- *Random jammer*: This jammer sends jamming signals toward the bands after selecting them randomly.
- *Intelligent jammer*: This jammer is able to leverage the environment's dynamics. Intelligent jamming occurs when a jammer chooses the band based on its experience and the assumption (with a high possibility) that an SU is switching to this band.

Techniques to tackle jamming:

Several defense approaches are available to protect CRNs against jamming. These approaches include detection of jammers, prevention, and mitigation. Once the SU receiver detects an attack, one of the spread spectrum methods, such as frequency hopping (FH), may be implemented. In this scenario, when the SU identifies a jamming danger, there is a quick transition to alternate to unjammed bands.

For instance, in [32], an FH approach-based traffic load is adopted. Nodes that use this scheme can change their channel hopping sequence based on their sending and receiving traffic. In addition, in [33], an anti-jamming hopping technique-based Tri- CH for CRNs has been adopted. The SU in this case jumps randomly between channels for a more secure system.

Machine learning (ML) approaches can also be utilized for detecting jamming attacks or maximizing security in the presence of jammers. To illustrate, in [34], an ML approach is utilized to avoid a jammer. The jamming and anti-jamming processes are both formulated based on the Markov game scheme. In [35], the authors design an anti-jamming scheme based on deep reinforcement learning (DRL). Specifically, a double deep Q-network (double DQN) is designed to model the confrontation between the CRN and the jammer, with the objective of maximizing the data rate of SUs. Moreover, in [36], a double DQN is also utilized to optimize two problems in the presence of a jammer and a relay, which are maximizing the throughput and the secrecy rate.

Game theory approaches are widely used for preventing jamming attacks in CRNs. In [37], the Stackelberg-based game theory framework is adopted to defend against jammers targeting SUs. In [38], first, the scenario of an SU that hops from one band to another as an anti-jamming game is studied. Then, based on analyzing the interaction between SUs and attackers, a channel hopping defense scheme is modeled using the Markov decision process approach. Finally, two learning schemes are adopted for SUs to gain knowledge of adversaries. Other game theories for anti jamming defense mechanisms have been investigated, which are based on the interaction between legitimate users and adversaries. For example, in [39] the system is modeled as a multi-stage game-based Dogfight and in [40] as a Markovian game model.

Finally, there are other techniques that can be used to overcome jamming attacks in CRNs without requiring extra infrastructure. In [41], an SU selects a channel with the lowest probability that the transmission delay of a packet is higher than a predefined delay requirement (packet-invalidity ratio). Hence, an SU will be assigned to the most secure channel for each transmission. In [42], the authors examine the case in which a malicious user analyzes the SU frequency's energy level and delivers jamming signals in its direction. In addition, they show that the FH technique does not mitigate the effect of jamming, and they instead propose two scenarios in which half- and full-duplex relays are involved to help the victim SU. The relay has the potential of multiplexing the victim's messages in its message to the destination.

13.3.3 Eavesdropping

CRNs encounter the potential of another threat, which is eavesdropping. Eavesdroppers seek to tap the private conversations between legitimate users. Both the PUs and the SUs networks may be targeted by eavesdroppers. This is predicated on the premise that while users broadcast their messages, users within the transmitter's coverage area may overhear them. In general, there are two sorts of eavesdroppers:

- *Active eavesdroppers*: It is presumed that these users access private communications without permission. Specifically, these users may be regarded as legitimate network users, while they are not permitted to access the channel and therefore cannot be trusted. Given this, the CSI of these eavesdroppers can be obtained. For example, pay-TV broadcast services where global CSI of the eavesdroppers is presumed to be provided at the transmitter [43].

- *Passive eavesdroppers*: These are the individuals who listen to information without broadcasting or delivering harmful signals to authorized receivers. These users are exclusively interested in accessing private information. In this instance, it is challenging to obtain the CSI of these eavesdroppers at the transmitter. Consequently, the transmitter typically transmits messages at a constant rate. This sort of eavesdropping is more frequent and more suitable to address in the literature.

Eavesdroppers may also be classified according to the manner in which they tap and handle the intercepted information as

- *Colluding eavesdroppers*: These eavesdroppers collaborate in tapping and analyzing the information. After intercepting the communications, the tapped messages are sent to a centralized processor for processing. Considering this, collaborating eavesdroppers are equivalent to a single eavesdropper with several antennas. Subsequently, one of the signals combining methods, such as maximal-ratio combining (MRC), may be applied [44].
- *Non-colluding eavesdroppers*: These individuals attempt to intercept communications individually. They cannot be regarded as a single eavesdropper in this instance. This is because it is probable that these eavesdroppers are placed in separate areas and do not operate together to intercept and analyze communications.

Due to the use of combining methods at the eavesdropper end and hence the high received SNR, colluding eavesdroppers are regarded to be more effective at intercepting communications. Therefore, additional security measures must be considered when eavesdroppers collaborate in the interception procedure [45]–[47].

Techniques to tackle eavesdropping

PLS techniques for wireless communications may prevent eavesdropping despite the absence of data encryption at the upper layer of the network. Consequently, it is preferable to use the physical characteristics of the radio channel to secure CRNs against eavesdroppers. Herein, a few of the most effective strategies for preventing eavesdropping are mentioned and discussed.

- *Cooperative jamming*: Apart from harmful jammers who attempt to interrupt the transmissions of legitimate users, another sort of jammer exists to support communication. This is termed friendly jamming or cooperative jamming (C_J). In this context, a legitimate jammer enhances the number of destructive jamming signals in the direction of the eavesdropper and limits it to legitimate users. In other cases, a friendly jammer may broadcast all the jamming signals to be received by all devices in the networks. However, it is commonly claimed that the legitimate receiver identifies the Pseudo-random sequence of the jamming signals and could therefore cancel them, causing no damage at the legitimate end [45]. This renders the main channel more robust and dependable than the wiretap channel, hence enhancing the confidentiality of the exchanged information. For instance, in [48], an SU-friendly jammer contributes to the improvement of the secrecy of PUs' communications by releasing jamming signals to the PU's destination and the eavesdropper. However, a nulling approach to prevent the PU receiver from being jammed is considered. In [49], SUs serve as a relay and a friendly jammer for the transmission of PUs in order to improve their energy efficiency and safety. In certain earlier studies, the collaboration between relaying and jamming has also been widely implemented. For example, in [50], two relays in an underlay CRN have been proposed. One of the relays is operating as a cooperative relay that forwards useful messages to the destination, while the other one sends jamming signals to mislead the eavesdropper. Additionally, in [51], an underlay cognitive satellite-terrestrial network is considered under the threat of eavesdroppers. A cooperative relay exists to forward the SU's messages to an end user. Another relay that operates as a cooperative jammer sends artificial noise (jamming signals) to harm the eavesdropper. In [50,51], it is presumed that the legitimate receivers recognize the Pseudo-random sequence of the artificial noise and hence can cancel them. However, this sequence is not recognized by the eavesdropper and thus jamming cannot be prevented. In certain scenarios, PUs might be under the threat of eavesdropping. For example, in [52], it is assumed that an SU acts as a friendly jammer, but it is not always trusted to perform this job, i.e., sending jamming signals to protect the PUs from eavesdroppers. Hence, a strategy based on the previously recorded behavior of SUs for the selection of the next friendly jammer is proposed. In exchange for this cooperation, once the PUs finish their transmissions and the channel is idle, SUs begin their transmissions.

- *Cooperative relaying*: As mentioned earlier in this chapter, from the perspective of PLS, to boost the safety of the transferred information, the conditions of the main channel should be enhanced, and/or the wiretap channel's conditions should deteriorate. In addition to cooperative jamming, it is also possible to use cooperative relaying to prevent eavesdropping. Deploying a relay at the main connection raises the link capacity, resulting in a high SNR at the destination. Consequently, the security of private communications transmitted over this channel is strengthened. To accomplish this, multiple protocols can be employed, such as amplify-and-forward (AF) and decode-and-forward (DF) schemes as they are well-known for their effectiveness in enhancing connection capacity. In addition, a relay selection strategy may be used to maximize the main link's capacity or SNR. There are several schemes for relay selection, including optimal relay selection, opportunist relay selection, sub-optimal relay selection, and partial relay selection. These fall under the max–min strategy umbrella. To illustrate, in [53], in a dual-hop underlay CRN, an optimum relay selection strategy is applied to select one of the available relays in order to maximize the secrecy rate with the presence of multiple eavesdroppers. In [54], four relay selection schemes are considered and contrasted. Furthermore, two-way relaying is a common cooperative relaying system that has been employed to enhance security and spectral efficiency. In [55], two-way relaying and jamming mechanisms have been implemented between SUs in order to enhance the security that has been compromised by a number of eavesdroppers.

- *Multiple antennas*: Multiple antennas can be employed at the transmitter, receiver, or relay (when a cooperative relaying strategy is implemented). Multiple antennas take benefit from the multiple copies of the signal, and as a result, either combining or selection algorithms may be used. To demonstrate, a transmitter antenna selection (TAS) scheme may be applied at the transmitter with the capability to select the antenna that produces the greatest received SNR at the legitimate receiver. Moreover, the MRC approach may be used to increase the quality of the received signal at the receiver. Realizing this, these approaches have been used extensively in the literature to enhance the PLS of the main link in CRNs. This is predicated on the notion that while these selections and combining approaches increase connection reliability, security is also raised. Examples of recent works that utilized the TAS schemes to maximize SNR and hence improve the security can be found in [56–60]. Moreover, in [61–65], the MRC technique has been adopted for maximizing the SNR at the legitimate receiver. Furthermore, various combining techniques, such as selection combining (SC) and generalized selection combining (GSC) have been utilized to improve the security of CRNs (see [64,66,67] for SC and [63,68–70] for GSC).
- *Beamforming*: Beamforming is one of the strategies employed in wireless communication networks to fulfill a variety of purposes. First, beamforming increases the SNR, which in turn increases the capacity of the transmission. This is because beamforming serves to focus signals toward the intended receiver, hence reducing multipath attenuation and interference. Second, since the signals are focused on a particular region, there is no interference created by the system. Specifically, when an antenna broadcasts in all directions, often known as an omni-directional antenna, this may lead to interference. Given these two factors, beamforming improves the security of the physical layer. Particularly, due to the fact that signals are sent in the direction of a particular receiver, it is very difficult for eavesdroppers to eavesdrop on conversations and extract information since they are required to be in the same physical vicinity as the intended receiver [71]. Beamforming has been actively exploited to enhance PLS for CRNs. In [72], for example, beamforming is employed to enhance the useful signals in the direction of the SU receiver and decrease interference at the PU receiver. In addition, beamforming may be used to broadcast artificial noise to eavesdroppers in order to decrease their interception capabilities, while preventing harmful signals to be transmitted to other authorized users, as in [73–75].

Certain enabling technologies may be coupled with PLS to enhance the security of CRNs further or to explore the PLS in a more realistic fashion. In the following sections, a few of these technologies and some recent developments in the field will be reviewed.

13.4 Energy harvesting for securing cognitive radio networks

Safeguarding the physical layer of CRNs against attacks requires energy. This is considered a challenge since many CRN scenarios presume energy-constrained devices, such as cognitive internet of things (IoT) devices and cognitive unmanned aerial vehicles (UAVs), and protecting these networks necessitates additional energy consumption. Using cooperative jamming and relaying to secure CRNs, for instance, costs energy in addition to the other energy-intensive activities that SUs and PUs execute.

Energy harvesting (EH) has a significant impact on PLS in addition to all the benefits it provides for enhancing the energy efficiency and lifetime of energy-constrained devices. Given this, all devices in CRNs may be equipped with the EH feature, allowing them to gather energy from a variety of sources to compensate for their energy consumption. This provides CRNs' users with additional opportunities to better secure networks. The employment of EH allows the network nodes to collect energy and store it, hence extending the battery life. Solar, wind, and radio frequency (RF) waves are all sources of energy that may be harvested via the EH process (see Figure 13.3). To provide energy to devices, EH converts AC electricity into DC current. This energy may be utilized for a multitude of applications, including enhancing energy efficiency and securing private information for CRN users. In addition, self-sustaining capabilities, reduced carbon emission, truly wireless nodes without battery replacement, and easy and quick deployment in any hazardous, hostile, or inaccessible environment are all benefits of EH.

To comprehend the significance of EH in enhancing PLS for CRNs, it is necessary to explore the two primary EH transmit and receiving systems, which will be explored in the next subsections.

13.4.1 Energy harvesting transmit schemes

On the transmitting side, there are several types of EH-transmitters as shown in Figure 13.4 and explained below [71]:

- *Wireless power transfer (WPT)*: In WPT, there is a transmission power station for charging devices through energy transfer. In [76], a clustered wireless sensor network (WSN) with an access point (AP) capable of producing power to charge all network nodes (sensors and cluster heads) was provided as an example of a WPT method. Given this, sensor nodes may conduct sensing, data processing, and communication using the collected energy.
- *Simultaneous wireless information and power transfer (SWIPT)*: RF signals are composed of both energy and information, which is why SWIPT is considered a feasible solution for energy shortages. Energy may be harvested from the received RF signal as two nodes share information. In contrast to the WPT, no additional infrastructure or costs are necessary for this sort of operation.

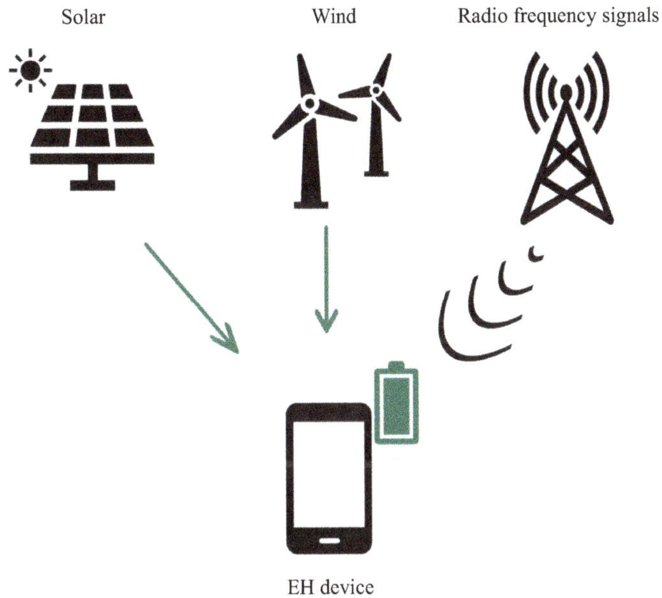

Figure 13.3 EH sources

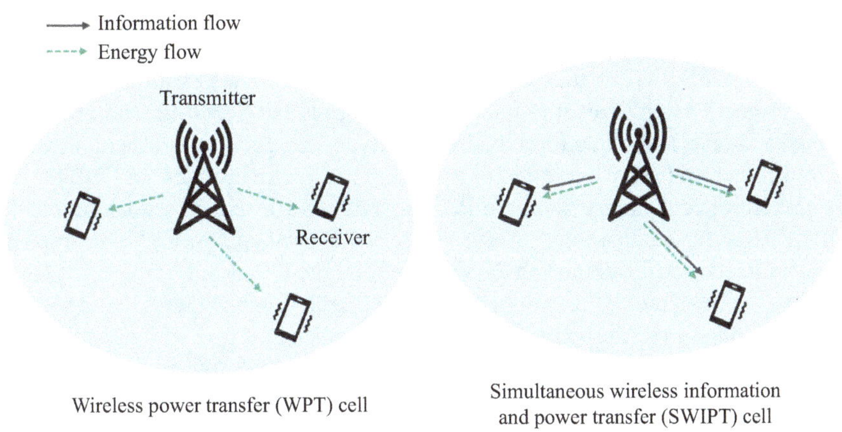

Figure 13.4 EH transmit schemes

13.4.2 Energy harvesting receivers

To be able to successfully benefit from the SWIPT scheme, the receiver should be designed with an energy harvester circuit to carry out either, power splitting (PS), time switching (TS), or antenna selection (AS) protocols as shown in Figure 13.5. They are defined as [71]:

- *PS protocol*: The power of the received signal is partitioned by the receiver using a power splitting factor (θ). Storage equipment like a battery or a capacitor will

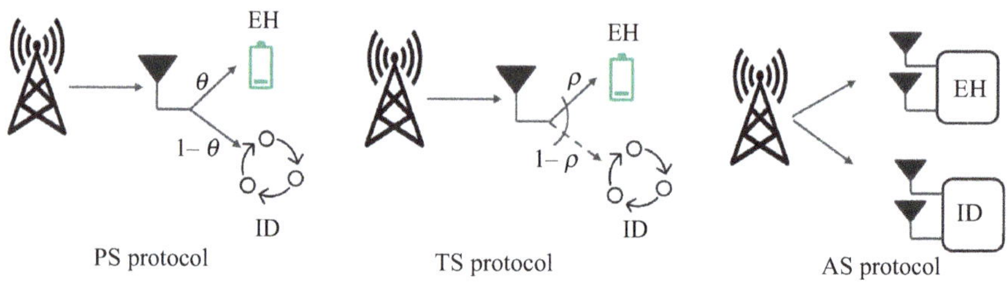

Figure 13.5 EH receiver schemes

be utilized to store the first portion of the power for EH purposes. The rest of the power is used for information decoding (ID).

- *TS protocol*: The available power is utilized, while the time is divided into two or more slots, with one slot devoted to EH and the remainder to ID. The switching is determined by the time switching factor (ρ).
- *AS protocol*: This design mounts a collection of antennas on the receiver, some of which are exploited for EH and others for ID.

13.4.3 Recent works

Several studies have applied EH to increase the energy efficiency of CRNs while investigating PLS. In [24], the SU relay gathers energy from the SU transmitter using the PS protocol and utilizes it to transport messages to the SU destination. Ref. [77] adopted a similar technique for a dual-hop CRN utilizing the TS protocol, in which the transmitter and relay harvested energy from ambient RF waves. In [78], the SU transmitter collects energy from the PU transmitter's RF transmissions via the TS protocol. Similar studies that have used TS or PS protocols in CRNs to improve energy efficiency while exploring PLS can be found in [79–81].

EH may be exploited for cooperative jamming in order to increase the security of CRNs against eavesdroppers. This is achieved when an SU or PU gathers energy for the purpose of creating and broadcasting jamming signals towards untrusted SUs or eavesdroppers [82–85]. In addition, it is frequently considered that the eavesdropper is collecting energy from legitimate users in CRNs in order to obtain the greatest eavesdropping capacity, as in [86]. In [83], a comparison is performed between the effects of an EH and a non-EH eavesdropper on the PLS of a CRN. Additionally, an external jammer can be employed to gather energy from either the transmitters of SUs or PUs in order to impair the wiretap channel by jamming, as described in [45,87,88]. To illustrate, in [45], an external SU jammer that harvests energy from the SU transmitter using the PS protocol is considered. This jammer utilizes the harvested energy to forward jamming signals toward the eavesdroppers. The SUs network is under the threat of multiple randomly distributed eavesdroppers. For this work, Figure 13.6 depicts the probability of non-zero secrecy capacity versus the interference threshold tolerable at the PU receiver (I_{th}). It can be noticed that as

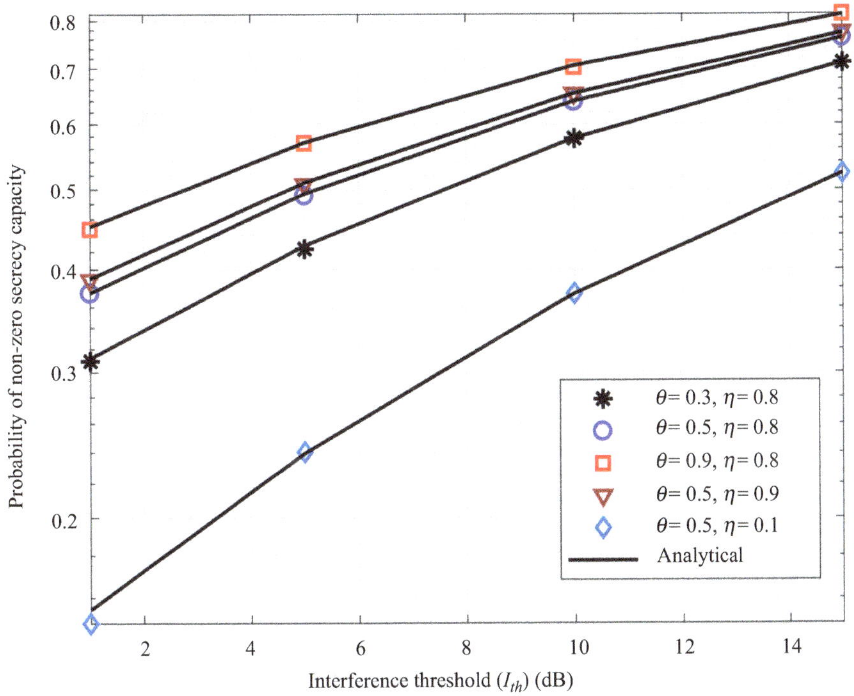

Figure 13.6 Improving PLS for CRNs via cooperative jamming-based EH [45]

the PS factor (θ) and the EH conversion efficiency coefficient (η) increase, implying a higher amount of energy harvested to impair the eavesdroppers, the security improves. This shows the effectiveness of the cooperative jamming-based EH in improving the PLS for CRNs.

13.5 Securing the physical layer of unmanned aerial vehicles-based CRNs

Unmanned aerial vehicles (UAVs), which are also known as drones, are aircraft without humans on board. UAVs have emerged recently due to their various applications, which range from military to civil applications due to their flexibility and fast mobility. These applications include video streaming, amateur photography and filming, people and weather monitoring, rescue and research, traffic control, cruise detection, and disaster recovery [89,90]. Hence, UAVs are critical components of every nation's airspace in terms of social, economic, and security.

UAVs provide promising wireless communication in the era of 5G and beyond. For instance, UAVs may act as user equipment, a flying base station, or a relay in flying ad-hoc networks. Depending on their application, the UAVs can be deployed as a high-altitude platform for long-term use or as a low-altitude platform for rapid and flexible deployment. However, one of the main challenges that UAVs face is spectrum scarcity. This is due to the fact that UAVs operate in unlicensed bands, which renders them competing with other devices operating in the same bands, such as smartphones, WIFI, and Bluetooth [89,91]. This makes the spectrum crowded and hence causes UAVs to face spectrum scarcity in the near future. To tackle this challenge, CRNs can be utilized to enable UAVs to access the licensed bands in 5G networks without causing harmful interference to licensed users. In 5G and beyond networks, UAVs equipped with CR technology can simultaneously access the wireless channel and perform many functions such as traffic surveillance, disaster management, and package delivery.

13.5.1 Challenges of UAVs-based CRNs

Several challenges arise with the integration of UAVs and CRNs in 5G and beyond that lead to serious security concerns. These challenges include [89]:

- *Spectrum mobility*: A critical aspect of UAVs-enabled CRN is spectrum mobility. That is, recent research has been focusing on stationary or low-mobility devices when studying spectrum handoffs. On the other hand, high mobility devices, such as UAVs lead to high spectrum mobility. That is, SUs operating as UAVs need to perform a high number of spectrum handoffs due to the appearance of PUs, which will negatively impact the UAVs' communication stability as it causes delays and interruptions in transmissions. Moreover, SUs might be under the threat of PUEAs while they are shifting from one band to another, which is another obstacle that UAVs may encounter in the context of security.
- *Response time*: Another important aspect to consider is response time and the delay associated with communication in UAVs. The processes performed by cognitive radio (CR) users, such as spectrum sensing and handover need to occur fast enough to avoid delays in time. This is due to the fact that the delay impacts the instant the UAVs begin functioning, which directly impacts the reliability of these UAVs and thereby the speed these UAVs communicate and deliver their messages. The delay in UAVs can be reduced via cooperative sensing and transmission in CRNs. However, from the perspective of the PLS, a reduction in reliability results in security concerns.
- *Channel modeling*: The fast change in the environment and the high-speed motion of aircraft cause a rapid change in the channel characteristics. These characteristics include different types of delays, fading, and Doppler spread, which impact spectral efficiency. This elevates the problem of channel modeling and the design of the system's physical layer. Orthogonal frequency division multiplexing (OFDM)-based systems are now being examined for aeronautical propagation environments as a potential technique for increasing spectrum efficiency. Nevertheless, legitimate users will experience variable levels of security as the received SNR fluctuates due to changes in channel characteristics.

- *Energy consumption*: In addition to the energy consumed by UAVs while covering large areas and performing long tasks, the spectrum sensing and handoff performed by UAVs in CRNs raise the problem of energy consumption. Indeed, SUs might pause their transmissions until the band is vacant instead of transferring to another band in order to save energy. To solve this issue, SUs can adopt the underlay mode in which there is no need to pause the transmissions as long as their transmission power is below the interference threshold at the PU receiver. However, security threats might arise here due to the continuous adaptation of the transmission power in the underlay mode.

It can be noticed from the previous challenges that communication security plays an important role in UAVs. In fact, UAVs-based CRNs are susceptible to different types of attacks, such as jamming, a global positioning system (GPS) spoofing (location spoofing) attack, and eavesdropping. Such security threats may lead the ground base station to lose connection with UAVs-based cognitive radios. This problem mainly arises in overcrowded or hostile areas. In addition, since these attacks impact the physical layer, PLS is a reliable approach for securing UAVs-enabled CRNs. In the following subsection, the main physical layer attacks that UAVs-based CRNs encounter and their well-known countermeasures are listed.

13.5.2 Attacks on the physical layer of UAVs-based CRNs and countermeasures

Some of the physical layer attacks are common between UAV-based CRN users and regular CRN users. In this subsection, the main types of attacks and countermeasures against them are listed.

- *Jamming*: Due to the shared and dynamic environment, one of the attacks that the physical layer of UAVs-based CRNs is susceptible to is jamming. Similar to CRNs, this threat degrades system performance since the jamming causes incorrect learning behaviors and incorrect judgment, in addition to creating a denial of service. In this case, energy consumption will occur due to re-transmissions of data. Specifically, the line-of-sight (LoS) connection between the UAV and the ground station exposes the users to terrestrial jamming threats [92]. In order to mitigate this danger and improve security, jamming must first be identified and avoided. Self-awareness (SA) is one strategy used to combat the jamming problem from the PLS standpoint. The user learns a hierarchical representation of the radio environment through SA. This is accomplished via a hierarchical dynamic Bayesian network (HDBN), which enables the radio to properly predict the status of the RF bands using Bayesian filters. Given this, SA can assist in identifying abnormalities produced by jamming attacks by using previous data. That is, whenever there is a divergence from the predictions, jamming operations are identified. Jamming detection in UAVs-based CRNs-based learning approaches is an intriguing topic of research since jamming is not always detectable owing to the emergence of intelligent jammers, such as in [92–95].

- *Location/ GPS spoofing*: One of the major physical layer attacks for cognitive UAVs is location/ GPS spoofing. This attack occurs when an adversary spoofs the SU receiver using a signal that is stronger than a genuine location satellite signal. As a result, a spoofing attacker may capture a UAV and control its traveling route. In addition, GPS spoofing attackers might hijack flying UAVs by modifying the transmission between two devices, which is a tactic known as a man-in-the-middle attack. The attacker will next force the UAV to land in an unapproved location and exploit its authorized network access [96]. In addition, GPS spoofing attacks may lead to a denial-of-service and severe interference to SUs and PUs. The spoofing operation begins with the transmission of signals that are synchronized with the authentic GPS signals detected by the target receiver. In fact, the attacker employs a GPS spoofing device, also known as a GPS simulator, to initiate a spoofing attack by transmitting counterfeit GPS signals to the target receiver. Then, the strength of the fake signals surpasses the strength of the real GPS signals. Hence, the victim's receiver will then lock onto the false signals and deviate from the actual ones. At this point, the receiver will be subjected to a spoofing attack [97].

  Countermeasures against location/ GPS spoofing attack:
  Several methods exist for detecting the location/ GPS spoofing attack, and some of the countermeasures are included below [97]:
  - *Self-checking*: It is one possible response to this attack that aids in detecting an irregularity in the GPS receiver by comparing the accuracy of the received GPS signals to the signals produced by fixed and well-known structures. For instance, the targeted receiver may examine the carrier's power to detect an unusual change.
  - *Smart antennas*: These are employed to identify the angle of arrival for each satellite signal. In this context, it is easier to identify spoofing since all fake signals are transmitted from the same location.
  - *Centralized detection scheme*: This approach is predicated on the premise that all SUs should register their locations in a database for database-driven CRNs. This database is intended to identify the dangers posed by spoofing. Thus, the system can ensure that every SU routinely updates its location information, and hence maintain a track of all SUs' positions. Then, an abnormal activity may be quickly identified if a large number of SUs travel abruptly to another location at a very high pace. In this instance, the CRN is deemed to be under spoofing operation.
  - *Environmental-radio-based location verification*: Herein, it is assumed that the SU pre-stores a radio environment map (REM) that includes the fingerprints of the signals of existing infrastructures. Then, without utilizing GPS, an SU estimates its position by matching the local signals' fingerprints to those stored in the REM. Therefore, this SU can avoid spoofing attacks.
  - *Peer location verification*: This strategy is based on the premise that each SU will emit an r-radius beacon signal containing its location to other nearby SUs to prevent collisions. Then, when an unauthenticated SU receives a

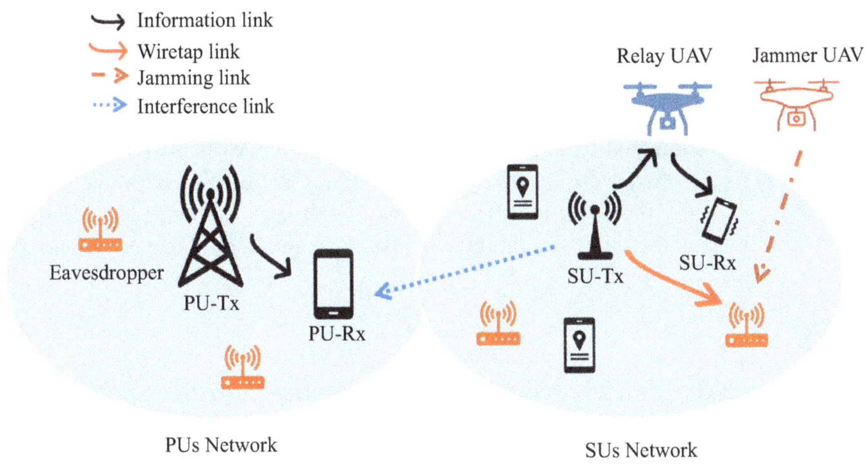

Figure 13.7 Illustration for methods to combat eavesdropping in UAVs-based CRNs

beacon signal, it verifies whether its GPS position is within a radius of r from the SU location. This SU will trust the received GPS position if it is within the specified radius; otherwise, it will indicate that it is under spoofing attack and remain silent.

- *Eavesdropping*: Due to the broadcast nature of cognitive UAV transmitters and the strong LoS connectivity between the UAV and the ground receivers, the transferred private messages are always susceptible to eavesdropping [98]. The main methods of combating eavesdropping for cognitive UAVs are shown in Figure 13.7 and are given in the following subsection.

Countermeasures against eavesdropping threat:

1. *Cooperative jamming*: UAVs can be employed as friendly jammers to enhance security by emitting artificial noise (jamming signals) to impair the wiretap channels of eavesdroppers. This is owing to the deployment flexibility, mobility, and robust LoS channels that UAVs provide [99]. Specifically, a UAV can reach an optimal location and altitude to aid SUs or PUs in transmitting jamming signals to eavesdroppers. It is worth mentioning that using conventional fixed jammers may result in a blockage, and as a consequence, their jamming effectiveness will be inferior to that of UAVs. Multiple studies, including [99–102], have addressed this topic. In addition, the trajectory of the UAV jammer may also be tuned to maximize the secrecy rate of the authorized users [98,103].
2. *Cooperative relaying*: When the direct route between the transmitter and the receiver is obstructed by several objects, resulting in an unstable connection, traditional fixed relays are used to enhance the link's performance. Regardless of the role of these relays, i.e., AF or DF relays, the relayed signals will be stronger, hence enhancing the SNR and security. Nonetheless,

fixed relays lack mobility and scalability qualities, which is problematic in densely obstructed areas. Consequently, UAVs have lately been used as relays owing to their ease of deployment and adaptability [104]. Deploying UAVs as relays further improve the PLS given their function in enhancing transmission quality. In [104], for instance, a DF UAV relay was deployed in CRNs to enhance the security of SUs. Similar UAV functions for CRNs may be seen in [105–107]. In [101,102], two UAVs are examined, one acting as a relay and the other as a jammer to reduce the quality of the eavesdropper's channel.

13.6 Cascaded fading channels and securing cognitive radio networks

Lately, cascaded fading channel paradigm has developed as a highly effective and realistic model for the transmission of signals. Figure 13.8 illustrates that the transmitted signal between the source (S) and the destination (D) is generated by several signals reflected from objects and scattered throughout the channel. These signals are represented by independent random variables that are not necessarily identically distributed. In the context of cascaded fading models, the received signal (Y_N) at D will be the multiplication of random variables X_i, for $i = 1, 2, \cdots, N$, with N is the cascade level as

$$Y_N = \prod_{i=1}^{N} X_i. \tag{13.4}$$

13.6.1 Applications of cascaded fading channels

Various types of communication systems employ cascaded fading channels to represent RF signal transmission, including mobile-to-mobile (M2M)/vehicle-to-vehicle (V2V) transmission channels [108], radio-frequency identification (RFID) pinhole channels [109,110], multi-hop relaying systems [2], and multiple-input-multiple-output (MIMO) keyhole communication systems [111]. The complete structure in multi-hop relaying systems may be represented using the cascaded fading channels model, in which the signal from the transmitter to the receiver is relayed from one

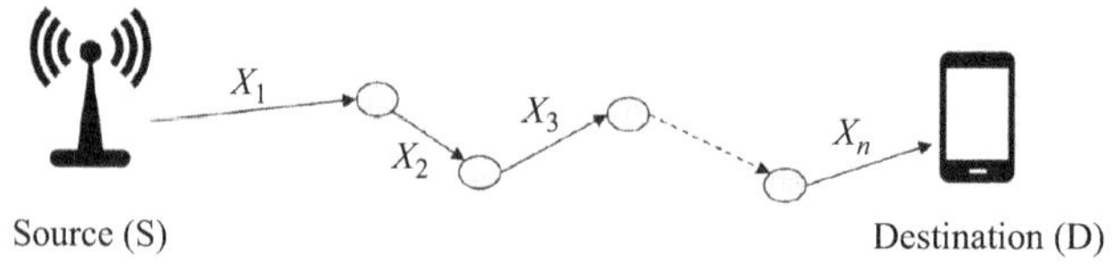

Figure 13.8 Cascaded channels modeling

node to another. Each node functions as a non-regenerative node. Additionally, the transmission in small keyholes within barriers can be represented by cascaded channels. These keyholes are also used to propagate the signal from the transmitter to the receiver, where each keyhole serves as a new source for the next one [112]. For instance, cascaded Rayleigh fading channels may be used to model inter-vehicular communication (IVC) channels, according to certain studies [110]. Keyhole channel models used in MIMO systems have been developed using double Rayleigh fading channels [112]. Double Rayleigh models are also employed to model M2M communication, and vehicular communications systems [113]. Furthermore, the connections for V2V communications were modeled using N$*$Nakagami-m fading channels [113].

13.6.2 Cascaded fading channels and PLS in CRNs

It has been broadly acknowledged in earlier research that multi-hop CRNs operate over classical fading channel models, such as the Rayleigh fading model. However, recent developments in cognitive vehicular networks (CVNs) and M2M communication have attracted attention. Traditional fading channel models, may not be suitable for simulating the propagation of signals in these networks [114]. This is due to the fact that they do not adequately simulate the signal transmission. Particularly, it is believed that the signal flows from the transmitter to the receiver without encountering any barriers along the way. In addition, when the network units are located in locations with a lot of scattering, an accurate technique is needed to describe the propagation of signals. In light of this, cascaded fading channels have been lately recommended for these types of networks.

Cascaded channel models have a significant influence on the security of communications, thus they must be used in the research and improvement of the PLS for CRNs and particularly for CVNs. In fact, in [3,44,45], PLS for a three-node wiretap system model is studied, in which the channels undergo cascaded κ–μ model. It was proved in these articles the importance of assuming cascaded channels that make them hard to be ignored when studying PLS. In [3], the secrecy outage probability has been examined assuming colluding eavesdroppers, as illustrated in Figure 13.9. SOP is investigated versus the average received SNR at the legitimate receiver Bob ($\bar{\gamma}_B$). κ and μ are the fading parameters for the main channel, while κ_e and μ_e correspond to the wiretap channel parameters. It is noticed that the PLS weakens as the main channel cascade level (N) grows. Moreover, when the amount of cascade for the wiretap channel (n_e) lowers, the shared information becomes less private. Thus, when n_e decreases, there are fewer obstructions in the route of the wiretap. This will enhance the eavesdropper's ability to intercept communications, hence reducing the security of the legitimate users' messages exchange.

In [22], a case study demonstrating the significance of considering cascaded channels for CRNs while boosting security was presented. Under the danger of an eavesdropper, an underlay CRN has been examined, where both channels, i.e. the main and wiretap channels, follow the cascaded Rayleigh model. In this case, it

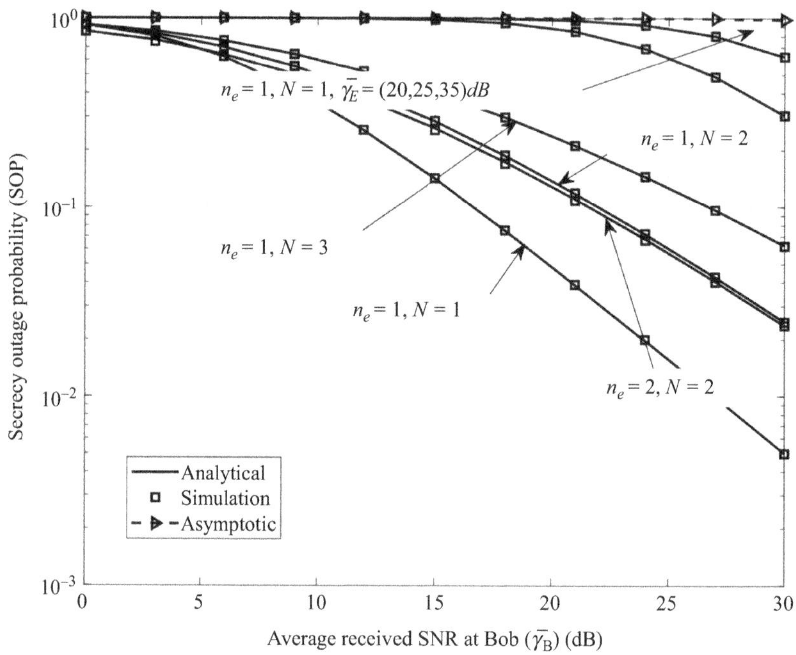

Figure 13.9 The lower bound of the secrecy outage probability. For the main channel: $\kappa = 0, \mu = 1$ and for the wiretap channel: $\kappa_e = 0, \mu_e = 1$ [3]

is assumed that the secondary network is situated in a heavily dispersed location. In addition, the probability of non-zero secrecy capacity has been studied for the considered system model to demonstrate the effect of the cascaded level of the main channel (N) as the SU receiver is equipped with multiple antennas (L_D). It is also presumed that the eavesdropper's channel cascade level is $n_e = 2$ and is equipped with multiple antennas (L_E). Figure 13.10 demonstrates that the security degrades as N rises. In other words, when N rises, more scattering is anticipated in the SUs' connection, resulting in more severe fading and less security.

13.6.3 Recent works

Exploring PLS and improving it for CRNs while considering cascaded channels has been an intriguing issue for addressing the aforementioned challenges. PLS for an underlay CRN over cascaded Rayleigh model and over cascaded κ–μ fading distribution has been studied in [115] and [116], respectively. The analyzed data and system model were adequate for characterizing CVNs. In [45,83], PLS for underlay CRNs has been investigated to further enhance the security of CRNs across cascaded channels. The improvement in PLS was made possible by a cooperative jamming-based

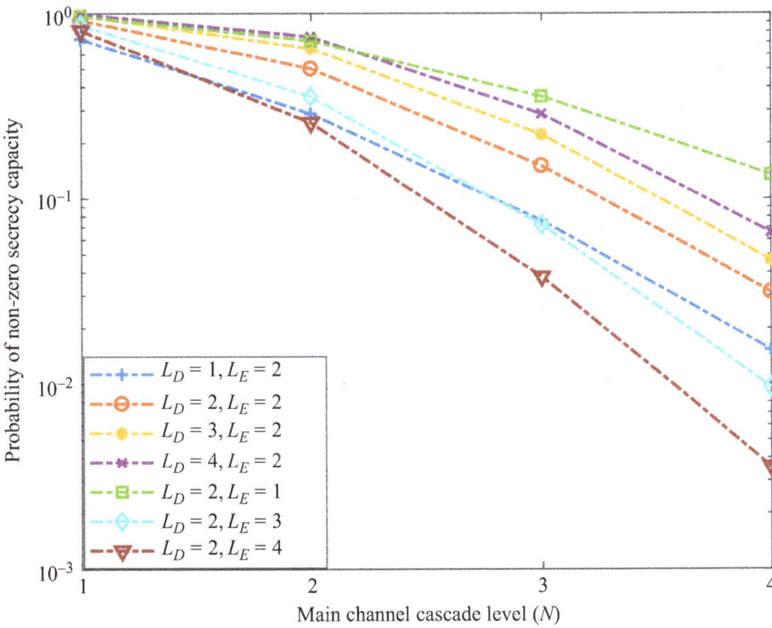

Figure 13.10 The probability of non-zero secrecy capacity versus the main channel cascade level (N) for the different number of antennas at SU-Rx and Eve. $n_e = 2$ [22]

EH approach. All of the aforementioned articles demonstrate that cascaded channels must be considered while analyzing or strengthening PLS for CVNs.

13.7 Conclusions

Upon observing the radio spectrum, it is discovered that spectrum utilization is concentrated in particular portions of the band. Moreover, due to the fixed frequency allocation, a significant amount of the spectrum is currently underutilized. Therefore, CRNs were suggested to address these issues by providing SUs with opportunistic access to the licensed bands. Due to the broadcast nature of CRNs, SUs and PUs are vulnerable to attacks that originate from inside and outside the network. Several types of attacks can impact users in CRNs, such as the PUEA, jamming, and eavesdropping. Since both SUs and PUs networks impact each other, the privacy of both of them should be protected from attacks. One possible approach to safeguard these networks is through the employment of the PLS approach. This chapter provides a summary of the most common threats to the physical layer of CRNs and discusses how to protect users against them. In addition, to further improve the PLS of CRNs, EH technique utilization is explored. UAVs-based CRNs and the main physical layer attacks and countermeasures are also presented. Furthermore, the impact of cascaded fading channels on the PLS of CRNs has been provided, proving the importance of considering them when studying the PLS of CRNs.

13.8 Future directions

Due to the significant achievements of combining PLS and CRNs, it becomes appealing to combine them with other emerging technologies for the development of future wireless networks. Combining PLS and CRNs with other technologies enables the attainment of a number of additional qualities, including optimal security and reduced complexity. In this section, we will discuss potential future research paths that should be pursued in the field of PLS and CRNs.

13.8.1 Cross-layer attacks

Prior research has mostly focused on protecting the physical layer of CRNs via PLS approach or the other upper network layers (e.g., network or MAC layers). Nevertheless, some attacks that target one network layer might affect other levels. These attacks are characterized as cross-layer threats. This may occur in several wireless network, and CRNs are no exception. Future research should concentrate more on the collaboration of PLS with other higher layers security approaches, such as cryptography, to guarantee that CRNs are protected to the greatest extent possible.

13.8.2 Machine learning algorithms

Machine learning (ML) techniques identify trends by interacting with the surroundings. CRNs become intelligent when they adapt and make decisions solely on the basis of their current and past acquired information [7]. Intelligent CRNs may make a choice to optimize the security of PUs and SUs from a PLS perspective, even in the absence of comprehensive and accurate previous knowledge from the environment. To optimize the security of CR users with the least complexity compared to other traditional and sophisticated optimization problems, ML techniques should be integrated with PLS.

13.8.3 Reflecting intelligent surfaces

A reflecting intelligent surface (RIS) is a type of uniform planner array constructed from low-cost passive reflecting components. Due to its potential to provide intelligent and adjustable wireless channel conditions, RISs have lately gained interest. The components of RISs may be regulated to tune the amplitude and phase of RF signals in order to raise or decrease signal strength in a certain direction [117]. In the context of PLS, this implies that it is feasible to enhance the SNR at the legitimate receiver and decrease it in all other directions in order to boost security. PLS and RISs for CRNs have been the focus of a limited amount of study, and additional investigation is urged to increase the privacy of information transferred between CR users.

13.8.4 Millimeter wave applications

Leveraging high-frequency communications in the millimeter wave (mmWave) band is one of the most potential 5G technologies. Due to the high frequency of mmWave, a larger bandwidth may be attained with a large number of antennas and highly directed broadcasts [118]. Combining PLS with mmWave would increase security for CRNs while delivering a higher data rate and reduced latency. Furthermore, the directional broadcasts and short wavelength of mmWave antennas may direct signals in the vicinity of authorized receivers, hence enhancing the PLS. Certain earlier research, including [119], has shown that PLS may be more effective when paired with mmWave rather than traditional microwave systems. Combining the advantages of mmWave, CRNs, and PLS would reach outstanding aims in terms of security and spectrum efficiency; however, more studies should be undertaken in this area since only a limited number of studies have been performed.

References

[1] Wang Q, Xu K, Ren K. Cooperative secret key generation from phase estimation in narrowband fading channels. *IEEE Journal on Selected Areas in Communications*. 2012;30(9):1666–1674.

[2] Ata SÖ. Secrecy performance analysis over cascaded fading channels. *IET Communications*. 2019;13(2):259–264.

[3] Tashman DH, Hamouda W. Cascaded κ-μ fading channels with colluding eavesdroppers: physical-layer security analysis. In: *2020 International Conference on Communications, Signal Processing, and their Applications* (ICCSPA); 2021. p. 1–6.

[4] Wyner AD. The wire-tap channel. *Bell Labs Technical Journal*. 1975;54(8): 1355–1387.

[5] Elkashlan M, Wang L, Duong TQ, *et al.* On the security of cognitive radio networks. *IEEE Transactions on Vehicular Technology*. 2015 Aug;64(8):3790–3795.

[6] Xiang Z, Yang W, Pan G, *et al.* Physical layer security in cognitive radio inspired NOMA network. *IEEE Journal of Selected Topics in Signal Processing*. 2019;13(3):700–714.

[7] Khalek NA, Hamouda W. From cognitive to intelligent secondary cooperative networks for the future Internet: design, advances, and challenges. *IEEE Network*. 2021;35(3):168–175.

[8] Khalek NA, Hamouda W. Learning-based cooperative spectrum sensing in hybrid underlay-interweave secondary networks. In: *GLOBECOM 2020 – 2020 IEEE Global Commun. Conf.*; 2020. p. 1–6.

[9] Moualeu JM, Ngatched TMN, Hamouda W, *et al.* Energy-efficient cooperative spectrum sensing and transmission in multi-channel cognitive radio networks. In: *2014 IEEE International Conference on Communications* (ICC); 2014. p. 4945–4950.

[10] Ali A, Hamouda W. Power-efficient wideband spectrum sensing for cognitive radio systems. *IEEE Transactions on Vehicular Technology*. 2018;67(4): 3269–3283.

[11] El Tanab M, Hamouda W. Resource allocation for underlay cognitive radio networks: a survey. *IEEE Communications Surveys & Tutorials*. 2017;19(2): 1249–1276.

[12] Ali A, Hamouda W. Low power wideband sensing for one-bit quantized cognitive radio systems. *IEEE Wireless Communications Letters*. 2016;5(1): 16–19.

[13] Ali A, Hamouda W. A novel spectrum monitoring algorithm for OFDM-based cognitive radio networks. In: *2015 IEEE Global Communications Conference* (GLOBECOM); 2015. p. 1–6.

[14] Khalek NA, Hamouda W. Intelligent spectrum sensing: an unsupervised learning approach based on dimensionality reduction. In: *ICC 2022 – IEEE Int. Conf. Commun.*; 2022. p. 171–176.

[15] Ali A, Hamouda W. Advances on spectrum sensing for cognitive radio networks: theory and applications. *IEEE Communications Surveys & Tutorials*. 2017;19(2):1277–1304.

[16] Elsaadany M, Hamouda W. Antenna selection for dual-hop cognitive radio networks: a multiple-relay scenario. *IEEE Transactions on Vehicular Technology*. 2017;66(8):6754–6763.

[17] El Tanab M, Hamouda W, Fahmy Y. On the distributed resource allocation of MIMO cognitive radio networks. In: *2015 IEEE Global Communications Conference* (GLOBECOM); 2015. p. 1–6.

[18] Elsaadany M, Hamouda W. Performance analysis of non-orthogonal AF relaying in cognitive radio networks. *IEEE Wireless Communications Letters*. 2015;4(4):373–376.

[19] El Tanab M, Hamouda W, Fahmy Y. Distributed opportunistic scheduling for MIMO underlay cognitive radio networks. *Wireless Communications and Mobile Computing*. 2016;16(15):2212–2224.

[20] Elsaadany M, Hamouda W. Enhancing the performance of amplify-and-forward cognitive relay networks: a multiple-relay scenario. In: *2015 IEEE Global Communications Conference* (GLOBECOM); 2015. p. 1–6.

[21] Khalek NA, Hamouda W. Unsupervised two-stage learning framework for cooperative spectrum sensing. In: *ICC 2021 – IEEE Int. Conf. Commun.*; 2021. p. 1–6.

[22] Tashman DH, Hamouda W. An overview and future directions on physical-layer security for cognitive radio networks. *IEEE Network*. 2021;35(3): 205–211.

[23] Ali A, Hamouda W. Spectrum monitoring using energy ratio algorithm for OFDM-based cognitive radio networks. *IEEE Transactions on Wireless Communications*. 2015;14(4):2257–2268.

[24] Bouabdellah M, El Bouanani F, Sofotasios PC, *et al.* Cooperative energy harvesting cognitive radio networks with spectrum sharing and security constraints. *IEEE Access*. 2019;7:173329–173343.

[25] Zou Y, Zhu J, Yang L, *et al.* Securing physical-layer communications for cognitive radio networks. *IEEE Communications Magazine.* 2015;53(9): 48–54.

[26] Yadav K, Dhar Roy S, Kundu S. Enhanced throughput performance under primary user emulation attack in cognitive radio networks by optimal threshold selection approach. In: *2018 2nd International Conference on Electronics, Materials Engineering & Nano-Technology* (IEMENTech); 2018. p. 1–6.

[27] Saber MJ, Sadough SMS. On optimal spectrum sensing strategy for cognitive radio systems under primary user emulation attack. In: *7'th International Symposium on Telecommunications* (IST'2014); 2014. p. 1113–1116.

[28] Chen R, Park JM. Ensuring trustworthy spectrum sensing in cognitive radio networks. In: *2006 1st IEEE Workshop on Networking Technologies for Software Defined Radio Networks*; 2006. p. 110–119.

[29] Karimi A, Taherpour A. Trusted throughput-efficient spectrum sensing against intelligent malicious behaviors in cognitive radio networks. In: *2018 9th International Symposium on Telecommunications* (IST); 2018. p. 357–362.

[30] Jin Z, Anand S, Subbalakshmi KP. Robust spectrum decision protocol against primary user emulation attacks in dynamic spectrum access networks. In: *2010 IEEE Global Telecommunications Conference GLOBECOM 2010*; 2010. p. 1–5.

[31] Nguyen-Thanh N, Ciblat P, Pham AT, *et al.* Surveillance strategies against primary user emulation attack in cognitive radio networks. *IEEE Transactions on Wireless Communications.* 2015;14(9):4981–4993.

[32] Chao CM, Lee WC, Wang CX, *et al.* A flexible anti-jamming channel hopping for cognitive radio networks. In: *2018 Sixth International Symposium on Computing and Networking Workshops* (CANDARW); 2018. p. 549–551.

[33] Chang GY, Wang SY, Liu YX. A jamming-resistant channel hopping scheme for cognitive radio networks. *IEEE Transactions on Wireless Communications.* 2017;16(10):6712–6725.

[34] Ibrahim K, Ng SX, Qureshi IM, *et al.* Anti-jamming game to combat intelligent jamming for cognitive radio networks. *IEEE Access.* 2021;9:137941–137956.

[35] Xu J, Lou H, Zhang W, *et al.* An intelligent anti-jamming scheme for cognitive radio based on deep reinforcement learning. *IEEE Access.* 2020;8: 202563–202572.

[36] Huang C, Chen G, Gong Y, *et al.* Joint buffer-aided hybrid-duplex relay selection and power allocation for secure cognitive networks with double deep Q-network. *IEEE Transactions on Cognitive Communications and Networking.* 2021;7(3):834–844.

[37] Ahmed IK, Fapojuwo AO. Stackelberg equilibria of an anti-jamming game in cooperative cognitive radio networks. *IEEE Transactions on Cognitive Communications and Networking.* 2018;4(1):121–134.

[38] Wu Y, Wang B, Liu KJR, *et al.* Anti-jamming games in multi-channel cognitive radio networks. *IEEE Journal on Selected Areas in Communications.* 2012;30(1):4–15.

[39] Li H, Han Z. Dogfight in spectrum: jamming and anti-jamming in multichannel cognitive radio systems. In: *GLOBECOM 2009 – 2009 IEEE Global Telecommunications Conference*; 2009. p. 1–6.

[40] Zhu Q, Li H, Han Z, *et al.* A stochastic game model for jamming in multichannel cognitive radio systems. In: *2010 IEEE International Conference on Communications*; 2010. p. 1–6.

[41] Bany Salameh HA, Almajali S, Ayyash M, *et al.* Spectrum assignment in cognitive radio networks for Internet-of-Things delay-sensitive applications under jamming attacks. *IEEE Internet of Things Journal.* 2018;5(3): 1904–1913.

[42] Chaudhary V, Jagadeesh H. Fast-forward mitigation schemes for cognitive adversary. *IEEE Transactions on Cognitive Communications and Networking*. 2021;7(4):1304–1319.

[43] Su R, Wang Y, Sun R. Destination-assisted jamming for physical-layer security in SWIPT cognitive radio systems. In: *2018 IEEE Wireless Commun. Networking Conf.* (WCNC); 2018. p. 1–6.

[44] Tashman DH, Hamouda W, Dayoub I. Secrecy analysis over cascaded κ-μ fading channels with multiple eavesdroppers. *IEEE Transactions on Vehicular Technology*. 2020;69(8):8433–8442.

[45] Tashman DH, Hamouda W, Moualeu JM. On securing cognitive radio networks-enabled SWIPT over cascaded κ-μ fading channels with multiple eavesdroppers. *IEEE Transactions on Vehicular Technology.* 2021;71(1):478–488.

[46] Tashman D, Hamouda W. Cascaded κ-μ fading channels with colluding and non-colluding eavesdroppers: physical-layer security analysis. *Future Internet*. 2021;13(8):205.

[47] Tashman DH, Hamouda W. Towards improving the security of cognitive radio networks-based energy harvesting. In: *ICC 2022 – IEEE International Conference on Communications*; 2022. p. 3436–3441.

[48] Sarikaya Y, Ercetin O, Gurbuz O. Control of cognitive networks with friendly jamming as a service. *IEEE Transactions on Cognitive Communications and Networking*. 2018;4(2):299–313.

[49] Liu X, Zheng K, Fu L, *et al.* Energy efficiency of secure cognitive radio networks with cooperative spectrum sharing. *IEEE Transactions on Mobile Computing*. 2019;18(2):305–318.

[50] Shah HA, Koo I. A novel physical layer security scheme in ofdm-based cognitive radio networks. *IEEE Access*. 2018;6:29486–29498.

[51] Bouabdellah M, Bouanani FE. A PHY layer security of a jamming-based underlay cognitive satellite-terrestrial network. *IEEE Transactions on Cognitive Communications and Networking*. 2021;7(4):1266–1279.

[52] Wen Y, Huo Y, Ma L, *et al.* A scheme for trustworthy friendly jammer selection in cooperative cognitive radio networks. *IEEE Transactions on Vehicular Technology*. 2019;68(4):3500–3512.

[53] Shokair M, Saad W, Ibraheem SM. Statistical analysis of a class of secure relay assisted cognitive radio networks. *China Communications*. 2018;15(12):174–189.

[54] Phi Son V, Nhat Binh L, Nguyen TT, *et al*. Physical layer security in cooperative cognitive radio networks with relay selection methods. In: *2021 International Conference on Advanced Technologies for Communications* (ATC); 2021. p. 295–300.

[55] Cao Z, Ji X, Wang J, *et al*. Security-reliability tradeoff analysis for underlay cognitive two-way relay networks. *IEEE Transactions on Wireless Communications*. 2019;18(12):6030–6042.

[56] Chetry S, Singh A. Physical layer security of outdated CSI based CRN. In: *2018 9th Int. Conf. Comput., Commun. Networking Technol.* (ICCCNT). Bangalore, India; 2018. p. 1–5.

[57] Honde V, M L, V NM, *et al*. Transmit antenna selection in MIMO cognitive radio networks. In: *2016 Online International Conference on Green Engineering and Technologies* (IC-GET); 2016. p. 1–6.

[58] Yan P, Zou Y, Zhu J. Transmit antenna selection to improve physical layer security for MIMO-CR systems. In: *2016 8th International Conference on Wireless Communications & Signal Processing* (WCSP); 2016. p. 1–4.

[59] Thakur S, Singh A. Secrecy performance analysis for underlay cognitive radio network with optimal antenna selection and generalized receiver selection. In: *2019 IEEE International Conference on Advanced Networks and Telecommunications Systems* (ANTS); 2019. p. 1–6.

[60] Khodeir MA, Alquran SM. Secrecy outage performance with EH and TAS for realistic underlay cognitive radio networks using MIMO systems. In: *2021 12th International Conference on Information and Communication Systems* (ICICS); 2021. p. 405–410.

[61] Shang Z, Zhang T, Wang X, *et al*. Secure transmission in cognitive relaying networks with full-duplex technique. In: *2020 IEEE 3rd International Conference on Electronics Technology* (ICET); 2020. p. 688–694.

[62] Zhao H, Liu H, Liu Y, *et al*. Physical layer security of maximal ratio combining in underlay cognitive radio unit over Rayleigh fading channels. In: 2015 *IEEE Int. Conf. Commun. Software Networks* (ICCSN). Chengdu, China; 2015. p. 201–205.

[63] Thakur S, Thakur A, Soni S, *et al*. On the performance of GSC/MRC based cognitive radio networks with multiple primary users. In: *2018 IEEE 13th International Conference on Industrial and Information Systems* (ICIIS); 2018. p. 251–257.

[64] Chen D, Cheng Y, Yang W, *et al*. Physical layer security in cognitive untrusted relay networks. *IEEE Access*. 2018;6:7055–7065.

[65] Chen Y, Zhang T, Qiao X, *et al*. Secure cognitive MIMO wiretap networks with different antenna transmission schemes. *IEEE Access*. 2021;9: 5779–5790.

[66] Shang Z, Zhang T, Liu Y, *et al*. Secrecy performance analysis of cognitive radio networks with full-duplex relaying. In: *2019 IEEE/CIC International Conference on Communications in China* (ICCC); 2019. p. 700–705.

[67] Shang Z, Zhang T, Cai Y, *et al.* Secure spectrum-sharing wiretap networks with full-duplex relaying. *IEEE Access*. 2019;7:181610–181625.

[68] Lei H, Zhang H, Ansari IS, *et al.* Secrecy outage performance for SIMO underlay cognitive radio systems with generalized selection combining over Nakagami-*m* channels. *IEEE Transactions on Vehicular Technology.* 2016;65(12):10126–10132.

[69] Lei H, Gao C, Ansari IS, *et al.* Secrecy outage performance of transmit antenna selection for MIMO underlay cognitive radio systems over Nakagami-*m* channels. *IEEE Transactions on Vehicular Technology*. 2017;66(3):2237–2250.

[70] Zhang T, Cai Y, Huang Y, *et al.* Secure full-duplex spectrum-sharing wiretap networks with different antenna reception schemes. *IEEE Transactions on Communications*. 2017;65(1):335–346.

[71] Alsaba Y, Rahim SKA, Leow CY. Beamforming in wireless energy harvesting communications systems: a survey. *IEEE Communications Surveys & Tutorials*. 2018;20(2):1329–1360.

[72] Tuan PV, Trung Duy T, Koo I. Multiuser MISO beamforming design for balancing the received powers in secure cognitive radio networks. In: *2018 IEEE Seventh International Conference on Communications and Electronics*(ICCE); 2018. p. 39–43.

[73] Zhu F, Yao M. Improving physical-layer security for CRNs using SINR-based cooperative beamforming. *IEEE Transactions on Vehicular Technology*. 2016;65(3):1835–1841.

[74] Nandan N, Majhi S. Physical layer security of full-duplex cognitive radio network using beamforming. In: *2019 International Conference on Internet of Things, Embedded Systems and Communications* (IINTEC); 2019. p. 100–105.

[75] Al-Nahari A, Geraci G, Al-Jamali M, *et al.* Beamforming with artificial noise for secure MISOME cognitive radio transmissions. *IEEE Transactions on Information Forensics and Security*. 2018;13(8):1875–1889.

[76] Choi H, Ron D, Sengly M, *et al.* Energy-neutral wireless sensor network based on SWIPT in wireless powered communication networks. In: *ICC 2020 – 2020 IEEE International Conference on Communications* (ICC); 2020. p. 1–7.

[77] Yan P, Zou Y, Ding X, *et al.* Energy-aware relay selection improves security-reliability tradeoff in energy harvesting cooperative cognitive radio systems. *IEEE Transactions on Vehicular Technology*. 2020;69(5): 5115–5128.

[78] Ding X, Zou Y, Zhang G, *et al.* The security-reliability tradeoff of multiuser scheduling-aided energy harvesting cognitive radio networks. *IEEE Transactions on Communications*. 2019;67(6):3890–3904.

[79] Li M, Yin H, Huang Y, *et al.* Physical layer security in overlay cognitive radio networks with energy harvesting. *IEEE Transactions on Vehicular Technology*. 2018;67(11):11274–11279.

[80] Maji P, Roy SD, Kundu S. Physical layer security with non-linear energy harvesting relay. In: *2019 10th International Conference on Computing, Communication and Networking Technologies* (ICCCNT); 2019. p. 1–6.

[81] Li C, Li G. Design of physical layer secure transmission in RF energy harvesting cognitive radio network. In: *2020 IEEE 6th International Conference on Computer and Communications* (ICCC); 2020. p. 212–219.

[82] Xu X, Bao J, Wang Y, *et al.* Cognitive radio primary network secure communication strategy based on energy harvesting and destination assistance. In: *2021 13th International Conference on Wireless Communications and Signal Processing* (WCSP); 2021. p. 1–5.

[83] Tashman DH, Hamouda W. Secrecy analysis for energy harvesting-enabled cognitive radio networks in cascaded fading channels. In: *ICC 2021 – IEEE International Conference on Communications*; 2021. p. 1–6.

[84] Khoshafa MH, Moualeu JM, Ngatched TMN, *et al.* On the performance of secure underlay cognitive radio networks with energy harvesting and dual-antenna selection. *IEEE Communications Letters*. 2021;25(6):1815–1819.

[85] Su R, Wang Y, Sun R. Secure cooperative transmission in cognitive AF relay systems with destination-aided jamming and energy harvesting. In: *2019 IEEE 30th Annual International Symposium on Personal, Indoor and Mobile Radio Communications* (PIMRC); 2019. p. 1–5.

[86] Tan R, Gao Y, He H, *et al.* Secrecy performance of cognitive radio sensor networks with an energy-harvesting based eavesdropper and imperfect CSI. In: 2018 *Asian Hardware Oriented Security and Trust Symposium* (AsianHOST); 2018. p. 80–85.

[87] Xu M, Jing T, Fan X, *et al.* Secure transmission solutions in energy harvesting enabled cooperative cognitive radio networks. In: *2018 IEEE Wireless Communications and Networking Conference* (WCNC); 2018. p. 1–6.

[88] Zhou F, Chu Z, Sun H, *et al.* Artificial noise aided secure cognitive beamforming for cooperative MISO-NOMA using SWIPT. *IEEE Journal on Selected Areas in Communications*. 2018;36(4):918–931.

[89] Santana GMD, Cristo RS, Dezan C, *et al.* Cognitive radio for UAV communications: opportunities and future challenges. In: *2018 International Conference on Unmanned Aircraft Systems* (ICUAS); 2018. p. 760–768.

[90] Jacob P, Sirigina RP, Madhukumar AS, *et al.* Cognitive radio for aeronautical communications: a survey. *IEEE Access*. 2016;4:3417–3443.

[91] Erdogan E, Altunbas I, Kabaoglu N, *et al.* A cognitive radio enabled RF/FSO communication model for aerial relay networks: possible configurations and opportunities. *IEEE Open Journal of Vehicular Technology*. 2021;2:45–53.

[92] Krayani A, Alam AS, Marcenaro L, *et al.* An emergent self-awareness module for physical layer security in cognitive UAV radios. *IEEE Transactions on Cognitive Communications and Networking*. 2022;8(2):888–906.

[93] Krayani A, Baydoun M, Marcenaro L, *et al.* Self-learning bayesian generative models for jammer detection in cognitive-UAV-radios. In: *GLOBECOM 2020 – 2020 IEEE Global Communications Conference*; 2020. p. 1–7.

[94] Mowla NI, Tran NH, Doh I, *et al.* Federated learning-based cognitive detection of jamming attack in flying ad-hoc network. *IEEE Access.* 2020;8: 4338–4350.

[95] Krayani A, Alam AS, Marcenaro L, *et al.* A novel resource allocation for anti-jamming in cognitive-UAVs: an active inference approach. *IEEE Communications Letters.* 2022;p. 1–1.

[96] Pigatto DF, Rodrigues M, de Carvalho Fontes JV, Pinto ASR, Smith J, Branco KRLJC. The internet of flying things, in *Internet of Things A to Z: Technologies and Applications* (p. 529–562), John Wiley & Sons, Ltd; 2018.

[97] Zeng K, Kondaji Ramesh S, Yang Y. Location spoofing attack and its countermeasures in database-driven cognitive radio networks. In: *2014 IEEE Conference on Communications and Network Security*; 2014. p. 202–210.

[98] Tao Z, Zhou F, Wang Y, *et al.* Resource allocation and trajectories design for UAV-assisted jamming cognitive UAV networks. i. 2022;19(5):206–217.

[99] He X, Li X, Ji H, *et al.* Resource allocation for secrecy rate optimization in UAV-assisted cognitive radio network. In: *2021 IEEE Wireless Communications and Networking Conference* (WCNC); 2021. p. 1–6.

[100] He X, Li X, Ji H, *et al.* Secure transmission based on marginal utility in UAV-assisted cognitive radio network. In: 2021 *IEEE International Conference on Communications Workshops* (ICC Workshops); 2021. p. 1–6.

[101] Wang Z, Guo J, Chen Z, *et al.* Robust secure UAV relay-assisted cognitive communications with resource allocation and cooperative jamming. *Journal of Communications and Networks.* 2022;24(2):139–153.

[102] Wu K, Wang D, Zhang R, *et al.* Secure cognitive communication via cooperative jamming. In: *2020 IEEE/CIC International Conference on Communications in China* (ICCC); 2020. p. 594–599.

[103] Nguyen PX, Nguyen HV, Nguyen VD, *et al.* UAV-enabled jamming noise for achieving secure communications in cognitive radio networks. In: *2019 16th IEEE Annual Consumer Communications & Networking Conference* (CCNC); 2019. p. 1–6.

[104] Shengnan C, Xiangdong J, Yixuan G, *et al.* Physical layer security communication of cognitive UAV mobile relay network. In: *2021 7th International Symposium on Mechatronics and Industrial Informatics* (ISMII); 2021. p. 267–271.

[105] Vo VN, Long NQ, Dang VH, *et al.* Physical layer security in cognitive radio networks for IoT using UAV with reconfigurable intelligent surfaces. In: *2021 18th International Joint Conference on Computer Science and Software Engineering* (JCSSE); 2021. p. 1–5.

[106] Liu N, Tang X, Xu D, *et al.* A learning approach towards secure cognitive networks with UAV relaying and active Jamming. In: *2021 13th International Conference on Wireless Communications and Signal Processing* (WCSP); 2021. p. 1–6.

[107] Chi-Nguyen D, Pathirana PN, Ding M, *et al.* Secrecy performance of the UAV enabled cognitive relay network. In: *2018 IEEE 3rd International*

Conference on Communication and Information Systems (ICCIS); 2018. p. 117–121.

[108] Ghareeb I, Tashman D. Statistical analysis of cascaded Rician fading channels. *International Journal of Electronics Letters*. 2020;8(1):46–59.

[109] Bekkali A, Zou S, Kadri A, *et al.* Performance analysis of passive UHF RFID systems under cascaded fading channels and interference effects. *IEEE Transaction on Wireless Communication*. 2015 March;14(3):1421–1433.

[110] Alghorani Y, Kaddoum G, Muhaidat S, *et al.* On the approximate analysis of energy detection over $n*$Rayleigh fading channels through cooperative spectrum sensing. *IEEE Wireless Communications Letters*. 2015 Aug;4(4): 413–416.

[111] Kaur A, Malhotra J. Cascade fading channel models for wireless communication—a survey. *International Journal of Computer Applications*. 2014;89(14):22–25.

[112] Peppas K, Lazarakis F, Alexandridis A, *et al.* Cascaded generalised-K fading channel. *IET Communication*. 2010;4(1):116–124.

[113] Boulogeorgos AA, Sofotasios PC, Selim B, *et al.* Effects of RF impairments in communications over cascaded fading channels. *IEEE Transaction on Communicaiton*. 2016 Nov;65(11):8878–8894.

[114] Lee J, Lee JH, Bahk S. Performance analysis for multi-hop cognitive radio networks over cascaded Rayleigh fading channels with imperfect channel state information. *IEEE Transactions on Vehicular Technology*. 2019;68(10):10335–10339.

[115] Tashman DH, Hamouda W. Physical-layer security for cognitive radio networks over cascaded Rayleigh fading channels. In: *GLOBECOM 2020 – 2020 IEEE Global Communications Conference*; 2020. p. 1–6.

[116] Tashman DH, Hamouda W. Physical-layer security on maximal ratio combining for SIMO cognitive radio networks over cascaded κ-μ fading channels. *IEEE Transactions on Cognitive Communications and Networking*. 2021;7(4):1244–1252.

[117] Almohamad A, Tahir AM, Al-Kababji A, *et al.* Smart and secure wireless communications via reflecting intelligent surfaces: a short survey. *IEEE Open Journal of the Communications Society*. 2020;1:1442–1456.

[118] Wu Y, Khisti A, Xiao C, *et al.* A survey of physical layer security techniques for 5G wireless networks and challenges ahead. *IEEE Journal on Selected Areas in Communications*. 2018;36(4):679–695.

[119] Wang L, Elkashlan M, Duong TQ, *et al.* Secure communication in cellular networks: the benefits of millimeter wave mobile broadband. In: *2014 IEEE 15th International Workshop on Signal Processing Advances in Wireless Communications* (SPAWC); 2014. p. 115–119.

Chapter 14

Machine learning for physical layer security

Mehmet Ali Aygül[1,2] and Hüseyin Arslan[3]

Due to the broadcast nature of the wireless medium, an illegitimate node can harm legitimate users with eavesdropping, jamming, and spoofing threats, as indicated in the previous chapters. These threats can be eliminated, detected, and prevented using physical layer security (PLS) approaches. However, providing PLS of the fifth generation and beyond (5GB) communication and sensing systems is challenging due to the requirements of these systems, such as heterogeneity, ultra density, and high mobility. This is especially the case when a large number of parameters are used simultaneously, and the problem or model itself is highly dimensional and complex. Machine learning (ML) and deep learning (DL) have recently emerged as potential tools for reducing the increasing complexity of wireless networks while simultaneously employing multiple parameters in multiple dimensions (improving the performance of wireless security systems). In this chapter, we first review the classic and recent ML and DL algorithms for PLS, the ongoing and anticipated issues, and the primary efforts made in this sector by leading industries and research entities. Then, we summarize the possible research direction for ML-based PLS.

14.1 Introduction

The number of wireless communication and sensing systems and their applications have increased dramatically in recent years. This presents new challenges (e.g., complexity and the requirement for more reliability) for wireless communication and sensing security. Artificial intelligence (AI)-based solutions can answer different kinds of challenges since they can optimize many parameters simultaneously in multiple dimensions with lower complexity. Therefore, AI-based wireless communications have attracted research interest from academia and relevant sectors as a potential solution to these problems.

Although AI is the generic name for learning-based algorithms, many works use the terms AI, ML, and DL interchangeably. The concept of AI is the creation of

[1]Department of Electronics and Communications Engineering, Istanbul Technical University, Turkey
[2]Department of Research & Development, Vestel, Turkey
[3]Department of Electrical and Electronics Engineering, Istanbul Medipol University, Turkey

intelligent machines. ML is a subset of AI that aids in the development of AI driven applications. DL is a subclass of ML which trains a model using a huge amount of data and advanced algorithms. In Figure 14.1, these three principles are depicted with a chronology.

For model development, the ML and DL-based algorithms alter parameters and adapt to their environment through experience. These developed models are capable of generating outputs from given inputs without the need for human assistance. The ML algorithm is able to observe the relationship between input and output due to the vast amount of data available and the repetitive interaction between exploitation and exploration. Intelligent decision-making and processing reduce excess resources and ensure their efficient use.

Without the loss of generality, Figure 14.2 illustrates the seven significant steps that constitute an ML (or DL) system. In the first step, data should be collected. Then, based on the algorithm and domain knowledge, feature extraction is made. There are wide varieties of ML models, which have their own advantages and disadvantages. Model selection is highly dependent on data samples, data qualities, performance requirements, dimensionality, complexity, accuracy, training time, etc. It is important to strike a balance between the general nature and the accuracy of the model. If the

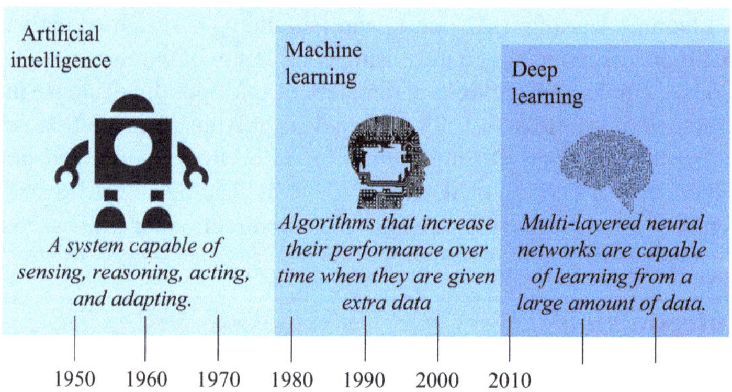

Figure 14.1 Fundamental properties of AI, ML, and DL

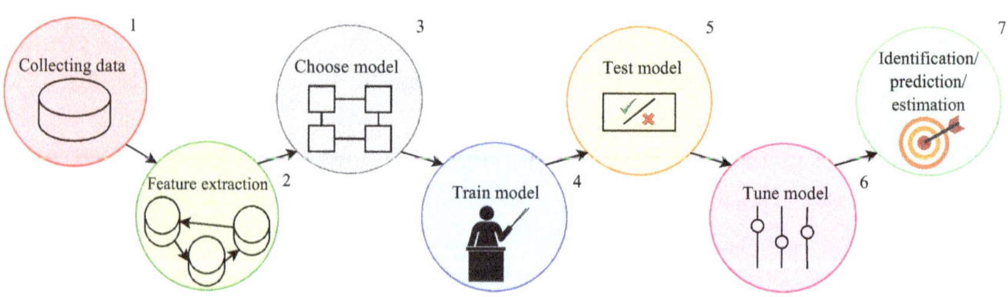

Figure 14.2 Simple ML processes

model is trained specifically to represent the training data, it can consider noise and other unnecessary details as features and lead to overfitting. On the other case, if the model skips an important/characteristic feature, it leads to underfitting. Generally, the number of parameters used is decreased to prevent overfitting and vice versa for underfitting. Finally, the trained ML algorithm will be able to identify, predict, or estimate the desired output.

14.2 ML algorithms

One of the most appealing aspects of ML is the wide range of algorithms available. According to the availability of data and training methods, they can be divided into four categories: supervised learning (SL), unsupervised learning (USL), semi-SL, and reinforcement learning (RL). When a large amount of prior data is available, SL, semi-SL, and USL ML approaches are appropriate, whereas RL techniques may be advantageous in the absence of past data. These four groups are detailed below. It is worth noting that hybrid techniques can be applied by combining two or more forenamed categories at the same time.

14.2.1 Supervised learning

In this category, there should be a dataset with some observations and the classes/labels of the observations. For instance, the observations can be received signal vector, and the output labels can be the decision of the legitimate/illegitimate user.

These models are trained on the labeled data and then used to forecast future events. The training procedure involves a training dataset with matching labels. Chosen learning algorithm generates a function through the training procedure to make predictions about unseen observations. After enough training, the model can provide targets for any new input. The learning algorithm can also compare its output to the ground truth output (the desired output that is wanted to be obtained by the learning algorithm) and detect faults so that it can adjust itself accordingly. There are two types of SL models that exist: regression and classification:

- *Regression*: In general, regression is the task of predicting a quantity, such as the number of illegitimate users.
- *Classification*: Classification algorithms determines between categories depending on the scenario. For example, a classification algorithm may decide whether a user is legitimate or illegitimate for a given input signal.

K-nearest neighbors, support vector machines (SVMs), decision trees, naive Bayes classifiers, neural networks (NNs), convolutional neural networks (CNNs), long short term memory (LSTM), and recurrent neural networks (RNNs) are the commonly used SL techniques. A decision tree-based algorithm is illustrated

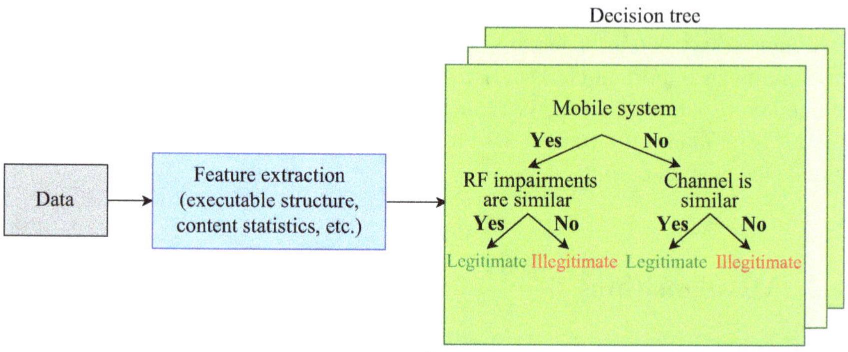

Figure 14.3 A decision tree example for PLS

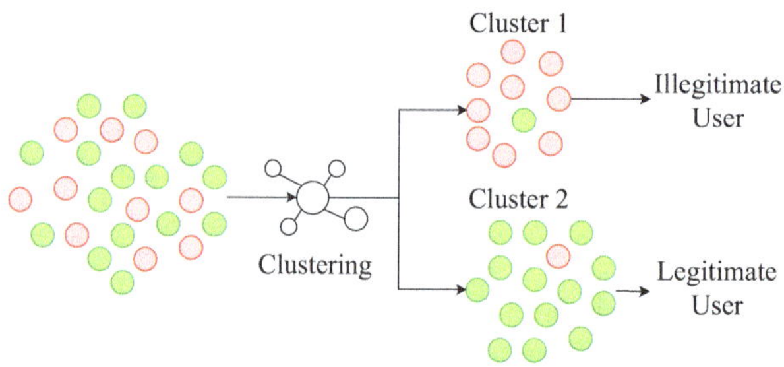

Figure 14.4 Clustering example for PLS

in Figure 14.3. For this technique, it is necessary to have wireless domain knowledge such as which impairments or what kind of channel properties can be exploited to identify the users' legitimacy.

14.2.2 Unsupervised learning

USL carries out analysis of unclassified/unlabeled datasets. It investigates how systems infer a function from unlabeled data to describe a hidden structure. The system does not anticipate the correct output; instead, it examines the data and can infer hidden structures from unlabeled data using datasets. Principal component analysis, autoencoder, K-means clustering, Gaussian mixer model, Bayesian networks, conditional random fields, self-organizing map, and generative adversarial network (GAN) are examples of USL algorithms. Figure 14.4 highlights an example of clustering for security. Here, we have legitimate and illegitimate user data samples and want to distinguish between them. Therefore, a clustering algorithm is used, and after the clustering algorithms, we will be able to distinguish them with some errors. Also, examples of these errors are illustrated in the figure.

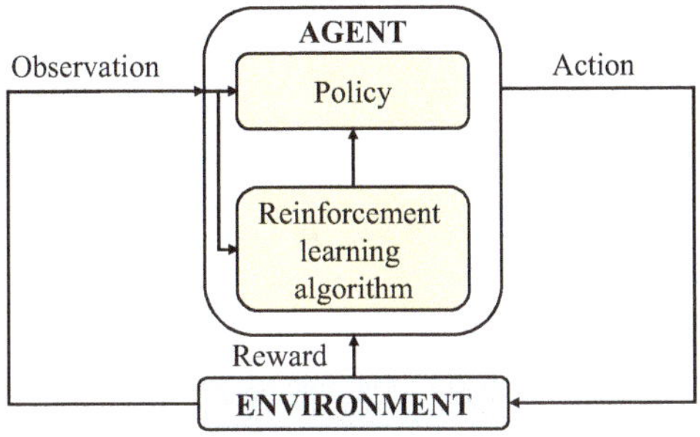

Figure 14.5 The basic description of the RL

14.2.3 Semi-supervised learning

Semi-SL is a cross between the SL and USL groups. Generally, it uses large amounts of unlabeled and small amounts of labeled data, combining the advantages of both SL and USL algorithms without the difficulties of finding a lot of labeled data. Therefore, it reduces the disadvantage of SL, which is having a large amount of labeled data. It can be used in several PLS scenarios where little amounts of labeled data exist (this can be caused by a lack of time for labeling, data availability, or security concerns). The semi-SL category includes transduction SVM, generative models, multiview algorithms, and graph-based methods.

14.2.4 Reinforcement learning

RL depends on rewarding desired behaviors and punishes the undesired ones. For this purpose, first, an estimation is made, and an output is obtained. Then, the error between desired and obtained output is compared. Based on the error, rewards and punishments are made. If the error is severe, the punishment is high, and the reward is low. If the error is low, the punishment is low, and the reward is high.

The essential elements of RL are trial and error search and delayed reward. This category of models enables the automatic determination of the optimal behavior in a given situation to get the best possible results. The model needs reward input to learn which action is best, which is known as *reinforcement signal*. RL category includes Q-learning, Alpha-Go, temporal difference learning, deep Q-learning, and temporal difference learning. The basic description of the RL is illustrated in Figure 14.5.

14.3 Deep learning algorithms

ML algorithms may be inadequate when it faces large amounts of high dimensional data. DL is a unique learning algorithm that deals with a large amount of multiple dimensional complex data thanks to its several hidden layers. As seen in Figure 14.6, DL employs many hidden layers, whereas ML only employs one hidden layer. Multiple hidden layers of DL algorithms allow for magnifying intrinsic data features while

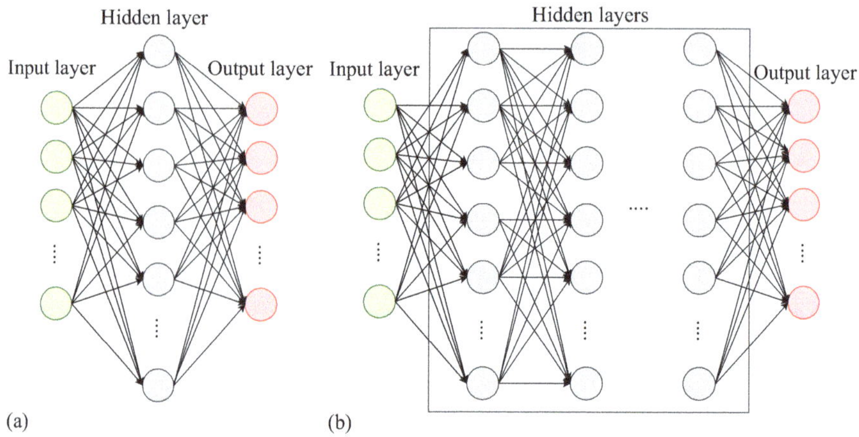

Figure 14.6 Illustrations of (a) ML and (b) DL algorithms

reducing unnecessary data at each layer. This approach is beneficial for complex issues in which multiple problems are attempted to be estimated at the same time, causing the system to become blind. Also, DL-based algorithms can gather information from raw data. As a result, these algorithms are potential candidates for solving complex problems. The deep neural networks (DNNs), CNN, and LSTM are well-known DL algorithms. These algorithms can be used to solve complex PLS problems.

14.4 Multi-task learning

In ML, we generally focus on optimizing for a specific parameter, such as a score on a business key performance indicator or a certain benchmark. To accomplish this, we typically train a single model or a group of models to perform the task. The models are then fine tuned and tweaked until their performance converges. While we may generally get adequate results in this manner, we ignore information that could help us perform even better than being laser focused on a single task. This information is explicitly derived from the training signals of multiple related tasks. We can improve the generalization of a model on the primary task by exchanging representations amongst associated tasks. This method is known as multi-task learning (MTL).

MTL has been successfully used in a wide range of ML applications, including natural language processing and speech recognition, as well as computer vision and image processing. MTL algorithms concern joint optimization of numerous loss functions and it arises to be promising for different security problems. For example, various types of attacks can happen simultaneously, and there can be relationships between them. MTL can be advantageous for this problem since it can solve multiple problems jointly by optimizing its weight based on different problems. Figure 14.7 depicts an example of MTL structure. In this figure, there are three different tasks.

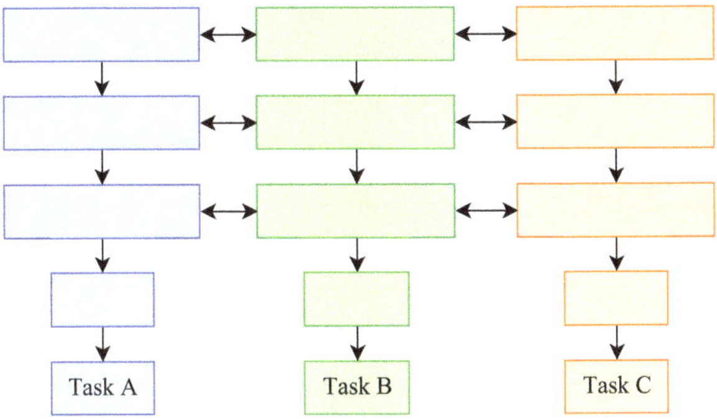

Figure 14.7 A usage of MTL for three related tasks

The first three layers learn these tasks together, while the last two layers learn separately.

14.5 Federated learning

A dataset that ML uses can include private datasets such as hospital and bank datasets. Therefore, privacy is a critical issue in ML-based solutions. These will be further discussed in the following parts. Federated learning (FL), also known as collaborative learning, trains an ML algorithm on multiple local datasets stored in local nodes without exchanging data samples explicitly. Since it does not exchange data samples explicitly, it can be considered safe. Here, the main idea is to train local models on local data samples and periodically exchange hyperparameters between these local nodes to produce a global model shared by all nodes. There are three types of FL: centralized, decentralized, and heterogeneous. These are further explained below.

- *Centralized learning*: A central server orchestrates the different steps of the algorithms and coordinates all of the participating nodes during the learning process in a centralized learning (CL) scenario. In CL, all possible nodes must send updates to a single entity, so there can be a server bottleneck. To prevent this, the server is responsible for node selection at the start of the training process and the aggregation of model updates received.
- *Decentralized learning*: The nodes are able to coordinate themselves in a decentralized learning context to obtain the global model. This configuration prevents single-point failures since model changes are solely shared across networked nodes without the orchestration of the central server. Still, the performance of the learning process may be influenced by the network topology. For blockchain-based FL, we refer [1].
- *Heterogeneous learning*: A growing number of diverse applications involve a significant number of heterogeneous clients. Most current FL techniques

presume that local models have the same global architecture as global ones. HeteroFL, a new FL framework, has been recently developed to address heterogeneous clients with a wide range of computation and communication capabilities. HeteroFL allows for training heterogeneous local models with constantly changing computation and non-i.i.d. data complexity while providing a single accurate global inference model.

14.6 Generative adversarial network

Ian Goodfellow and his colleagues created the GAN technique of ML [2]. The main concept of a GAN is built on *indirect* training via a discriminator, which is another NN that determines how *realistic* an input is. This means that the generator is trained to deceive the discriminator instead of minimizing the distance to a certain image. Therefore, the model is able to learn in a USL fashion.

GANs can create synthetic data based on their training experience with actual data. An example of it is illustrated in Figure 14.8. In this figure, random noise is used to generate a dataset that is similar to a real dataset. Its procedures are as follows. First, random noise is given as an input to the GAN generator, and the first dataset is generated; then generated data is compared with the real dataset in the discriminator, and the weights of the generator are iteratively updated. These processes are repeated based on the quality of the generated data and complexity until the generated data is similar to the real dataset. As a result, this created data, which a user would not normally present in training, is used to investigate a wide range of prospective assaults, including previously unknown threats. This makes the

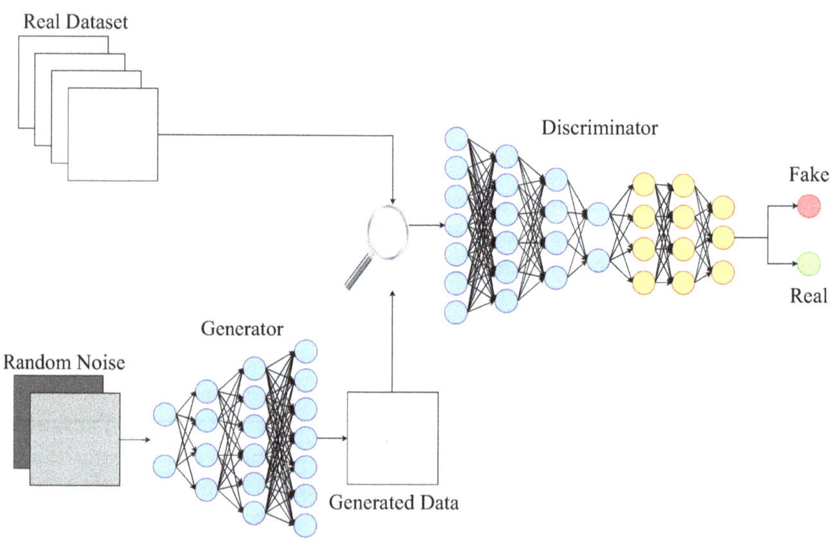

Figure 14.8 An example of GAN for PLS

GANs methodology extremely useful in a system security context where new data risks emerge regularly. GANs appear to be ideal for training a NN to simulate similar attacks based on a known attack. GANs allow for the isolation and prediction of future attacks, allowing a designer to construct stronger security measures to prevent those attacks before they even occur in the thoughts of the attackers. It can also be exploited by hackers to generate new threats. For example, the generated dataset can be used to spoof legitimate users. We further discuss GAN from both security and attack perspectives below.

14.6.1 Generative adversarial networks in security defenses

In the literature, GAN is used to obscure sensitive information, cyber intrusion and malware detection, secure image steganography, and security analysis. These are detailed below. Here note that in PHY, there are only a few works for the usage of GAN, but it should be investigated. GANs in upper layer security attacks and defence techniques can be investigated to find new frameworks and algorithms in PLS.

- *Obscuring sensitive information*: Companies and government agencies frequently hold highly sensitive material that is not accessible to researchers. Hospitals, for example, have a lot of sensitive patient data. Banks also store statements and financial details in a secure manner. If the secure data were to be shared with academicians or analysts, it might provide valuable insights and aid in future studies. According to [3], a well-trained GAN can generate new data that can be used to represent the original data. As a result, the original data can be kept safe, and the GAN generated data will most likely reflect the same trends and insights as the original data.
- *Cyber intrusion and malware detection*: Cyber intrusions are used to assault and compromise a computer by compromising the security system or making the operating environment insecure. Cyber intrusions have a variety of consequences, including unauthorized information release, tampering, and data destruction. An intrusion detection system (IDS) monitors the network and detects any malicious activity, and alerts the user. A GAN-based model could be an excellent choice for developing an IDS. By learning the features of regular data, GANs are employed in intrusion detection. GANs have also been used by researchers to detect malware [4]. In the literature, new security tactics and algorithms against various threats are proposed using GANs. An example is called Defense-GAN, which is trained to model the distribution of unperturbed pictures [5]. This GAN can be used to defend against various attack methods while also improving existing defense strategies. The majority of existing defense methods are tailored to the specific attack models. By exploiting the generative potential of GANs, defense GAN may be deployed with any classifier and on any assault.
- *Secure image steganography*: Steganography is a method of disguising secret information within an image. The image used to disguise the data is referred to as the cover, while the image with the embedded hidden message is referred to as the stego image. Unlike cryptography, steganography attempts to hide

the message's presence. Secure image steganography can be done with GANs. GANs have become an effective security measure and a popular study area because steganography also deals with images.
- *Security analysis*: GANs can assist in determining whether the various security components are met simultaneously. Integrity, confidentiality, and availability are examples of these security components.

14.6.2 Generative adversarial networks in security attacks

GANs can also be used from an attack perspective; breaking ciphers, password guessing, and malware generation and attacks against intrusion detection systems.

- *Breaking (cracking) ciphers*: Ciphertexts are encoded or encrypted messages that hide patterns in order to ensure data integrity. Encryption is the process of transforming raw data into ciphertexts using encryption algorithms in order to prevent unwanted access to the data. *Symmetric encryption* is a type of encryption in which the sender and receiver encrypt and decode the message's data using the same key. However, NNs have evolved to be able to decrypt data based on their mutual outputs. This has given rise to a new field known as *neural cryptography*, which integrates NNs with cryptography and cryptanalysis. The output weights of two NNs can be synchronized to generate secret keys.
- *Password guessing*: Password guessing is a type of brute force attack used to guess a password. GANs can be used to conduct and enhance password guessing to increase the quality of the attack. GANs can learn from the data distribution of the billions of leaked passwords and then uses that information to create higher-quality password guesses.
- *Malware generation and attacks against intrusion detection systems*: GANs have been used to improve IDSs, but, one drawback of ML is that it may also be used to produce malware that defeats an IDS. By training on input noise and malware instances, this approach generated adversarial malware examples, which were then supplied to a replacement NN detection system alongside legitimate ones [6]. The GAN is trained to minimize malicious properties in order to bypass the substitute detector. As soon as the IDS gets a good grasp of the assault, the probability distribution can be altered, making it unpredictable. Malware attacks have been limited by the lack of properly labeled datasets. To solve this problem, a data augmentation technique is presented for generating additional tagged datasets from existing malware datasets.

14.7 Interpretable ML

Interpretable ML allows humans to understand the reasons for the solutions. It contrasts with the concept of the "black-box" ML. Interpretable ML promises to improve users' performance by refining their mental models and eliminating misconceptions about ML solutions. Interpretable ML's goal is to explain what has been done, what

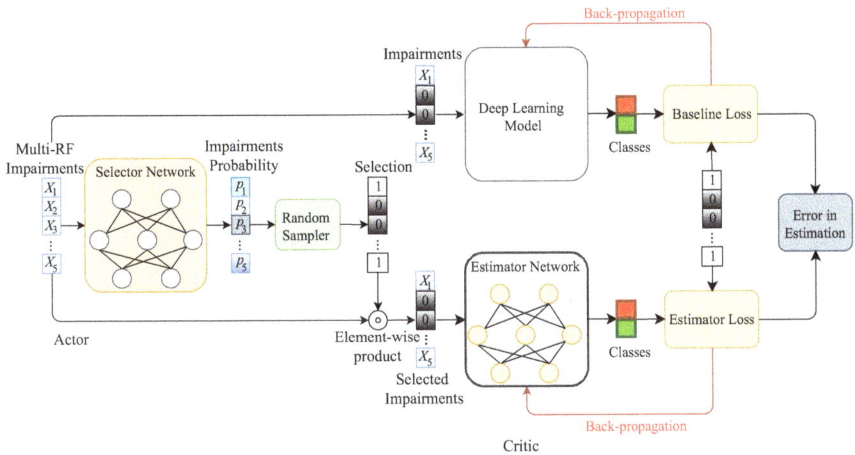

Figure 14.9 An example for interpretable ML for PLS.

is being done now, and what will be done next, as well as to reveal the knowledge on which the actions are based.

White-box ML algorithms and black-box ML algorithms are two types of ML algorithms. White-box ML models produce results that are intelligible to domain professionals. On the other hand, black-box models are complicated to explain, and even domain specialists struggle to comprehend them. Transparency, interpretability, and explainability are concepts used for interpretable ML algorithms. Transparency is granted if the procedures that extract model parameters from training data and create labels from testing data can be described. Interpretability is the ability to understand the ML model and provide the underlying reason of its decision. Explainability is widely acknowledged as vital. It is defined as the collection of features of the interpretable domain which have led to the production of a decision for a particular example. If algorithms meet these criteria, they can be used to explain, track, and verify decisions, improve algorithms, and investigate new facts.

With a white-box ML method, which is interpretable in and of itself, it is possible to produce a high accuracy outcome. This is especially important for sectors like security, where understanding judgments and establishing trust in algorithms are critical. An example of interpretable ML for security is illustrated in Figure 14.9. In this figure, an ML algorithm is used with radio frequency (RF) impairments to learn whether the user is legitimate or illegitimate, and RL is used to learn which impairments decide the decision of the ML algorithm.

14.8 Privacy protection in ML

Although ML is critical for future PLS solutions, it also comes with new security threats. Privacy threats are examples of these threats. In this section, we explain

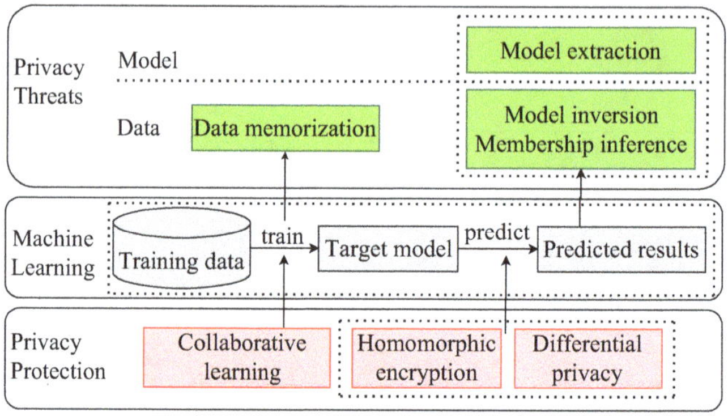

Figure 14.10 The framework of ML privacy threats and protection

privacy threats on models and data, and privacy protection techniques. Also, the framework of ML privacy threats and protection is given in Figure 14.10.

14.8.1 Privacy threats

One of the most serious threats to data collecting and analysis is privacy concerns. Data-driven ML algorithms always create a privacy concern since this data can be obtained by an unauthorized user. We describe privacy threats to models and data in specific scenarios in this section, as well as briefly discuss defense techniques.

14.8.1.1 Privacy threats on model

The model is publicly accessible through the application programming interface on the ML as a service platform. On the other hand, hyperparameters/architectures and parameters/functions that use unique learning techniques can be considered private, and they are hidden from illegitimate users. However, there are attack techniques for learning these, which will be further discussed below.

- *Parameters/functions*: Ref. [7] showed how to extract target models with high accuracy via assaults on NNs, decision trees, and logistic regression. They also suggested a new path-finding technique for extracting decision trees that uses confidence values as quasi-path identifiers. Then, the Jacobian matrix-based data augmentation strategy is presented to steal a substitute model and generate highly transferrable adversarial cases [8]. Samples near the target model's decision boundary can be captured by determining the substitute model's sensitivity to training points.
- *Hyperparameters/architectures*: Wang *et al.* [9] created a scenario in which hyperparameters are stolen. Model parameters are, intuitively, the minimal value of the objective function on the training set; for example, gradients vectors are close to zero vector. Therefore, the hyperparameters, model parameters, and

training set connections can be expressed as a linear system of equations. Finally, the optimal hyperparameter is solved using the least square approach. To reverse engineer black-box NNs, Oh *et al.* [10] developed *model attributes*. Static query and dynamic input creation are two attack strategies. The static query approach teaches the meta classifier the mapping relationship between the model's output combination on a set of queries and the attributes of the meta classifier. Internal information is obtained via the dynamic input creation method via carefully planned queries and predictions.

14.8.1.2 Privacy threats on data

- *Data memorization attacks*: Data owners frequently use third-party original code to generate high-accuracy models without truly understanding what this code does or whether it leaks sensitive data. Song *et al.* [11] used the third-party untrusted algorithm in the trusted isolation environment to implement the data memorization attack. The training set information is encoded in the model without impacting its accuracy significantly. Finally, the attacker obtains a white-box model or black-box access in order to extract the encoded data. There are three attack approaches in a white-box scenario: directly encoding information in the least significant bits of parameters, forcing parameters to be significantly associated with sensitive information, and encoding information into signs. The final two are accomplished by introducing malicious regularizers. Direct and sign encoding can retrieve information based on parameters and continuous signs, respectively, in the decoding step. For decoding, the correlation technique employs a linear transformation to map parameters back to feature space. The original training set is supplemented with synthetic inputs whose labels are encoded with sensitive information in a black-box scenario.
- *Model inversion attacks*: Researchers normally have access to datasets, but the models developed from them are made public. In this regard, Fredrikson *et al.* [12] developed a model inversion attack against the warfarin model to investigate the extent to which the model leaks data privacy. The attacker builds an inference system based on basic population information, marginal distributions, and steady doses that predict individual genotypes using the correlation between attributes.

  The conventional model inversion attack is suitable for inferring sensitive properties from tiny collections. Following that, Fredrikson *et al.* [13] grew more confident in the broadened the attack. In the case of face recognition, the attacker is aware of the label and anticipates reconstructing the image linked with it. Although no pixel value is supplied, it can be considered an optimization challenge, unlike earlier attacks. The attacker generates the loss function through the objective class function label in the model inversion attack, i.e., using gradient descent to iteratively move the input along the negative direction and taking the local minimum value as a representative sample of the label. The image labeled with high confidence by the model is intuitively similar to the original input and can be considered an average training data sample. On the other hand, it may be applied primarily for cases with identical training samples and cannot challenge sophisticated deep models effectively. Yang *et al.* [14] trained the inversion model with general data and trimmed prediction vectors so that it could reconstruct samples from partial outcomes. The inversion model reconstructs the input using probabilities and forms an autoencoder-like architecture when combined with the target model.

- *Membership inference attacks*: In the ML field, Shokri *et al.* [15] introduced membership privacy. Through black-box access to the model, which quantifies the leaking of membership information produced by prediction, the attacker can establish if the known data record is in the training set. A classification problem is the most common membership inference attack. The attackers use shadow models to replicate the target model in this attack. Afterwards, they compute the prediction vectors y_{in} and y_{out} for the training and testing sets of shadow models, respectively, and create records labeled "in" and "out," where "in" represents members of the training set and "out" represents members of the testing set. Finally, given some prediction vectors, the attack model is trained to infer membership properties.

14.8.2 Privacy protection

14.8.2.1 Collaborative learning

In CL, data must be shared with third parties that may contain sensitive information that was mistakenly recorded and is no longer under personal control. To address these issues, distributed learning is used to strike a compromise between utility and privacy, as previously stated. The model architecture is agreed upon by all participants in this manner. Participants must obtain the most recent parameters from the server before each session of local training. The model is then trained on the local dataset using stochastic gradient descent. Finally, to update the parameter server, gradients are selectively shared. This eliminates the need to exchange personal information and takes advantage of other participant models via parameters, preventing overfitting. Google researchers proposed FL, a comparable distributed architecture that is utilized for local model updates of mobile phone end users, thus addressing data sharing privacy concerns.

14.8.2.2 Homomorphic encryption

Homomorphic encryption was proposed by Rivest *et al.* [16], in which the encrypted data can be calculated without being decrypted. Early encryption methods were only partially homomorphic, meaning they were multiplicative or additive homomorphic. Gentry proposed the fully homomorphic encryption system based on the ideal lattice, and it showed a wide application potential in the cloud environment [17].

14.8.2.3 Differential privacy

As one of the most advanced privacy protection technologies, differential privacy [18] minimizes the risk of inferring unique person information in statistical searches.

For example, assume a hospital frequently discloses cancer patient statistics for data analysis. In a given period, only $\mathscr{A}$ was tested, resulting in a statistical change. To determine whether $\mathscr{A}$ has cancer, a differential attack can be utilized. Differential privacy is used to protect against different types of threats. If the odds of producing the same query result are close to each other on neighboring datasets (just one record is different), the privacy of this record is secure.

14.9 Prediction of security attacks

Most of the security techniques look at whether there is an attack in the current system or not. In other words, they detect the attack after the attack occurs, and the techniques that predict future attacks are very limited. Many of the studies predicting future systems are only aimed at detecting attacks in the near future. Few studies in state of art suggest some techniques to determine whether specific attacks will occur in the future. However, these techniques are very limited, and when more than one attack occurs at the same time, the success of these techniques decreases.

A study is not available in the state of the art that can detect attacks in different layers of open systems interconnection (OSI) with a single model. However, a technique used in any layer, for example, in the network layer, can also be used for other layers without the need for much change. For example, time series (indicating whether the attack will occur within the specified time period or which attack may occur) are generally used in these studies. These studies can be carried out over long periods of time; however, the predictive performance for a long time will not be successful enough for the near future.

A method can be developed ensuring that attacks are made to any at least one OSI layer and/or to any at least one subcategory of said OSI layers in wireless communication systems can be predicted. Also, it can be possible to propose an algorithm that predicts the attack that may occur in all layers and in which layer it will occur. Thus, it is ensured that measures are taken in advance with the estimation/determination made. Wireless communication systems can be used safely in daily life in this way. Also, this method can be continuously improved with its learnable structure, which can determine whether it is necessary to take an additional measure to prevent the attack to be made and in which layer the attack will be. In addition, the method can provide information about the location of the attack and whether the attack will be resolved if no additional measures are taken, that is to say, whether there is a need to take additional measures for that future attack. These additional features will increase the predictive success of other problems both when learning by the machine and by enriching the dataset. In addition, attacks in the OSI layer can be estimated in combination with the security attack prediction technique. In addition, the method can determine the attacks in each OSI layer and even the subcategories in that layer. An example of attack prediction is given in Figure 14.11.

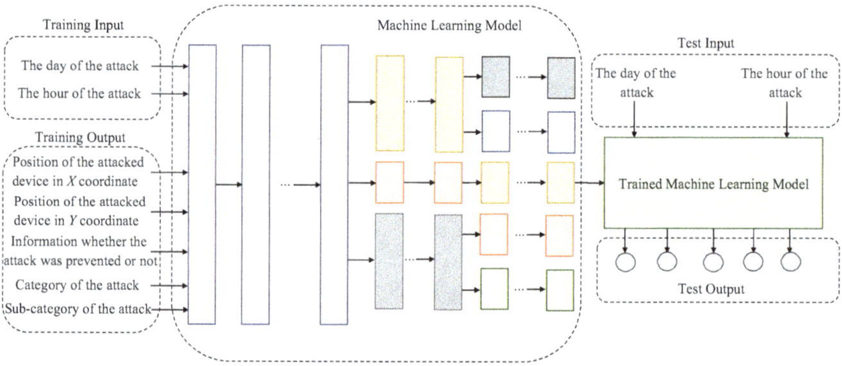

Figure 14.11 An example for attack prediction

14.10 Selected use cases of ML for physical layer security

In this part, we highlight some use cases that are new in the literature and they can give future research ideas to readers.

14.10.1 Signal relation-based physical layer authentication

This section introduces the extension of the work in [19]. The corresponding method is illustrated in Figure 14.12(a) and (b) as a training and testing stage, respectively. In this figure, the transmitted data is generated, and the effect of the RF impairments is added to the signal. Afterwards, the signal is passed through a wireless channel, and noise is added. On the receiver side, received signal symbols are obtained. For the detection of received signal symbols, two different solutions are developed in [19]. These are classical estimation and an ML-based method. In the classical approach, the effect of the channel and RF impairments are estimated. On the other hand, in the ML approach, ML is trained based on the received signal and the information on whether the signal is coming from legitimate or illegitimate users. In the testing stage, the received signal is fed directly to the trained ML algorithm. Again, channel and RF impairments are estimated for the classical approaches, and based on a hypothesis test [20] it is decided whether the user is legitimate.

14.10.2 Multiple radio frequency impairments

Several works on the physical layer authentication (PLA) use the RF impairments, such as frequency offset and phase noise, for user authentication [21]. In these works, the ML-based methods have been preferred recently [22]. However, these methods classify the user as legitimate or illegitimate according to the RF impairment characteristics of the received signal rather than directly detecting the RF impairments. For example, these methods use the received signal as the input and the user information is legitimate or illegitimate as the output and train their network according

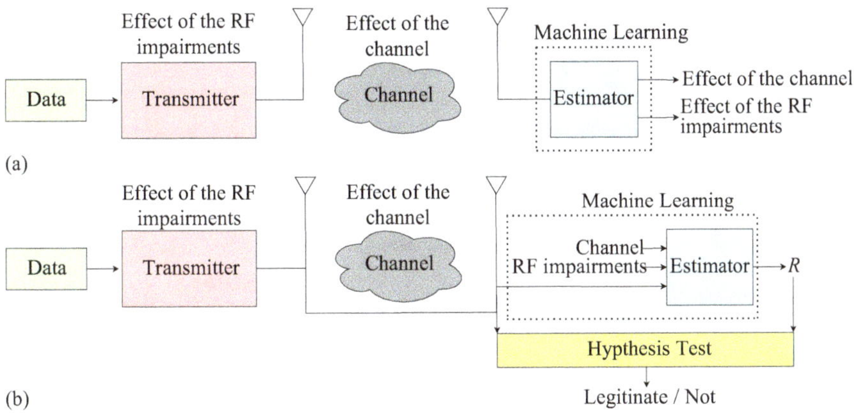

Figure 14.12 A signal relation-based PHY authentication method: (a) training and (b) testing stages

to these datasets. However, these methods are not interested in the estimation of RF impairments.

PLA studies train their models in the training stage according to the legitimate signal. After that, they decide if the signal is legitimate or illegitimate according to the received signal in the testing stage. However, this is disadvantageous in the sense that they cannot provide the information on which impairment they give the decision that the signal is illegitimate. In other words, the machine gives the decision to the user if it is legitimate or not, as a black-box, without estimating the RF impairments. Therefore, this reduces the reliability. On the other hand, if the RF impairments are estimated correctly, then estimated values can be used to learn whether the user is legitimate or not. Here the problem was estimating RF impairments truly when multiple RF impairments exist simultaneously. Ref. [23] can estimate multiple RF impairments jointly, and estimated impairments can be used to decide whether the user is legitimate or not. Also, user authentication can be made based on the hardware distortions, which indicate RF impairments. Identification of hardware distortions is made in [24]. An example of some of the RF impairments and corresponding hardware components are illustrated in Figure 14.13, where DAC and PA stand for digital to analog converter and power amplifier, respectively. As seen from this figure, the RF impairments effects are exclusive, and they can be used for PLA. However, instead of using ML for direct authentication, more solid decisions can be made using domain knowledge such as impairments characteristics, as explained earlier. Also, after the impairments are estimated, another ML model can be used as a thresholding mechanism to decide whether the user is legitimate or not.

14.10.3 Cognitive radio security

Cognitive radio (CR) allows primary users (PUs) and secondary users (SUs) to share spectrum. More specifically, it allows SUs to take advantage of empty spectrum

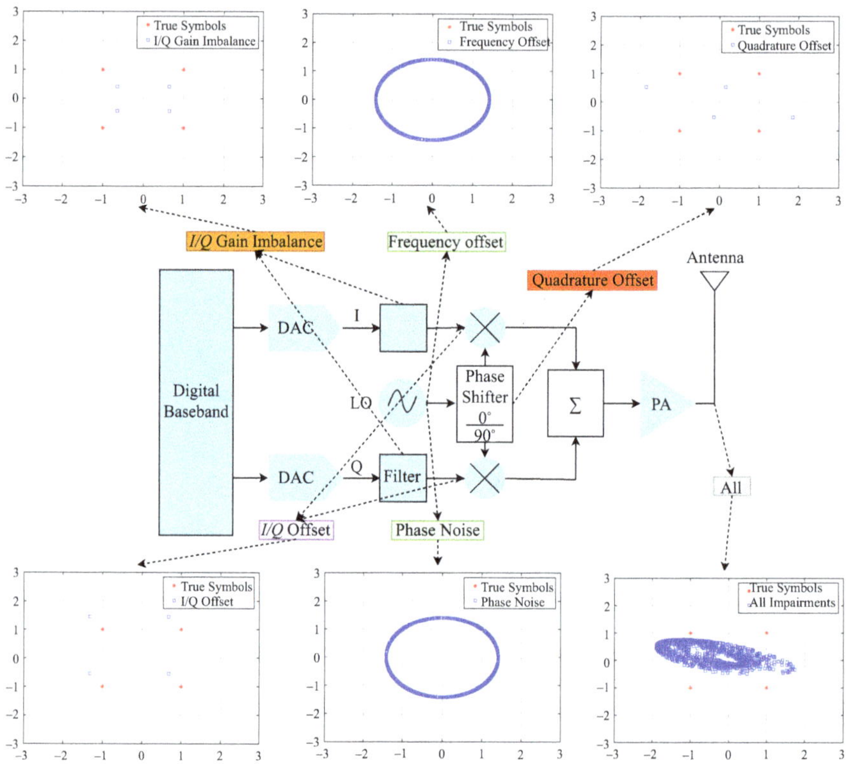

Figure 14.13 RF impairments in a simple receiver

bands while avoiding hurting PUs by following these steps: (i) assessing whether or not the channel is occupied, (ii) selecting the best part of the spectrum depending on their quality of service requirements, (iii) collaborating with other users to gain access to the spectrum, and (iv) exiting the channel when a PU begins to broadcast its data [25].

Many security vulnerabilities are inherently present in CR. Eavesdropping, jamming attacks, and primary user emulation attacks (PUEAs) are examples of CR security threats. PUEAs and jamming attacks are the most important of these because they can obstruct spectrum exploitation by producing false spectrum occupancy alarms. A primary user emulator can simulate the transmission characteristics of a legitimate PU, and a jammer can produce deliberate interference. The impact of attacks leads to a false conclusion about spectrum occupancy in both circumstances.

Recent literature uses ML algorithms to detect PUEAs and jamming attacks. This is built on training a machine to recognize such attacks using features extracted from received signals. SVM-based techniques regard attack detection as a classification process [26]. NN algorithms with energy detection and cyclostationary features have been applied more recently. In this case, energy detection is used first to detect

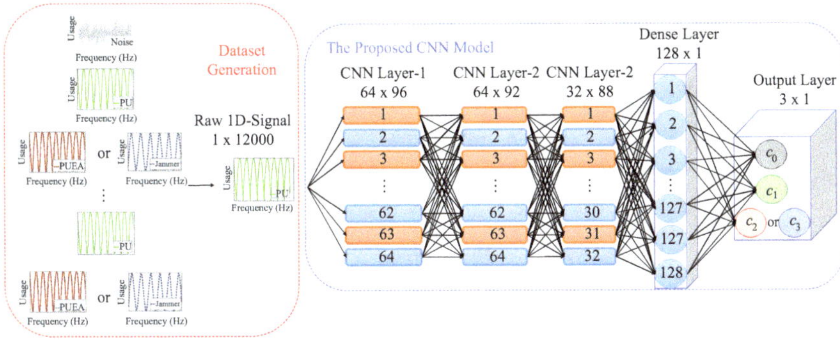

Figure 14.14 An example of CR security

the presence of a user. Then, to discriminate between legitimate users and attackers, cyclostationary traits are used. Similarly, RF fingerprinting [27] is utilized to establish RF fingerprint profiles for each transmitter for PUEA detection. Also, Ref. [28] employs sparse coding convergence patterns as feature extraction and uses ML-based classification to detect such assaults.

ML-based attack detection approaches, despite their apparent success, all have the issue of requiring multiple phases of feature extraction. A DL-based approach is proposed in [29] to detect the PUEA and jamming attack without explicit feature extraction. A CNN architecture is designed and optimized to do this. The algorithm is depicted in Figure 14.14 [29].

14.10.4 Internet of Things security

The IoT has recently become a hot study research area since it connects multiple sensors and items to interact directly with one another without human involvement. The requirements for large-scale IoT deployment are fast expanding, and security is a big concern. Any enterprise trying to secure IoT devices more scalable and efficiently with automation and aberrant behavior detection may benefit from ML. In a multiple-layered strategy to identify hidden data patterns, ML delivers promising results in detecting large-scale threats in IoT with computational efficiencies [30]. Traditional ML algorithms may not deliver state-of-the-art approaches, efficiency, or improved performance results, but DL can. Therefore, several papers implement unique DL techniques such as CNN and LSTM for IoT security.

14.11 Performance metrics

The effectiveness of security techniques can be measured using a variety of performance metrics. Classic performance metrics, which are used to evaluate classical security solutions, are employed in eavesdropping and this type of security threat. To evaluate the performance of the algorithms in the authentication context, accuracy

Table 14.1 Formulas of the performance metrics

Performance metric	α	π	ψ	F_1-score
Formula	$\frac{\xi+\sigma}{\xi+\sigma+\upsilon+\mu}$	$\frac{\xi}{\xi+\upsilon}$	$\frac{\xi}{\xi+\mu}$	$2\times\frac{\pi\times\psi}{\pi+\psi}$

(α), precision (π), recall (ψ), F_1-score, true positive rate (*TPR*), and false positive rate (*FPR*) performance metrics and receiver operating characteristic (ROC) curves, which are common in ML, are used.

The accuracy measure is defined as the ratio of correct predicted observations to total observations. The precision measure is the ratio of positive expected observations to all predicted positive observations. The recall is the proportion of accurately predicted positive observations to the total number of observations in the actual class. Finally, the harmonic average of precision and recall is used to get the overall accuracy of classifier models F_1-score. These performance measures are listed in Table 14.1, where ξ, σ, υ, and μ represent the number of true positive, true negative, false positive, and false negative results, respectively.

A signal is considered to be true positive (*TP*) if it belongs to class *i* and is appropriately classified to that class. It is referred to as a false negative (*FN*) if it's incorrectly identified as belonging to a separate class *j*. The signal is counted as false positive (*FP*) if it does not belong to class *i* and is incorrectly categorized as such. Finally, it is a true negative (*TN*) if it does not truly belong to *i* but is labeled as belonging to *i*. To this end, the true positive rate (*TPR*) or recall can be calculated using the formula $TPR = TP/(TP + FN)$, whereas the false positive rate (*FPR*) can be calculated using the formula $FPR = FP/(FP + TN)$. The ability of a classifier to discriminate between distinct classes is demonstrated by ROC curves and area under the ROC (area under the ROC (AUROC)) curve values. A probabilistic curve with a *TPR* on the vertical axis and a *FPR* on the horizontal axis is called a ROC. The *TPR* should ideally be 1, and the *FPR* should be 0. In general, the better the performance, the closer the ROC curve is to the top left corner. Similarly, the AUROC curve with higher values indicates better performance.

14.12 Computation

ML algorithms are widely used to reduce the computational complexity of problems. Still, some ML algorithms can add extra complexity, especially in the training stage. Also, when the number of hidden layers increases, computational complexity further increases. For example, the data is subjected to a substantial mathematical calculation in ML. In DL, this is multiplied by a significant margin. This task is beyond the capabilities of the ordinary central processing unit (CPU) architecture. The graphic processing unit (GPU) was created for gaming, but it has proven to be the optimal solution for DL thanks to its capability of parallel working. There is heavy research and development made to allow more efficient hardware solutions. Therefore, it is

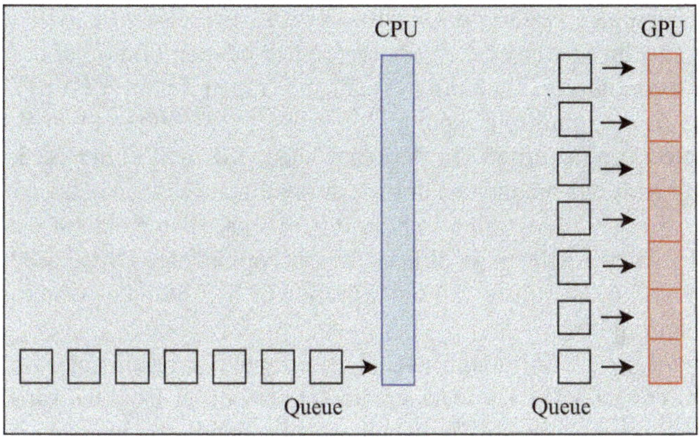

Figure 14.15 CPU versus GPU in terms of computation

expected to be a more efficient solution in the future. In Figure 14.15, you can see an illustration of CPU versus GPU for computing. The figure shows that GPU can work in parallel.

14.13 Open challenges and future directions

Although the advantages of ML in several use cases of PLS, there are still open challenges for practical ML approaches. This section highlights these challenges with future directions. Also, several use cases for ML in security are given.

- *Standard*: Although ML algorithms are recommended in many sources for PHY security, their standard work is very limited. This reduces the use of these algorithms in daily life. In this direction, it is necessary to carry out standard studies for ML security in PHY. Although there are standard activities for ML in Third Generation Partnership Project (3GPP) and Wi-Fi (Institute of Electrical and Electronics Engineers (IEEE) 802.11 AIML TIG study group), they are quite limited.
- *Theoretical analysis*: Theoretical analysis of ML for security is still in the early stage and it should be studied for reliable results.
- *Cost*: ML algorithms, particularly DL algorithms, are expensive and can consume a lot of energy and computational resources. In addition, training time is a restriction for DL models. Also, as the size of the training dataset grows, training time increases dramatically. Therefore, time efficient ML techniques should be researched. Also, more efficient hardware solutions such as advanced GPU can be developed in the future.
- *Hyperparameter optimization*: It is difficult to find optimum hyperparameters for a given ML model, environment, and application. Therefore, there should be a standard for it.

- *5GB technologies collaboration*: To serve all use cases of 5GB [31], future 5GB architecture requires the coexistence and cooperation of all traditional and modern technologies, including broadband transmission, LTE/LTE-Advanced systems, massive multiple-input multiple-output (MIMO), network slicing, and mobile cloud engineering. On the other hand, ML will either participate independently with one on one technology or intelligently manage the entire system. Orchestration and integration in hybrid networks, slice isolation method, security, and privacy are the main difficulties that sophisticated ML-based algorithms must overcome. Therefore, the orchestration of ML-based algorithms should be studied thoroughly.
- *Adversarial attack*: Although ML is an effective solution for PLS, it creates particular new attacks. These adversarial attacks differ from the various security threats that researchers are familiar with. As a result, the first step in defending against them is comprehending the various sorts of adversarial attacks and ML algorithms. Therefore, all the attacks should be surveyed in detail.
- *Supporting different services*: Telecommunication operators used the integrated services differentiated services models before 5GB technology, but 5GB technology introduces enhanced mobile broadband (eMBB), ultra-reliable low-latency communication (uRLLC), and massive machine-type communication (mMTC). Thus, 5GB is projected to meet the bandwidth and latency needs of various vertical applications and services. Therefore, the future network must be able to support new business models based on heterogeneously oriented services, as well as deliver services in all use case scenarios. Thus, we should perform in-depth and thorough research to identify real-time issues for all 5GB use cases, as well as embedded ML-based optimum solutions for the future era.
- *Roaming security and parameter optimization*: Updates on the user's security parameters are critical while traveling from one network to another. Especially in the 5GB densification scenario, these parameters should be optimized jointly. Still, the problem is delivering these services to different subscribers in diverse places and even optimizing these parameters for the specific scenario. ML can be used to solve this problem thanks to its ability to optimize multiple parameters jointly.
- *Using ML algorithms for sensing*: In recent years, using ML techniques to learn and adapt to the wireless environment has become increasingly common. However, their usage for security sensing is limited, and it should be studied. In addition to that, the protection of sensing data should be studied.
- *Using ML algorithms for IoT*: The topic of ML security solutions for IoT devices has risen to prominence in recent years, making today's researchers interested in contributing more to it. However, IoT applications require low complexity algorithms, which ML can provide. Therefore, ML should be addressed for security.
- *Using blind signal analyses for security*: Blind signal analysis comprehensively studied with ML algorithms [32]. These techniques can be used for both PHY security and attack perspectives.

- *CR security attacks*: As the number of users rises, the risk of large-scale events generating significant changes in spectrum utilization, either accidentally or maliciously, increases. The 5GB system must prevent significant swings in spectrum consumption and be responsive to them while maintaining an acceptable level of performance. CR technology is used to optimize the spectrum. However, there are also attacks on the CR. In these cases, ML-based algorithms can be used to learn the environments and recommend optimal outputs by detecting CR security attacks.
- *Security and privacy challenges in 5GB networks*: Different security concerns and privacy policy requirements can be included in 5GB networks by providing diverse services, multiple network slices, and resource sharing for different verticals [33]. As a result, complex research difficulties are raised and handled, considering the impact of one slice on another, efficient coordination mechanisms, and the impact of complete network systems, especially in multiple domain infrastructures. These issues can be met using ML-based algorithms that maintain network performance.
- *Using complexity of an ML algorithm for eavesdropping*: Although the complexity of ML is disadvantageous, this complexity can be used to create mechanisms that will be used for security, such as artificial channels, RF impairments, and signals for eavesdropping. Thus, the eavesdropper will face this extra complexity that ML brings.
- *Using ML algorithms for joint security threats*: ML can be used to estimate, predict, take precautions, compensate, etc. for joint problems. Also, attacks can happen in several layers of OSI simultaneously. An ML algorithm can be used for this problem since it is difficult to build relationships between different layers and ML is promising for this kind of problem.
- *Security levels*: In a security architecture, there are service-driven limits. However, these protections are not always essential in current cellular specifications. Therefore, the security levels of the applications and environment should be detected. ML can be used for this purpose thanks to its ability to understand different requirements.
- *Mobility management in future networks*: 5GB communication systems are expected to support high-speed devices and their security is still open to research. ML can be used for that since it has several algorithms which can respond quickly in the testing stage.

14.14 Conclusion

The nature, structure, and operational circumstances of 5GB communication and sensing systems are different for different scenarios, environments, and applications. Depending on the technology used, security in the future wireless network will be drastically different. Approaches based on ML have been shown to be potential solutions for meeting the various requirements of emerging wireless networks.

ML can do this thanks to a variety of algorithms that have been designed to operate well in a variety of situations. In addition, ML allows you to work with multiple dimensional data and many parameters while maintaining high performance and simplicity. However, there are still significant obstacles to its successful use for security. It even introduces additional security threats that are not present in traditional procedures. Therefore, there are several research gaps to be studied in ML-based security.

Acknowledgement

This work was supported in part by the Scientific and Technological Research Council of Turkey (TUBITAK) under Grant no. 119E433.

References

[1] Pokhrel SR, Choi J. Federated learning with blockchain for autonomous vehicles: analysis and design challenges. *IEEE Transactions on Communications*. 2020;68(8):4734–4746.

[2] Goodfellow I, Pouget-Abadie J, Mirza M, *et al.* Generative adversarial nets. *Advances in Neural Information Processing Systems*. 2014;27: 2672–2680.

[3] WAIR Development. Obscuring and analyzing sensitive information with generative adversarial networks. *World Wide Technology Technical Report*; 2019.

[4] Kim JY, Bu SJ, Cho SB. Malware detection using deep transferred generative adversarial networks. In: *Proc. Int. Conf. Neural Inf. Process*. Springer; 2017. p. 556–564.

[5] Samangouei P, Kabkab M, Chellappa R. Defense-GAN: protecting classifiers against adversarial attacks using generative models. arXiv preprint arXiv:180506605. 2018;.

[6] Hu W, Tan Y. Generating adversarial malware examples for black-box attacks based on GAN. arXiv preprint arXiv:170205983. 2017;.

[7] Tramèr F, Zhang F, Juels A, *et al*. Stealing machine learning models via prediction APIs. In: *USENIX Security Symposium* (USENIX Security 16); 2016. p. 601–618.

[8] Papernot N, McDaniel P, Goodfellow I, *et al.* Practical black-box attacks against machine learning. In: *Proc. ACM Asia Conf. Comput. Commun. Sec.*; 2017. p. 506–519.

[9] Wang B, Gong NZ. Stealing hyperparameters in machine learning. In: *2018 IEEE Symposium on Security and Privacy* (SP); 2018. p. 36–52.

[10] Oh SJ, Schiele B, Fritz M. Towards reverse-engineering black-box neural networks. In: *Explainable AI: Interpreting, Explaining and Visualizing Deep Learning*. New York, NY: Springer; 2019. p. 121–144.

[11] Song C, Ristenpart T, Shmatikov V. Machine learning models that remember too much. In: *Proceedings of the ACM SIGSAC Conference on computer and communications security*; 2017. p. 587–601.

[12] Fredrikson M, Lantz E, Jha S, *et al.* Privacy in pharmacogenetics: an end-to-end case study of personalized warfarin dosing. In: *23rd USENIX Security Symposium* (USENIX Security 14); 2014. p. 17–32.

[13] Fredrikson M, Jha S, Ristenpart T. Model inversion attacks that exploit confidence information and basic countermeasures. In: *Proceedings of the 22nd ACM SIGSAC Conference on Computer and Communications Security*; 2015. p. 1322–1333.

[14] Yang Z, Zhang J, Chang EC, *et al.* Neural network inversion in adversarial setting via background knowledge alignment. In: *Proceedings of the ACM SIGSAC Conference on Computer and Communications Security*; 2019. p. 225–240.

[15] Shokri R, Stronati M, Song C, *et al.* Membership inference attacks against machine learning models. In: *IEEE Symposium on Security and Privacy* (SP). New York, NY: IEEE; 2017. p. 3–18.

[16] Rivest RL, Adleman L, Dertouzos ML, *et al.* On data banks and privacy homomorphisms. *Foundations of Secure Computation*. 1978;4(11):169–180.

[17] Gentry C. Fully homomorphic encryption using ideal lattices. In: *Proc. 41 annual ACM Symp. Theory Computing*; 2009. p. 169–178.

[18] Dwork C, McSherry F, Nissim K, *et al.* Calibrating noise to sensitivity in private data analysis. In: *Theory of Cryptography Conference*. New York, NY: Springer; 2006. p. 265–284.

[19] Aygül MA, Büyükçorak S, da Costa DB, *et al.* Signal relation-based physical layer authentication. In: *IEEE International Conference on Communications* (ICC). New York, NY: IEEE; 2020. p. 1–6.

[20] Pei C, Zhang N, Shen XS, *et al.* Channel-based physical layer authentication. In: *IEEE Global Communications Conference* (Globecom). New York, NY: IEEE; 2014. p. 4114–4119.

[21] Xie N, Li Z, Tan H. A survey of physical-layer authentication in wireless communications. *IEEE Communications Surveys & Tutorials*. 2020;23(1):282–310.

[22] Jagannath A, Jagannath J, Kumar PSPV. A comprehensive survey on radio frequency (RF) fingerprinting: traditional approaches, deep learning, and open challenges. arXiv preprint arXiv:220100680. 2022;.

[23] Aygül MA, Memisoglu E, Arslan H. Joint estimation of multiple RF impairments using deep multi-task learning. arXiv preprint arXiv:210914321. 2021;.

[24] Aygül MA, Memisoglu E, Cirpan HA, *et al.* Identification of distorted RF components via deep multi-task learning. arXiv preprint arXiv:220701707. 2022;.

[25] Yucek T, Arslan H. A survey of spectrum sensing algorithms for cognitive radio applications. *IEEE Communications Surveys & Tutorials*. 2009;11(1):116–130.

[26] Arul Selvi S, Sundararajan M. SVM based two level authentication for primary user emulation attack detection. *Indian Journal of Science and Technology*. 2016;9.

[27] Rehman SU, Sowerby KW, Coghill C. Radio-frequency fingerprinting for mitigating primary user emulation attack in low-end cognitive radios. *IEEE Institution of Engineering and Technology (IET) Communication*. 2014;8(8):1274–1284.

[28] Furqan HM, Aygül MA, Nazzal M, *et al*. Primary user emulation and jamming attack detection in cognitive radio via sparse coding. *EURASIP Journal on Wireless Communications and Networking*. 2020;2020(1):1–19.

[29] Aygül MA, Furqan HM, Nazzal M, *et al*. Deep learning-assisted detection of PUE and jamming attacks in cognitive radio systems. In: *2020 IEEE 92nd Vehicular Technology Conference* (VTC2020-Fall). New York, NY: IEEE; 2020. p. 1–5.

[30] Ahmad R, Alsmadi I. Machine learning approaches to IoT security: a systematic literature review. *Internet of Things*. 2021;14:100365.

[31] Yuan G. Key technologies and analysis of computer-based 5G mobile communication network. In: *Journal of Physics: Conference Series*. vol. 4. Bristol: IOP Publishing; 2021. p. 042001.

[32] Aygül MA, Naeem A, Arslan H. Blind signal analysis. In: *Wireless Communication Signals: A Laboratory-based Approach*. New York, NY: Wiley; 2021;p. 355–381.

[33] Ziani A, Medouri A. A survey of security and privacy for 5G networks. In: *Emerging Trends in ict for Sustainable Development*. New York, NY: Springer; 2021. p. 201–208.

Chapter 15

Communications network security using quantum physics

A.S. Atilla Hasekioğlu[1] *and Orkun Hasekioğlu*[2]

Applications of quantum technologies are rapidly progressing in communications systems implementations. In particular, quantum key distribution (QKD) and quantum random number generation have become key applications for cryptography and physical layer security. Unlike classical cryptography and network security, in QKD, security is ensured by the laws of physics and is, in principle, unbreachable by an interceptor. In particular, peculiar properties and postulates of quantum physics, such as superposition, entanglement, and quantum measurement uncertainty, are exploited to achieve system security. Quantum random number generator (QRNG) is also an important component in the implementation of QKD protocols. Likewise, in QRNG, true randomness (unpredictability using past data and uniformity of statistical distribution) is guaranteed by the laws of quantum physics. In this study, we review prominent aspects of QKD and QRNG, related standards, protocols, and hardware components. Moreover, a comprehensive review of the current QKD protocols and quantum random number generation methods is also presented.

15.1 Introduction

Secure communications based on the principles of quantum systems are one of the pioneering applications of quantum technologies. Unlike classical communications systems, the security achieved by quantum technology applications is inherent, ensured by the laws of physics, and has proven to be unbreachable from information theoretic point of view [1]. In this sense, quantum technologies open up an unprecedented window of possibilities that will never be feasible by classical communications alone. The laws of quantum physics that are commonly observed in small-scale particles, such as atoms, and subatomic particles like photons, are more fundamental than classical physics. In the sense that classical equations of motion (Newtonian mechanics) can be derived from principles of quantum physics as stated

[1]Informatics and Information Security Research Center (BILGEM), The Scientific and Technological Research Council of Türkiye (TUBITAK), Turkey
[2]Scientific and Technological Research Council of Türkiye (TUBITAK), Turkey

by the correspondence principle. As Planck's constant approaches zero, the quantum mechanics formulation converges to the Newtonian mechanics' formulation of the equations of motion.

In our daily experience, we are accustomed to working with large objects. Experience with photons, electrons, or atoms, on the other hand, is unintuitive and may not be easily internalized. In quantum secure communications and quantum computation, two such quantum postulates are exploited: superposition and entanglement. In a general sense, we will refer to quantum communication and computation systems as quantum information systems.

In quantum information systems, the fundamental unit of information is referred to as a qubit (quantum bit) as opposed to a bit in classical communications and computation. A qubit is in a superposition state of the natural basis states $|0>$ and $|1>$, pronounced ket 0 and ket 1, which is called the Dirac notation. A qubit can be in a complex linear superposition of these basis states. However, when measured, one can only obtain one of these basis states. This measurement process is referred to as the collapse of the superposition state. Whether one measures the $|0>$ or $|1>$, the state is probabilistic with a distribution determined by the magnitude square of the associated complex coefficient. The complex coefficient itself is called the amplitude of the corresponding basis. Information is encoded in the quantum states of the qubits. In almost all quantum communication applications, a qubit is a photon prepared in a certain superposition state.

The topic of quantum secure communications deserves a separate volume in its own right. Inevitably, in this chapter, we will limit our discussions to secure quantum key distribution, the most common, mature, and well-established quantum physics application in communications. The secure quantum key distribution networks (QKDN) with QRNGs and related protocols will be in scope.

The field of quantum secure communications is mature with well-established theoretical models and commercial applications. Standardization efforts have been underway for the past few years. Standards development organizations, such as IEEE, ITU, ETSI, and ISO, have formed active working groups for this purpose.

Quantum key distribution (QKD) is one of the relatively few commercially available applications of quantum technologies to free-space and fiber-optic communications systems and networks. In this section, we present an overview of various QKD communications protocols, their characteristics, and related commercialization efforts. The emphasis will be on the communications security protocols and standardization aspects of such protocols. As much as possible, we will follow the emerging terminology and definitions that have recently been introduced and concurred upon by the standards development organizations.

A communications protocol is a list of actions performed by communicating parties to ensure a reliable and safe exchange of information. QKDN uses a large variety of such protocols to ensure the secure distribution of keys that are used for the encryption and decryption of messages in a symmetric key encryption communications network.

In Figure 15.1, the layers of protocols involved in a QKDN are depicted, as defined by the standards development organizations [2,3]. In this review, the

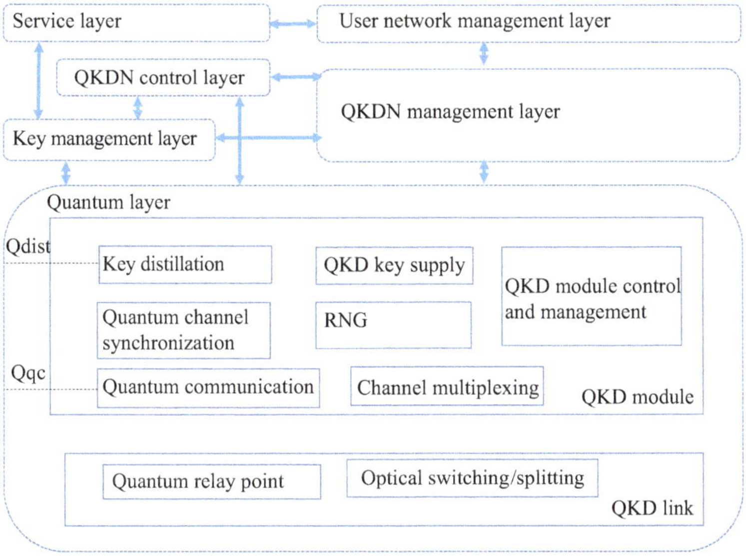

Figure 15.1 Layers of protocols of a QKDN, as depicted by standards development organizations

emphasis will be on the quantum layer box, where quantum-related technologies, e.g., QKD, QRNG, and key distillation, are implemented. The remaining boxes are mostly similar to any classical secure key distribution network.

A QKD protocol is implemented in the QKD module. The Qqc interface of the quantum communication function box and the Qdist interface of the key distillation function box in Figure 15.1 are called the quantum channel and classical channel, respectively. A quantum channel is a medium where qubits are transmitted. The medium can be free space or a fiber-optic channel. The classical channel is the medium where classical bits of information are transmitted. Classical information exchange is needed to support the reliable operation of the communications protocols.

In Figure 15.1, the random number generator (RNG) function box provides the randomness required for the quantum key generation, transmission, and distillation functions. Distillation is part of the QKD post-processing which will be discussed later.

The QKD module control and management function maintain the overall management of the other auxiliary functions of the QKDN. Detailed functions of the QKDN protocol layers are discussed in the standards documents, specifically in [4].

The quantum communication function is responsible for the actual physical production of the qubits, including quantum state preparation, information encoding, quantum state transmission, and quantum state measurement.

The other auxiliary functions in the quantum layer depicted in Figure 15.1 are quantum channel synchronization, channel multiplexing for the QKD module and

QKD link, optical switching/splitting, and quantum relay link, which are referred to as the transport functions. They consist of the physical layer components that realize the QKD protocols. Some of these functions are optional and implementation-dependent.

In this review, we will not cover the details of these auxiliary functions, except for mentioning their relations with the other functions in the quantum layer. The channel synchronization function supplies the clock and is responsible for timing synchronization. The channel multiplexing function is implementation dependent. When needed, it is used for wavelength division multiplexing of both classical and quantum channels.

In the QKD link module, the quantum channel and synchronization signal are switched through the switching/splitting function. The implementation-dependent quantum relay function is an untrusted intermediate function that is used in certain protocols like measurement device independent (MDI) and twin field (TF) to extend the transmission distance. These protocols are discussed later.

15.2 QKD system description and components

Before describing QKD network protocols and layers, we would like to give an overview of a generic QKD system with its components to present the overall description of a QKDN. Many variations of QKDN applications have been implemented. However, in physical implementations, basic components are similar, and fundamentally, they all use principles of quantum physics to ensure the secure distribution of shared keys.

In QKD applications, non-orthogonal quantum basis states of photons (e.g., polarization or phase state) or entangled photon pairs are used. These protocols make use of the quantum physical principle that it is not possible to faithfully decode such quantum states without a priori information on the encoding basis.

Figure 15.2 depicts a general-purpose QKD system [5], consisting of a transmitter (Alice) and a receiver (Bob). Depending on the particular implementation, the photon source may be a single photon generator, an entangled pair of photons, or a laser pulse. In some applications, the photon source is an attenuated laser pulse

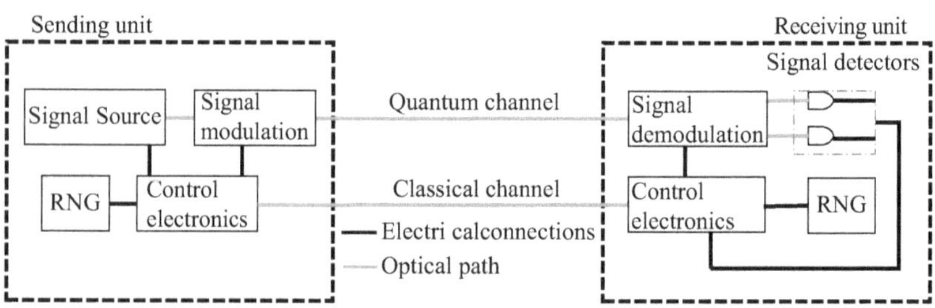

Figure 15.2 A general purpose QKD system

such that single distinguishable photons are generated and encoded. The modulator encodes the qubit state, typically in the phase or polarization of the photon source. The randomness required is supplied through a RNG that may again be based on quantum physical principles (QRNG) or a classical random number generator.

Similarly, the receiver, Bob, consists of a demodulator, a control unit, an RNG, and photon detectors. Note that the transmitter and receiver are connected through two channels: a quantum channel and a classical channel. They may be separate optical channels or a single frequency or time-division multiplexed channel. The quantum channel is used to transmit the photonic qubits, and the classical channel is used to exchange classical information necessary to conduct the QKD protocol, such as synchronization, post-processing, raw key sifting, and privacy amplification that will be discussed later.

Figure 15.3 describes such a QKD receiver and a transmitter using an Asymmetric Mach Zender Interforemeter (AMZI) [5,6], as an example. Here, the photon source is an attenuated laser pulse. The laser diode, intensity modulator, and attenuator constitute the single-photon source. The AMZI phase modulates the photons in a superposition state. The randomness of the phase is ensured by the RNG. The upper single-mode fiber is the quantum channel over which the qubit photons are transmitted. The lower single-mode fiber is the classical channel where protocol management and clock signals are transmitted.

Figure 15.4 depicts a receiver and transmitter QKD pair based on entangled photon sources [5,7]. Entangled pair of photons are polarization encoded, at 810 nm and 1550 nm. 810 nm photon is encoded at one of the four polarization states (horizontal, vertical, 45 deg, and 135 deg). The polarizing beam splitters (PBS) classify the photons based on their states. The Avalanche photo diode (APD) photodetectors detect the polarization state of the entangled qubit photons. The entangled qubits are transmitted via the wavelength division multiplexer (WDM) through the upper single mode fiber quantum channel.

More recently, in 2016, satellite-based free space QKD was achieved with the Micius satellite with an entangled photon pair generation payload [8].

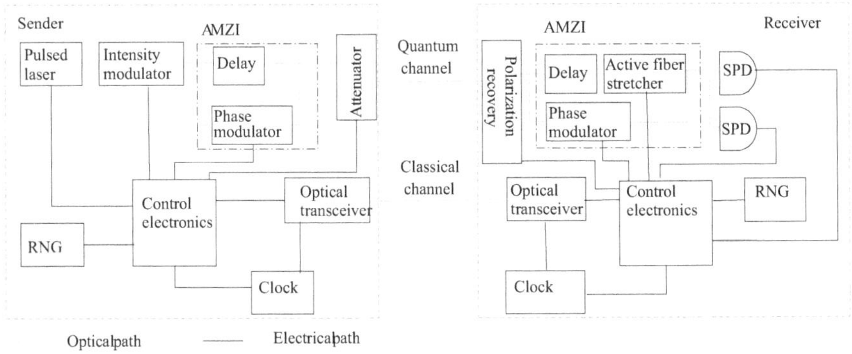

Figure 15.3 Schematic of a QKD system using AMZI

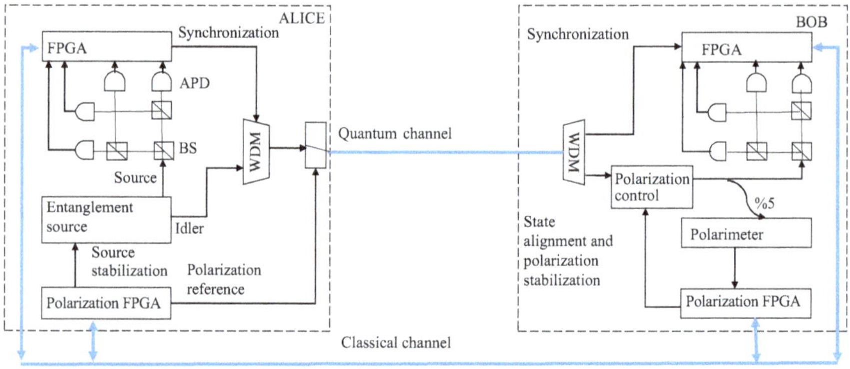

Figure 15.4 Schematic of an entanglement-based QKD system

This experiment has proven the applicability of entanglement-based QKD through the atmosphere and free space.

15.2.1 QKD photon sources and detectors

15.2.1.1 Single photon detector

A single photon detector converts the photons hitting the detector to an electrical signal in the output. Important specifications are detection repetition rate, detection probability, spectral response, dark count probability, after pulse probability, dead time, recovery time, signal jitter, and photon number resolution depth. Single photon avalanche photodiodes (SPAD), using InGaAs are common. They have single photon sensitivity over a wide range of wavelengths (900–1700 nm) which is suitable for fiber optic QKD systems.

Superconducting nanowire single-photon detectors (SNSPD) work at temperatures as low as a few degrees of Kelvin, requiring He-cooled cryogenics [9,10], as opposed to the SPADs working at room temperature or temperatures achievable by thermoelectric cooling. Because of fast switching and recovery durations, SNSPDs have low timing jitter and less than 10 ns reset time.

15.2.1.2 Single photon source

Typically in a QKD, quantum information is encoded on the photons in the form of phase, polarization, or angular momentum that is recovered at the receiver. Ideally, one would like to use a photon source that generates a single photon when triggered (on demand). However, in practice, there is also a probability of generating no photons or more than a single photon when triggered. Physical single photon generators are based on excited quantum dots, defects in diamonds, single ions, or atoms. Another approach, as we have seen, is to attenuate a laser beam to a point where individual photons are generated with a Poisson distribution.

Entangled photon pairs are another commonly used resource in QKDN. Most commonly, entangled photon pairs are produced through an optical process called spontaneous parametric down conversion (SPDC). A pump photon hitting a properly designed crystal with nonlinear optical properties, such as Beta-Barium Borate (BBO) and Lithium Niobate, emits two photons at a lower frequency such that the total energy and total momentum are conserved. Detecting one of the photons indicates the emittance of the second. In this sense, this is referred to as the heralded single photon generation.

In entangled photon pair generation using SPDC, the crystal may be optically excited either with a pulsed or continuous wave laser. This is a second-order nonlinear optical process. The incoming photon is called the pump photon, while the two outgoing photons are called signal and idler photons.

Some of the most relevant parameters of single photon sources are optical pulse repetition rate, photon number probability distribution, mean photon number per pulse, power stability, second-order correlation function, frequency, spectral line width, timing jitter, and temporal pulse shape.

15.3 QKDN protocols standardization

A communications protocol can be defined as a procedure applied by the communicating parties to achieve a specific goal. A typical QKDN involves various layers of protocols, apart from the QKD function, mostly borrowed from conventional communication networks.

The QKDN protocol enables symmetric key distribution, ensuring secure generation and transmission of keys to the communicating parties. This is typically accomplished by dedicated hardware to prepare and transmit the raw key and post-processing software to process and distill the raw key. The QKDN protocols and related standardization efforts attempt to define the rules and steps involved in such a protocol so that secure key transmission is achieved.

QKD is one of the most important commercially available applications of quantum technologies in communications. It allows two communicating parties, Alice and Bob, to share a secret key over a quantum channel. QKD establishes key exchange that is in many cases proven to be information theoretically secure.

The first QKD protocol was proposed by Charles Bennett and Gilles Brassard in 1984 [11]. Many other QKD protocols have been invented afterward, some of them commercialized. However, standardization efforts have only recently started [12]. Besides establishing a common framework among the equipment suppliers, an important purpose of standardization is to identify and remedy possible security loopholes and breaches.

Classical public key and symmetric key cryptographic protocols have been standardized [13–15]. Similar standards for QKD protocols are underway within various industry working groups, such as IEEE, ISO, and ITU.

In this study, we review some of the most prominent QKD protocols that are, in our view, commercially promising. There are a number of review papers discussing QKD protocols in detail [16,17] and more recently [18–20]. Here, we emphasize important aspects and parameters of these protocols that are especially relevant in communications security and standardization.

15.4 QKDN quantum layer protocols

As shown in Figure 15.1 and explained in [3], standardization document, a QKDN protocol application consists of Key management, QKDN control, QKDN management, QKDN service, and User network management layers. Apart from the quantum layer that is unique to QKDN, the other protocol layers or their equivalents exist in conventional key distribution networks. The quantum layer is responsible for implementing the QKD protocols and establishing symmetric keys through point-to-point QKD links.

The quantum layer implements the QKD protocols that establish symmetric keys between the trusted nodes. In the quantum layer, there are auxiliary functions, such as quantum channel synchronization, and channel multiplexing. The functional description of the details of the protocol, layers depicted in Figure 15.1, is explained in [3].

Here, we focus on what is referred to as the quantum layer by the standardization organizations. This layer consists of the QKD module, QKD link module, and interfaces to other layers. The QKD protocols and key distillation are implemented in the QKD module. The randomness is achieved through a quantum or classical random number generator.

Here, we will limit our discussions to the core protocols in the quantum layer of the QKDN, i.e., QKD protocols, which will include the quantum communication and key distillation function boxes, that are within the QKD module. QRNG is discussed in another section.

15.4.1 Performance parameters of QKD

The QKD protocols proposed since the invention of BB84 are all motivated to improve key generation rate, key distribution distance, simplify hardware implementation, and the security level in practice.

Although the QKD protocols differ in the details of the physical implementation, the overall information flow is very similar. Here, we first summarize the overall information flow in a QKD protocol and then review examples of the physical implementations of QKD systems.

15.4.2 Overview of the information flow

A typical QKD protocol consists of two major parts:

- The generation and transmission of the qubits representing the raw key, communicated through the quantum channel.
- Classical post-processing of the raw key that uses communication through the classical channel.

Every QKD protocol contains this general procedure, although the details vary depending on the protocol. Generation and transmission of the raw key data exploit two important properties of qubits: superposition and entanglement, which we will explain next.

15.4.2.1 Qubit generation and transmission

Superposition use case: Alice encodes a string of classical random bits into a set of non-orthogonal quantum states of photons. Alice, then, transmits the quantum states (qubits) to Bob using the quantum communications channel.

Bob receives the qubits and performs a measurement on them, producing another classical random bit sequence, called the raw key . The random bit sequence encoded by Alice and the random bit sequence produced by Bob are correlated but different random sequences. Repeating this process, Alice can transmit a long sequence of raw key data to Bob.

As an example, in BB84 protocol (Figure 15.5), the QKD transmitter produces single photons and encodes a stream of bits on the transmitted photons (qubits). One of the four different polarization states is chosen randomly for each photon.

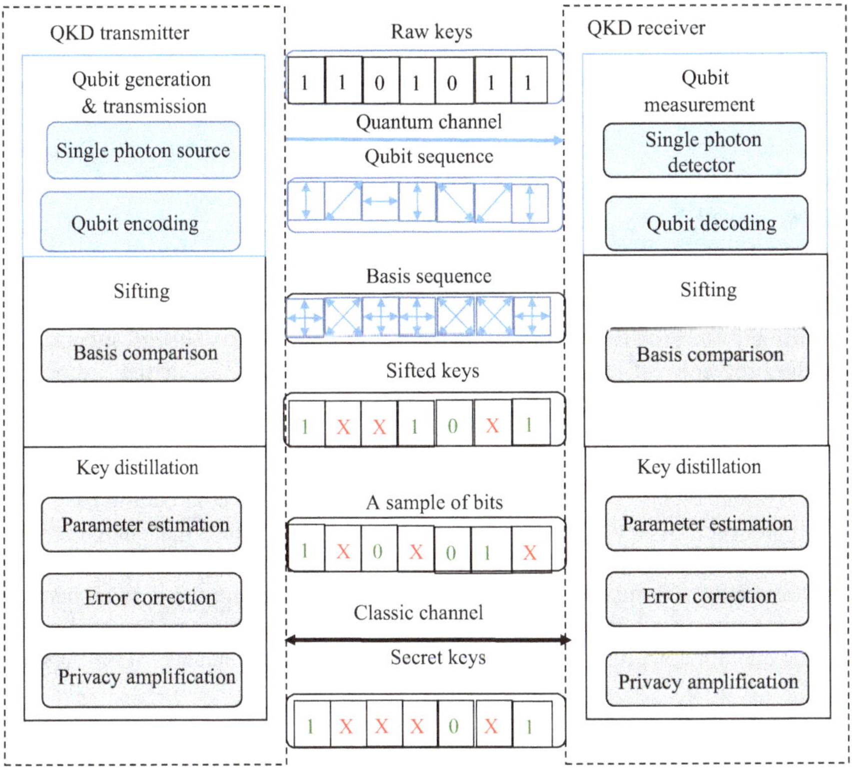

Figure 15.5 Steps for BB84 protocol

The 0 bit is encoded into a horizontal or 135° diagonal state and the 1 bit is encoded into a vertical or 45° diagonal state. The QKD transmitter sends the qubits to the QKD receiver through a quantum channel. The QKD receiver measures the incoming qubits by selecting one of the two bases (standard or diagonal) randomly and records the outcomes and the selected bases [21].

Entanglement use case: Unlike classical data, the quantum states representing bits of classical bits cannot be copied, as a consequence of the no-cloning theorem. This property of the qubits is the fundamental reason behind the security of QKD. In classical key distribution, an eavesdropper (Eve) can always intercept classical data, modify it, and re-transmit it, as in the well-known man-in-the-middle security attack.

In addition to the encoding in the superposition states of the qubits method, raw key data transmission is also possible using the entanglement property of qubits. This involves generating entangled photon pairs and Bell-state measurements for verification of non-interception of the qubits by an eavesdropper during transmission.

Both superposition and entanglement raw key data generation have been demonstrated and have commercial applications.

15.4.2.2 Classical post processing

After the exchange of the raw key data, the raw key is in the classical bit form and the classical post-processing takes place. The purpose of classical post-processing is to convert the qubits received and measured by Bob to a shared key suitable for symmetric key encryption and decryption. The post-processing part of the QKD protocol requires the transmission of classical information between Alice and Bob over a classical channel.

Classical post-processing of the received qubits, usually involves several stages: sifting, parameter estimation, error correction, and privacy amplification, which we explain below.

Sifting is the process of eliminating the parts of the raw key that are inconsistent with the protocol and known to be indeterminate by the rules of the protocol to be the same with Alice and Bob. The resulting reduced key obtained from the raw key is called the sifted key. As in the BB84 protocol (Figure 15.5), at this stage, the QKD receiver broadcasts the bases that have been used for decoding via the classical public channel. The QKD transmitter compares the polarization basis state of each qubit sent with the broadcast basis and retains the qubits that share the same encoding/decoding basis as the sifted raw key.

Afterwards, a portion of the sifted key is used to estimate the channel parameters, such as the ratio of the identical bits to the non-identical bits between Alice and Bob, the quantum bit error rate (QBER). This also gives an indication of the presence of Eve, undesired interception, and noise.

An error-correction scheme agreed upon between Alice and Bob may be implemented to correct some of the errors. In the privacy amplification stage, part of the error-corrected key is discarded to eliminate the information that may be obtained by Eve.

15.4.3 Types of QKD protocols

Different QKD protocols are implemented with variations in the raw qubit generation and post-processing stages, described above. We will first describe examples of such variations and demonstrate them through different QKD protocols.

One may classify QKD protocols based on the following:

- Preparation and measurement of qubits: Use of superposition or entanglement properties of qubits.
- Trusted measurement device independent QKD protocols or device dependent protocols.
- One-way (unidirectional) or two-way (bidirectional) qubit transmission.
- Qubit encoding and decoding technique used, such as discrete-variable (DVQKD) or continuous-variable (CVQKD).

In DVQKD, the dimensionality of the encoded information on the photon is finite and discrete. The information can be encoded in the polarization, discrete phase, or the discrete-time bin of single photons. The encoded photons received by Bob are detected using single photon detectors (SPD). Well-known examples of DVQKD are BB84 [11], E91 [1], B92 [22], the Six-State Protocol [23], BBM92 [24], SARG04 [25], Coherent-One-Way (COW) protocol [26,27], Differential Phase Shift (DPS) [28,29], Round Robin Differential Phase Shift (RRDPS) [30,31], Twin-Field (TF) protocol [32,33], Discrete Variable Measurement Device Independent (DV-MDI) [34,35], and Device Independent QKD (DI-QKD) [36].

On the other hand, in continuous variable CVQKD protocols, Alice encodes again a discrete variable string of raw key data on the quadrature components (X and P or position and momentum) of the electromagnetic field of the photons (laser) in the optical phase space. Bob, after receiving the encoded photons, uses coherent homodyne or heterodyne detection to recover the encoded information by Alice. Notice that, unlike DVQKD where one uses single photon generators and detectors, in the case of CVQKD one typically uses lasers (coherent quantum state of light) as photon sources and homodyne detectors to recover the raw bit values encoded in the quadrature components of the laser light. The acronym continuous variable (CV) stems from the fact that the measurement output of the detectors is a continuous signal.

In heterodyne detection, one measures both quadratures simultaneously. This causes additional noise due to the uncertainty relation. On the other hand, a higher key rate is possible using heterodyne detection.

Well-known examples of CVQKD are Gaussian modulation-based CV protocol [37], discrete modulation-based CV protocol [38–40], and continuous variable measurement device independent (CV-MDI) protocol [41], which will be explained later.

15.4.4 Security of QKD protocols

The main objective of a QKD protocol is to transmit a secure key from Alice to Bob (transmitter to receiver) that cannot be intercepted or compromised through outside attacks by Eve. The QKD protocol, therefore, consists of well-defined steps to achieve this goal, as structured by the standardization working groups, such as [42]. It is possible to prove the security of such protocols using information theoretical arguments. However, in practical implementations, information theoretical security may not hold due to various application-related security loopholes. Therefore, the protocol standards documentation needs to address the physical implementation-related aspects as well. There are intense ongoing academic research, commercialization efforts, and standardization activity to ensure security in practical implementations.

The security of the QKD protocols can be theoretically proven through information theoretical arguments and laws of quantum physics. These proofs usually assume no limitations on the computational capabilities and perfect physical device implementations. Other simplifying assumptions may include, authenticated classical channel, i.e., Eve can intercept but cannot change the classical messages which are usually assumed public in most protocols, in addition, random number generators used to implement the protocol are truly random so that they are computationally unpredictable, device hardware and software implementations exactly mimic the definition of the protocol in the standards. These conditions may not be fulfilled in real-life applications causing security levels to be over-estimated in many cases. For further discussions on security proofs of QKD algorithms, see [43–47].

15.4.4.1 Implementation security

The common assumption in the security analysis in cryptography is based on Kerckhoffs' principle which states that Eve has access to the detailed description and design of the cryptographic devices that are used by Alice and Bob. This assumption implies device-independent QKD protocols.

The security of the QKD protocol needs to be extended to the entire layers of protocols as discussed earlier. The added complexity of the specific QKD protocol implementation can lead to loopholes for several side-channel attacks, such as time shift attacks, blinding attacks, detector saturation attacks, spatial mode mismatch attacks, and Trojan horse attacks. The standards documentation [48,49] that are work-in-progress attempt to address such practical security aspects.

In a DVQKD blinding attack [50,51], SPDs are illuminated with a comparatively higher power rather than the expected single photons. This causes detectors to be saturated and insensitive to single photons. This can enable Eve to manipulate the SPD signals. Similarly, CVQKD systems can also be subject to the blinding attack [52,53], where Eve manipulates the homodyne detector by high-power laser illumination. The blinding attack can be eliminated by verifying the incident illumination power or monitoring the operating mode of the detector to be in pulsed mode rather than linear response mode. There are other means of security breaching, where Eve can interfere by sending photons through the QKD link ports of the transmitter or the receiver since they are open ports to the outside world [20,54].

15.4.5 QKD protocols

15.4.5.1 Discrete variable QKD protocols (DVQKD)

As explained earlier, in the DVQKD protocol implementations, the photons are detected using SPD.

DVQKD protocols are based on the quantum states of single photons, such as vertical and horizontal polarization. State measurement is done using SPDs. Below, we cite a few prominent examples of DVQKD:

- BB84, B94, and other versions of BB84 protocols: BB84 protocol [11] is the most widely studied and implemented QKD protocol that uses the superposition property of qubits. The physical implementation examples of the BB84 protocol are realized in fiber optic-based QKD networks, such as, the DARPA QKD network [55], SECOQC QKD network [56], and Tokyo QKD network [57] as well as satellite-based wireless QKD networks [58]. B92 protocol [22] is a version of the BB84 protocol where two non-orthogonal qubits are transmitted instead of four. The six-state protocol [23] is another version of the BB84 protocol that uses six discrete quantum states on three orthogonal bases instead of four states as in the original BB84. The implementation is comparatively more sophisticated but enables higher rate of key data transmission.
- SARG04 protocol: The SARG04 protocol [25], demonstrated in the Tokyo QKD network [57], is also derived from the BB84 protocol with modified encoding/decoding schemes, enabling robustness to the photon number splitting (PNS) attacks.
- Ekert (E91) protocol and variations: The E91 protocol usually referred to as the Ekert protocol, first proposed in [1], was the first QKD protocol using the entanglement property of qubits. Eavesdropping and other types of information interception are detected by performing Bell inequality tests on the qubits received by Bob. Violation of the Bell inequality tests indicates eavesdropping and raw qubits are retransmitted until the data passes the Bell inequality tests. One of the research challenges is to generate high data rate entangled photon pairs and to perform coincidence analysis on the received entangled pairs of photons. The coincidence analysis ensures the detected photon pairs by Alice and Bob belong to the same entangled pair of photons rather than random uncorrelated pairs of photons.

  The BBM92 protocol again proposed by Charles Bennett [24] is yet another version of the BB84 protocol that uses entanglement instead of the superposition principle as in the original BB84. Similar to the Ekert protocol, Bell inequality analysis is used to detect eavesdropping. Implementation examples are described in the SECOQC QKD network [56] and Tokyo QKD network [57] test facilities.
- COW protocol: The COW QKD protocol is first introduced in [27,59] with a detailed description in [60]. Information bits are encoded in time bins, through a string of weak coherent light pulses from a continuous laser as it is shown in Figure 15.6. The COW protocol demonstrations include TUBITAK BILGEM and SECOQC QKD testbeds [56].

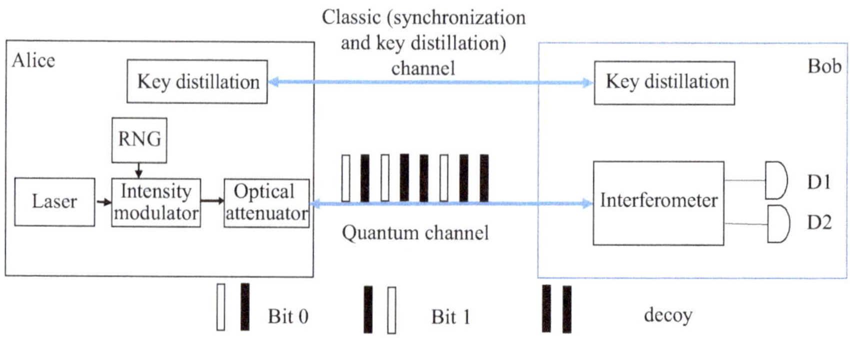

Figure 15.6 Schematic of the coherent one-way (COW) QKD protocol

- DPS protocol: The DPS protocol [28,29] demonstrated in [57] uses a string of coherent photon pulses, encoding the information bits in the relative phase between nearby pulses. The protocol has similar features to the COW protocol. DPS QKD is a distributed-phase-reference QKD protocol similar to COW, easy to implement, and resistant against PNS attacks.
- MDI-QKD protocol: The MDI-QKD protocols ([34,35]) attempt to eliminate the security issues stemming from the measurement device where Eve may have tempered with or breached Bob's receiver.

Commercialization status for DVQKD

Commercialization of many of the DVQKD protocols is reported in [61]. Some examples of commercialization of BB84 and its variations include QuantumCtek and Qasky (China), KT Corp. (Korea), NEC, NTT and Toshiba (Japan), QRate (Russia), Toshiba (the UK and Japan), MagiQ and BBN Raytheon (USA). The photon entanglement-based implementation (E91 QKD) is done by the Austrian Institute of Technology. There are many other commercialization examples, such as ID Quantique, QUBITEKK (USA), and S-Fifteen Instruments. SpeQtral will apply the BBM92 protocol on satellite key distribution.

15.4.5.2 CVQKD protocols implementation examples

The second major class of QKD protocols is the CVQKD. In quantum field theories, specifically in quantum electrodynamics (QED), photons are quantized excitations of electromagnetic fields. Electromagnetic fields have quadrature components commonly referred to as X and P or position and momentum components in the optical phase space. In CVQKD, the information is encoded in the X and P quadrature components of the electromagnetic field radiation by Alice. Bob uses homodyne or heterodyne detection to decode the information in the received laser light. Similar to the DVQKD, based on this principle, variations of CVKQD protocols are introduced. We will mention some of the prominent such protocols.

- CVQKD based on coherent states of photons: The GG02, Gaussian modulated coherent state (GMCS) protocol introduced by Grosshans and Grangier

[37] in 2002, is widely used and analyzed. In GG02, Alice encodes information on the quadrature components of coherent photon states of laser light. Bob uses homodyne detection to decode the information. One advantage of GG02 is its implementation using widespread optical communications equipment and parts. Various improvements are introduced to reduce losses and extend transmission lengths, such as reverse reconciliation and post-selection methods.

One version of GG02 (no switching GMCS) that uses heterodyne detection, sometimes referred to as conjugated detection exists. Heterodyne detection enables simultaneous measurement of both position and momentum quadrature components. Measurement and sifting stages are eliminated in this scheme which makes passive and simpler equipment feasible. GG02 is the most common CVQKD protocol implemented using conventional fiber networks in distances reaching to 100 km [62–67].

The QKD protocol layer passes relevant protocol parameters on the status of the quantum channel (noise, key rate, etc.) and the QKD module (shot noise variance, raw and secret key rates, etc.) to the higher layers.

- Unidimensional CVQKD (UD CVQKD): Refs. [68,69] introduced UD CVQKD protocol in which Alice uses a single quadrature modulation and Bob uses a randomly switched homodyne detection. UD CVQKD protocols are easier to implement, as they use only one modulator for a single quadrature. The security analysis of coherent-state UD CVQKD has been performed by [70,71].
- MDI-CVQKD: The MDI-CVQKD protocol [34,35] reduces the side channel attack vulnerabilities of the measurement devices of QKD systems. In this protocol, instead of Alice or Bob, measurements are done in an intermediary station without any security oversight. There are no existing implementations of this protocol. MDI-CVQKD is suitable for extended distances in metropolitan areas [40,41].
- Discrete modulation coherent state (DMCS): The DMCS protocol uses two coherent states to encode a binary string [72,73]. More recently, multiple coherent state modulation versions (three, four, and any) have also been suggested [74–76] were proposed. Theoretical security proof of DMCS has been studied in [40,77].

Classical post-processing in CVQKD

Following the transmission of qubits through the quantum communication function, classical post-processing on the raw key data is needed to obtain the final key shared by Alice and Bob that are identical. The post-processing steps may be broadly called authentication and information exchange. Although the details may vary among different CVQKD protocols, the main post-processing stages may be summarized as follows:

Alice and Bob first authenticate their identities by using a predetermined symmetric key. Authentication is followed by the information exchange between Alice and Bob to reconcile the integrity of the final key.

Authentication:

- Alice and Bob have a predetermined shared key. Alice transmits the message, hash, and key index. Then Alice waits for Bob's confirmation.
- Bob validates the certification transmitted by Alice and transmits acknowledgment and identification.
- When Alice receives the acknowledgment from Bob, she sends a final acknowledgment to Bob.

Authentication ensures the mutual identity validation of Alice and Bob.

The authentication is followed by the information exchange, whereby Alice and Bob use frame and data formats as specified by the protocol. The integrity of the data frames is checked by hash algorithms. If the key reconciliation is successful, Alice transmits a success flag to Bob.

15.4.5.3 Commercialization status for CVQKD

Detailed information on various CVQKD products can be found in [61]. There are several commercial implementations of CVQKD, starting from the first CVQKD product using GG02 in 2012 by SeQureNet. The raw data rate is 1 MHz with a maximum transmission distance of up to 80 km [78]. In addition, XT Quantech Co. has introduced a no switching CV-QKD version in 2018 with a raw data rate of 10 MHz [79]. Quintessence Labs [80] and Huawei (Germany) have introduced CVQKD products based on the discrete modulation protocol.

15.5 Quantum random number generators

Random numbers are the essence of various information and communications technology applications, such as statistics, gaming, banking, and cryptography. In these many applications, pseudo random numbers generators (PRNG) are preferred because they can be produced at high rates. They are commonly used in Monte Carlo simulations and various programming languages.

PRNGs are based on deterministic algorithms and are not reliable enough to create a sufficient level of security for some applications. A true random number generator (TRNG), on the other hand, uses natural phenomena to create random numbers, such as atmospheric noise, and thermal noise.

The third category of random number generators is the QRNG. The main advantage of a QRNG is that the statistical independence of the random numbers is guaranteed by the laws of physics, providing a high-quality entropy and randomness source. QRNG, thus, produces an independent identically distributed sequence of random numbers with enhanced cryptographic security. Examples of such sources are the inter-decay times of radioisotopes, the time difference between individual photons in a weak laser source, etc.

As discussed before, QKDN protocols and quantum cryptography also need a reliable randomness source. QKD systems are open to attacks, if the state preparation bases and the raw key bits are not chosen in a truly random way, as it has been shown for the BB84 protocol [81,82].

Random bit sequences are expected to have two main properties: The bit sequences have to be uniformly distributed and statistically unpredictable from past data. Uniform distribution of random bit sequences can be achieved by using several methods. Although the output of the QRNG entropy sources is unpredictable, most of the time the probability distribution of the bit sequence is not uniform. For some applications such as encryption, the probability distribution is required to be close to uniform. To achieve uniformity, the output sequence is post-processed by using randomness extractors [83]. QKD protocols, such as BB84, can be regarded as a mechanism for the secure distribution of random number sequences between two parties.

In secure random number generation and post-processing, external attacks such as side-channel attacks, and attacks using quantum computers have to be taken into account. For example, as a randomness extractor, Trevisan's extractor [84] which is proven to be secure against quantum attacks can be employed. For other quantum computation resistant extractors, check [85,86].

There are many quantum entropy sources based on shot noise [87], which is generated from quantum effects due to the granularity of the flow of quantum particles, e.g., photons in a coherent state, tunneling of particles in a finite potential, etc.

There are several types of optical quantum random number generators, for example, generators based on branching paths with beam splitters [88], inter-arrival time of photons in a coherent state, quantum vacuum fluctuations [89,90], phase noise of lasers, spontaneous photon emission [91], etc. For detailed descriptions and comparisons of these generators, check [83].

15.5.1 Types of QRNGs

15.5.1.1 Trusted QRNGs

Trusted QRNG assumes that the sub-parts including the entropy source and the measurement apparatus used to extract randomness are well characterized and trusted. The theory behind the trusted QRNG is usually simple and its implementation is suitable for practical applications.

To prevent the output random bits to be under the control of an adversary, the sub-parts of the device should be trusted which means that, for example, the photon sources such as lasers and the single-photon detectors should not be controlled or affected by the adversary. This is why this type of QRNG device is called a trusted device.

There are two types of trusted QRNGs: trusted non-optical QRNGs and trusted optical QRNGs.

- Trusted non-optical QRNGs
  - Radioactive decay: Radioactive decay of particles is a quantum phenomenon used to generate random numbers and can be explained by the uncertainty principle. The decay of radioisotopes is independent and identically distributed (i.i.d.) Poisson process and inter-decay time are exponentially distributed.

  - Thermal noise (Johnson noise) and shot noise: The noise present in electronic circuits, thermal noise, is also a source of entropy to extract randomness. Thermal noise is due to the random motion of the electrons in the circuit elements like resistors or diodes. The discontinuous jumps in the current due to the quantum fluctuations in the motion of electrons have a quantum origin and are called shot noise.
- Trusted optical QRNG
  - Single photon detection:
    * Measuring qubit: Randomness is generated by measuring qubits in a superposition state. Photon states can be in a superposition of possible polarization states or a superposition of possible paths.
    * Measuring temporal arrivals: QRNGs generate randomness from the arrival time of photons, such as a weak laser source, an LED, or a single-photon source. They are similar to the QRNGs based on radioactive decay.
  - Macroscopic photo-detection: Classical quantities, like amplitude or phase, are measured to generate random numbers in this type of QRNGs.
    * Vacuum noise: The zero-point vacuum fluctuation of the electromagnetic field is the source of randomness in this type of QRNGs The amplitude and phase quadratures of the field can be measured repeatedly to generate random numbers.
    * Amplified spontaneous emission: In this type of QRNGs, the phase noise of the amplified spontaneous emission is the source of randomness.

15.5.1.2 Self-testing QRNGs

The random number generation methods described so far assume trust in the QRNG devices. However, in practice, the devices may be defective or vulnerable to an adversary.

By focusing only on the input-output statistics, not the device details of the QRNG, testing can be done by comparing the output distribution against the distributions dictated by the physical laws. In entanglement-based systems, self-testing QRNGs measure statistical correlations by observing a Bell inequality violation. Observation of a Bell inequality violation implies entanglement and quantum randomness in the measurement of the qubit state. Some types of self-testing QRNGs include random number expansion and randomness amplification.

15.5.1.3 Semi-self-testing QRNGs

In practice, some parts of the random number generation device may be well characterized than other parts and one can trust only some parts of the device. The security level of such a device is between the trusted and fully self-testing devices. While trusted QRNGs provide a high data rate with low cost and low security and self-testing QRNGs with high security with low data rate, semi-self testing QRNGs represent a trade-off between these two types of QRNGs. Source device-independent and measurement device-independent are different models of QRNGs. For more information about the types of QRNGs, see [92].

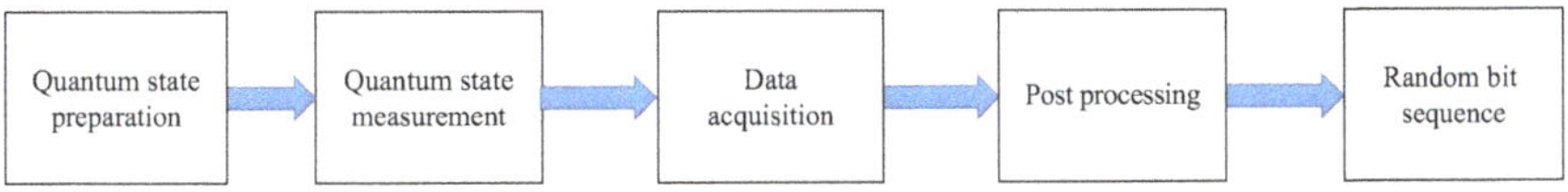

Figure 15.7 Steps for quantum random number generation

15.5.2 Steps in quantum random bit generation

Generating a random number sequence from quantum noise (entropy source) can be described in four steps [86,93] as shown in Figure 15.7:

1. Quantum state preparation: In this step, quantum superposition or entanglement states are produced. Photons produced by a weak laser source are an example of a quantum state. The quantum source can be optical or non-optical.
2. Quantum state measurement: Measurement basis is applied to the states derived from the quantum state entropy preparation. The combination of Step 1 and Step 2 composes quantum randomness or entropy source.
3. Raw data acquisition: This step is to generate raw data from quantum state measurement results. A digitization step is needed if the quantum state measurement results are analog.
4. Post-processing: Post-processing algorithms such as randomness extractors are applied to have uniformly distributed data for certain applications.

To be secure against quantum computer attacks, the randomness extractor is required to be a quantum-proof extractor. Currently, only two types of quantum-proof randomness extractors exist: Trevisan's extractor [84,86] and a two-universal hashing extractor [85,86]. For other extractor methods, check recommendations for random number generators (RNGs) [94–96].

15.5.3 QRNG based on vacuum fluctuations

To show the steps for quantum random number generation, as an example, we present the QRNG based on vacuum fluctuations [97]. In this type of QRNG, there are two main parts:

- Optical part (quantum entropy source).
- Post-processing part (processing unit to extract random bits from the entropy source).

Figure 15.8 schematically shows the setup for QRNG. The coherent laser is propagated through the BS and splits into two paths. Each detector PD1 and PD2 detects optical input signals and converts them to currents, i1 and i2. Subtracting the two output signals from each other results in the cancellation of common mode classical noise. This type of setup is also known as a balanced homodyne detector [98]. The homodyne setup is very beneficial to getting rid of the classical noise and observing the shot noise dominantly.

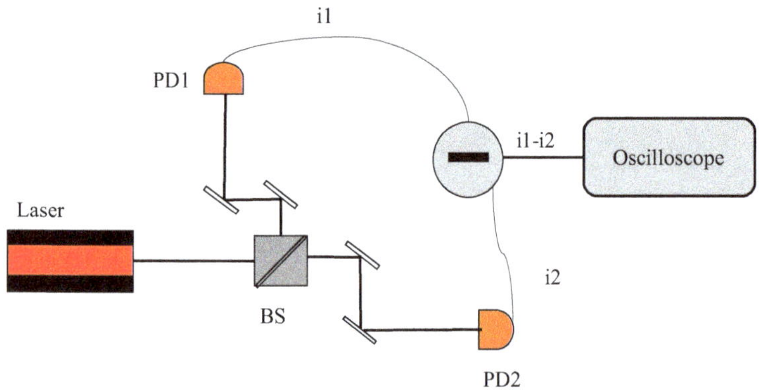

Figure 15.8 Schematic of the quantum random number generator. A beam splitter (BS) splits the beam with the central wavelength of 1,550 nm, onto two photodetectors (PD1, PD2) to create photocurrents i1 and i2. The current difference between i1 and i2 originates from the shot noise. The difference signal is obtained, sampled, and stored at the oscilloscope

The difference output between the outputs of two photo-detectors forms a quantum entropy source based on shot noise. Up to this point, Step 1 and Step 2 are realized. After the difference signal is obtained, sampled, and stored at the oscilloscope, Step 4 (post-processing) is applied to the collected data.

15.6 Conclusion and future direction

In this chapter, the well-established applications of quantum technologies to physical layer security, namely QKDNs and QRNGs, are explained. Standardization efforts for these applications are being pursued, and commercial products are available. As such, they are expected to be a part of the next-generation communications networks. Nevertheless, there is an ongoing research to further increase the key distribution data rate and the level of security, while decreasing the cost. In the near future, it is expected that through quantum entanglement distribution networks (quantum internet), though it is in its infancy, other applications such as distributed quantum computing, quantum communications using entangled qubits, clock synchronization, secure identification, and quantum sensor networks will be realized [99].

References

[1] Ekert A. Quantum cryptography based on Bell's theorem. *Physical Review Letters*. 1991;67(6):661–663.

[2] ITU-T. Quantum key distribution network protocols: quantum layer; 2021. Technical Report ITU-T FG QIT4N D2.3-part 1.
[3] ITU-T. Quantum key distribution networks: functional architecture; 2020. Recommendation ITU-T X.1710.
[4] ITU-T. Quantum key distribution network (QKDN) protocols: key management layer, QKDN control layer, and QKDN management layer; 2021. Technical Report ITU-T FG QIT4N D2.3-part 2.
[5] ETSI. Quantum Key Distribution (QKD); components and internal interfaces; 2018. ETSI GR QKD 003 V2.1.1 (2018-03).
[6] Dynes J, Yuan Z, Sharpe A, *et al.* Practical quantum key distribution over 60 hours at an optical fiber distance of 20 km using weak and vacuum decoy pulses for enhanced security. *Optics Express*. 2007;15(13):8465–8471.
[7] Treiber A, Poppe A, Hentschel M, *et al.* A fully automated entanglement-based quantum cryptography system for telecom fiber networks. *New Journal of Physics*. 2009;11:045013.
[8] Yin J, Cao Y, Li YH, *et al.* Satellite-to-ground entanglement-based quantum key distribution. *Physical Review Letters*. 2017;119(20):200501.
[9] Natarajan CM, Tanner MG, Hadfield RH. Superconducting nanowire single-photon detectors – physics and applications. *Superconductor Science and Technology*. 2012;25:063001.
[10] Dauler EA, Grein ME, Kerman AJ, *et al.* Review of superconducting nanowire single-photon detector system design options and demonstrated performance. *Optical Engineering*. 2014;53(8):081907.
[11] Bennett C, Brassard G. Quantum cryptography: public key distribution and coin tossing. In: *Proceedings of IEEE International Conference on Computers, Systems and Signal Processing*, Dec. 1984.
[12] Hasekioglu ASA, Hasekioğlu O. Quantum key distribution and quantum networks standardization efforts. In: *2020 28th Signal Processing and Communications Applications Conference* (SIU); 2020. p. 1–4.
[13] Federal Information Processing Standards. Announcing the ADVANCED ENCRYPTION STANDARD (AES); 2001. Publication 197.
[14] ISO/IEC. Information technology – security techniques – encryption algorithms. Part 3: Block Ciphers; 2010. ISO/IEC 10833-3:2010.
[15] Moriarty K, Kaliski B, Jonsson J, *et al.* PKCS# 1: RSA cryptography specifications version 2.2; 2016.
[16] Gisin N, Ribordy G, Tittel W, *et al.* Quantum cryptography. *Reviews of Modern Physics*. 2002;74(1):145.
[17] Scarani V, Bechmann-Pasquinucci H, Cerf NJ, *et al.* The security of practical quantum key distribution. *Reviews of Modern Physics*. 2009;81(3):1301–1350.
[18] Diamanti E, Lo HK, Qi B, *et al.* Practical challenges in quantum key distribution. *NPJ Quantum Information*. 2016;2(1):1–12.
[19] Pirandola S, Andersen UL, Banchi L, *et al.* Advances in quantum cryptography. *Advances in Optics and Photonics*. 2020;12(4):1012–1236.

[20] Xu F, Ma X, Zhang Q, *et al.* Secure quantum key distribution with realistic devices. *Reviews of Modern Physics*. 2020;92(2):025002.
[21] Cao Y, Zhao Y, Wang J, *et al.* Cost-efficient quantum key distribution (QKD) over WDM networks. *Journal of Optical Communications and Networking*. 2019;11(6):285–298.
[22] Bennett CH. Quantum cryptography using any two nonorthogonal states. *Physical Review Letters*. 1992;68(21):3121.
[23] Bruß D. Optimal eavesdropping in quantum cryptography with six states. *Physical Review Letters*. 1998;81(14):3018.
[24] Bennett CH, Brassard G, Mermin ND. Quantum cryptography without Bell's theorem. *Physical Review Letters*. 1992;68(5):557.
[25] Scarani V, Acín A, Ribordy G, *et al.* Quantum cryptography protocols robust against photon number splitting attacks for weak laser pulse implementations. *Physical Review Letters*. 2004;92(5):057901.
[26] Gisin N, Ribordy G, Zbinden H, *et al.* Towards practical and fast quantum cryptography. arXiv preprint quant-ph/0411022. 2004.
[27] Stucki D, Brunner N, Gisin N, *et al.* Fast and simple one-way quantum key distribution. *Applied Physics Letters*. 2005;87(19).
[28] Inoue K, Waks E, Yamamoto Y. Differential phase shift quantum key distribution. *Physical Review Letters*. 2002;89(3):037902.
[29] Inoue K, Waks E, Yamamoto Y. Differential-phase-shift quantum key distribution using coherent light. *Physical Review Letters*. 2003;68(2):022317.
[30] Sasaki T, Yamamoto Y, Koashi M. Practical quantum key distribution protocol without monitoring signal disturbance. *Nature*. 2014;509(7501):475–478.
[31] Zhang Z, Yuan X, Cao Z, *et al.* Practical round-robin differential-phase-shift quantum key distribution. *New Journal of Physics*. 2017;19(3):033013.
[32] Lucamarini M, Yuan ZL, Dynes JF, *et al.* Overcoming the rate distance limit of quantum key distribution without quantum repeaters. *Nature*. 2018;557 (7705):400–403.
[33] Ma X, Zeng P, Zhou H. Phase-matching quantum key distribution. *Physical Review X*. 2018;8(3):031043.
[34] Lo HK, Curty M, Qi B. Measurement-device-independent quantum key distribution. *Physical Review Letters*. 2012;108(13):130503.
[35] Braunstein SL, Pirandola S. Side-channel-free quantum key distribution. *Physical Review Letters*. 2012;108(13):13052.
[36] Acín A, Brunner N, Gisin N, *et al.* Device-independent security of quantum cryptography against collective attacks. *Physical Review Letters*. 2007; 98(23):230501.
[37] Grosshans F, Grangier P. Continuous variable quantum cryptography using coherent states. *Physical Review Letters*. 2002;88(5):057902.
[38] Silberhorn C, Ralph T, Lütkenhaus N, *et al.* Continuous variable quantum cryptography: beating the 3 dB loss limit. *Physical Review Letters*. 2002;89(16):167901.
[39] Ralph TC. Continuous variable quantum cryptography. *Physical Review A*. 1999;61:010303.

[40] Lin J, Upadhyaya T, Lütkenhaus N. Asymptotic security analysis of discrete-modulated continuous-variable quantum key distribution. *Physical Review X.* 2019;9(4):041064.

[41] Pirandola S, Ottaviani C, Spedalieri G, *et al.* High-rate measurement-device-independent quantum cryptography. *Nature Photonics.* 2015;9:397–402.

[42] ITU-T. Security framework for quantum key distribution networks; 2020. Recommendation ITU-T X.1710.

[43] Ben-Or M, Mayers D. General security definition and composability for quantum & classical protocols. arXiv preprint quant-ph/0409062. 2004.

[44] Renner R. Security of quantum key distribution. *International Journal of Quantum Information.* 2008;6(1):1–127.

[45] Müller-Quade J, Renner R. Composability in quantum cryptography. *New Journal of Physics.* 2009;11:085006.

[46] Portmann C, Renner R. Security in quantum cryptography. arXiv:210200021 v2 [quant-ph]. 2021.

[47] ETSI. Quantum key distribution (QKD); Security proofs; 2010. Group Specification ETSI GS QKD 005 V1.1.1.

[48] ISO/IEC. Security requirements, test and evaluation methods for quantum key distribution. ISO/IEC 23837 Committee Draft 2.

[49] ETSI. QKD Common Criteria Protection Profile for QKD; Draft Group Specification ETSI DGS/QKD-016-PP, Early draft V.0.5.3.

[50] Gerhardt I, Liu Q, Lamas-Linares A, *et al.* Full-field implementation of a perfect eavesdropper on a quantum cryptography system. *Nature Communications.* 2011;2(349).

[51] Lydersen L, Wiechers C, Wittmann C, *et al.* Hacking commercial quantum cryptography systems by tailored bright illumination. *Nature Photonics.* 2010;4:686–689.

[52] Qin H, Kumar R, Alleaume R. Quantum hacking: saturation attack on practical continuous-variable quantum key distribution. *Physical Review A.* 2016;94(1):012325.

[53] Qin H, Kumar R, Makarov V, *et al.* Homodyne-detector-blinding attack in continuous-variable quantum key distribution. *Physical Review A.* 2018; 98(1):012312.

[54] ETSI. Implementation Security of Quantum Cryptography; 2018. ETSI White Paper No. 27.

[55] Elliott C, Pearson D, Troxel G. Quantum cryptography in practice. In: *Proceedings of the 2003 Conference on Applications, Technologies, Architectures, and Protocols for Computer Communications* (SIGCOMM '03); 2003. p. 227–238.

[56] Poppe A, Peev M, Maurhart O. Outline of the SECOQC quantum-key-distribution network in Vienna. *International Journal of Quantum Information.* 2008;6(2):209–218.

[57] Sasaki M, Fujiwra M, Ishizuka H, *et al.* Tokyo QKD network and the evolution to secure photonic network. In: *CLEO:2011 Laser Applications to Photonic Applications*, OSA Technical Digest paper JTuC1. 2011.

[58] Liao SK, Cai WQ, Liu WY, *et al.* Satellite-to-ground quantum key distribution. *Nature*. 2017;549(7670):43–47.

[59] Stucki D, Walenta N, Vannel F, *et al.* High rate, long-distance quantum key distribution over 250 km of ultra low loss fibres. *New Journal of Physics*. 2009;11(7):075003.

[60] Walenta N. Concepts, components and implementations for quantum key distribution over optical fibers. PhD thesis; 2013. doi: 10.13140/2.1.1345.8880.

[61] ITU-T. Quantum information technology for networks standardization outlook and technology maturity: quantum key distribution network; 2021. ITU-T Technical Report.

[62] Fossier S, Diamanti E, Debuisschert T, *et al.* Field test of a continuous-variable quantum key distribution prototype. *New Journal of Physics*. 2009;11(4):045023.

[63] Alléaume R, Bouda J, Branciard C, *et al.* SECOQC white paper on quantum key distribution and cryptography (2007). arXiv preprint quant-ph/0701168. 2007.

[64] Lodewyck J, Bloch M, Garcia-Patron R, *et al.* Quantum key distribution over 25 km with an all-fiber continuous-variable system. *Physical Review A*. 2007;976(4):042305.

[65] Huang D, Lin D, Wang C, *et al.* Continuous-variable quantum key distribution with 1 Mbps secure key rate. *Optics Express*. 2015;23:17511.

[66] Huang D, Lin D, Wang C, *et al.* Long-distance continuous-variable quantum key distribution by controlling excess noise. *Scientific Reports*. 2016;6:19201.

[67] Wang C, Huang D, Huang P, *et al.* 25 MHz clock continuous-variable quantum key distribution system over 50 km fiber channel. *Scientific Reports*. 2015;5:14607.

[68] Usenko VC, Grosshans F. Unidimensional continuous-variable quantum key distribution. *Physical Review A*. 2015;92(6):062337.

[69] Gehring T, Jacobsen CS, Andersen UL. Single-quadrature continuous-variable quantum key distribution. *Quantum Information & Computation*. 2016;16(13-14):1081–1095.

[70] Wang P, Wang X, Li J, *et al.* Finite-size analysis of unidimensional continuous-variable quantum key distribution under realistic conditions. *Optics Express*. 2017;25(23):27995–28009.

[71] Liao Q, Guo Y, Xie C, *et al.* Composable security of unidimensional continuous-variable quantum key distribution. *Quantum Information Processing*. 2018;17(113).

[72] Zhao YB, Heid M, Rigas J, *et al.* Asymptotic security of binary modulated continuous-variable quantum key distribution under collective attacks. *Physical Review A*. 2009;79:012307.

[73] Leverrier A, Grangier P. Unconditional security proof of long distance continuous-variable quantum key distribution with discrete modulation. *Physical Review Letters*. 2009;102(18):1–4.

[74] Leverrier A, Grangier P. Continuous-variable quantum key distribution protocols with a discrete modulation. arXiv preprint arXiv:10024083. 2010.

[75] Bradler K, Weedbrook C. Security proof of continuous-variable quantum key distribution using three coherent states. *Physical Review A*. 2018;97(2):022310.

[76] Papanastasiou P, Lupo C, Weedbrook C, *et al*. Quantum key distribution with phase encoded coherent states: asymptotic security analysis in thermal-loss channels. *Physical Review A*. 2018;98:012340.

[77] Ghorai S, Grangier P, Diamanti E, *et al*. Asymptotic security of continuous-variable quantum key distribution with a discrete modulation. *Physical Review X*. 2019;9(2):021059.

[78] Jouguet P, Kunz-Jacques S, Leverrier A, *et al*. Experimental demonstration of long-distance continuous-variable quantum key distribution. *Nature Photonics*. 2013;7(5):378–381.

[79] Huang D, Huang P, Li H, *et al*. Field demonstration of a continuous-variable quantum key distribution network. *Optics Letters*. 2016;41:3511–3514.

[80] Weedbrook C, Pirandola S, García-PatróN R, *et al*. Gaussian quantum information. *Reviews of Modern Physics*. 2012;84(2):621–669.

[81] Bouda J, Pivoluska M, Plesch M, *et al*. Weak randomness seriously limits the security of quantum key distribution. *Physical Review A*. 2012;86(6):062308.

[82] Li HW, Yin ZQ, Wang S, *et al*. Randomness determines practical security of BB84 quantum key distribution. *Scientific Reports*. 2015;5:16200.

[83] Herrero-Collantes M, Garcia-Escartin JC. Quantum random number generators. *Reviews of Modern Physics*. 2017;89(1):015004.

[84] Trevisan L. Extractors and pseudorandom generators. *JACM*. 2001;48: 860–879.

[85] König R, Renner R. Sampling of min-entropy relative to quantum knowledge. *IEEE Transactions on Information Theory*. 2011;57:4760–4787.

[86] Ma X, Xu F, Tan X, *et al*. Postprocessing for quantum random-number generators: entropy evaluation and randomness extraction. *Physical Review A*. 2013;87:062327.

[87] Landauer R. Solid-state shot noise. *Physical Review B*. 1993;47:16427–16432.

[88] Nie YQ, Zhang HF, Zhang Z, *et al*. Practical and fast quantum random number generation based on photon arrival time relative to external reference. *Applied Physics Letters*. 2014;104:051110.

[89] Shi Y, Chng B, Kurtsiefer C. Random numbers from vacuum fluctuations. *Applied Physics Letters*. 2016;109(4):041101.

[90] Dandasi A, Ozel H, Hasekioglu O, *et al*. Effect of photon statistics on vacuum fluctuations based QRNG. *Journal of Optics*. 2021;23(6):065201. Available from: https://doi.org/10.1088/2040-8986/abd9dd.

[91] Li X, Cohen AB, Murphy TE, *et al*. Scalable parallel physical random number generator based on a superluminescent LED. *Optics Letters*. 2011;36(6): 1020–1022.

[92] Mannalath V, Mishra S, Pathak A. A comprehensive review of quantum random number generators: concepts, classification and the origin of randomness; 2022. ArXiv:2203.00261v1 [quant-ph].

[93] ITU-T. Recommendation ITU-T X.1702: quantum noise random number generator architecture; 2019. Rec. ITUT X.1702 (11/2019).
[94] NIST. Recommendation for the Entropy Sources Used for Random Bit Generation; 2018. NIST SP 800-90B.
[95] BSI. A proposal for: functionality classes for random number generators; 2011. BSI AIS20/AIS31.
[96] ISO/IEC. Information technology – Security techniques – Random bit generation; 2011. ISO/IEC 18031:2011.
[97] Hasekioğlu A, Çinkaya Yilmaz, Acar, *et al.* Silicon Photonics for Quantum Fibre Networks (SQUARE) QRNG Report; 2022. QuantERA 2017 program, Project No. 117F289.
[98] Jakeman E, Oliver C, Pike E. Optical homodyne detection. *Advances in Physics*. 1975;24(3):349–405.
[99] Wehner S, Elkouss D, Hanson R. Quantum internet: a vision for the road ahead. *Science*. 2018;362(6412).

Index

www.ingramcontent.com/pod-product-compliance
Lightning Source LLC
LaVergne TN
LVHW011837200926
848484LV00002B/364